# PREALGEBRA
## THIRD EDITION

# STUDENT'S
# SOLUTIONS MANUAL

## Judith A. Penna

# Marvin L. Bittinger
*Indiana University—Purdue University at Indianapolis*

# David J. Ellenbogen
*Community College of Vermont*

 **ADDISON-WESLEY**

An imprint of Addison Wesley Longman, Inc.

Reading, Massachusetts • Menlo Park, California • New York • Harlow, England
Don Mills, Ontario • Sydney • Mexico City • Madrid • Amsterdam

Reproduced by Addison Wesley Longman from camera-ready copy supplied by the author.

Copyright © 2000 Addison Wesley Longman, Inc.

ISBN: 0-201-34026-7

2 3 4 5 6 7 8 9 10  PHC  02 01 00

# Table of Contents

# Chapter 1

# Operations on the Whole Numbers

## Exercise Set 1.1

**1.** 5742 = 5 thousands + 7 hundreds + 4 tens + 2 ones

**3.** 27,342 = 2 ten thousands + 7 thousands + 3 hundreds + 4 tens + 2 ones

**5.** 5609 = 5 thousands + 6 hundreds + 0 tens + 9 ones, or 5 thousands + 6 hundreds + 9 ones

**7.** 2300 = 2 thousands + 3 hundreds + 0 tens + 0 ones, or 2 thousands + 3 hundreds

**9.** 2 thousands + 4 hundreds + 7 tens + 5 ones = 2475

**11.** 6 ten thousands + 8 thousands + 9 hundreds + 3 tens + 9 ones = 68,939

**13.** 7 thousands + 3 hundreds + 0 tens + 4 ones = 7304

**15.** 1 thousand + 9 ones = 1009

**17.** A word name for 85 is eighty-five.

**19.**

88,000

Eighty-eight thousand

**21.**

123,765

One hundred twenty-three thousand,
seven hundred sixty-five

**23.**

7, 754, 211, 577

Seven billion,
seven hundred fifty-four million,
two hundred eleven thousand,
five hundred seventy-seven

**25.**

1, 867,000

One million,
eight hundred sixty-seven thousand

**27.**

1, 583, 141,000

One billion,
five hundred eighty-three million,
one hundred forty-one thousand

**29.**

Two million,
two hundred thirty-three thousand,
eight hundred twelve

Standard notation is   2, 233, 812.

**31.**

Eight billion

Standard notation is   8,000,000,000.

**33.**

Nine trillion,
four hundred sixty billion,

Standard notation is   9,460,000,000,000.

**35.**

Two million,
nine hundred seventy-four thousand,
six hundred

Standard notation is   2, 974, 600.

**37.** 2 3 [5] , 8 8 8

The digit 5 means 5 thousands.

**39.** 4 8 8, [5] 2 6

The digit 5 means 5 hundreds.

**41.** 8 9, [3] 0 2

The digit 3 tells the number of hundreds.

**43.** 8 9, 3 [0] 2

The digit 0 tells the number of tens.

**45.** ◈

**47.** All digits are 9's. Answers may vary. For an 8-digit read-out, for example, it would be 99,999,999. This number has three periods.

## Exercise Set 1.2

**1.**

| 7 e-mail messages Tuesday | 8 e-mail messages Wednesday | 15 e-mail messages altogether |
|---|---|---|
| 8 | + 7 | = 15 |

**3.**

| One parcel contains 500 acres. | The other parcel contains 300 acres. | The total purchase is 800 acres. |
|---|---|---|
| 500 acres + | 300 acres | = 800 acres |

**5.** 14 mi + 13 mi + 8 mi + 10 mi + 47 mi + 22 mi = 114 mi

**7.** 18 in. + 8 in. + 18 in. + 8 in. = 52 in.

**9.** 325 ft + 325 ft + 325 ft + 325 ft = 1300 ft

**11.**
```
   3 6 4
 +   2 3
   3 8 7
```
Add ones, add tens, then add hundreds.

**13.**
```
   1
  1706
+ 3482
  5188
```
Add ones: We get 8.  Add tens: We get 8 tens.  Add hundreds: We get 11 hundreds, or 1 thousand + 1 hundred. Write 1 in the hundreds column and 1 above the thousands.  Add thousands: We get 5 thousands.

**15.**
```
   1
   86
+  78
  164
```
Add ones: We get 14 ones, or 1 ten + 4 ones. Write 4 in the ones column and 1 above the tens. Add tens: We get 16 tens.

**17.**
```
   1
   99
+   1
  100
```
Add ones: We get 10 ones, or 1 ten + 0 ones. Write 0 in the ones column and 1 above the tens. Add tens: We get 10 tens.

**19.**
```
   11
   789
+  111
   900
```
Add ones: We get 10 ones, or 1 ten + 0 ones. Write 0 in the ones column and 1 above the tens. Add tens: We get 10 tens, or 1 hundred + 0 tens. Write 0 in the tens column and 1 above the hundreds.  Add hundreds: We get 9 hundreds.

**21.**
```
   1
   909
+  101
  1010
```
Add ones: We get 10 ones, or 1 ten + 0 ones. Write 0 in the ones column and 1 above the tens. Add tens: We get 1 ten. Add hundreds: We get 10 hundreds.

**23.**
```
    1
   8113
+   390
   8503
```
Add ones: We get 3. Add tens: We get 10 tens, or 1 hundred + 0 tens. Write 0 in the tens column and 1 above the hundreds. Add hundreds: We get 5. Add thousands: We get 8.

**25.**
```
     1
    356
+ 4910
   5266
```
Add ones: We get 6. Add tens: We get 6. Add hundreds: We get 12 hundreds, or 1 thousand + 2 hundreds. Write 2 in the hundreds column and 1 above the thousands. Add thousands: We get 5.

**27.**
```
   121
   3870
     92
      7
+   497
   4466
```
Add ones: We get 16 ones, or 1 ten + 6 ones. Write 6 in the ones column and 1 above the tens. Add tens: We get 26 tens, or 2 hundreds + 6 tens. Write 6 in the tens column and 2 above the hundreds. Add hundreds: We get 14 hundreds, or 1 thousand + 4 hundreds. Write 4 in the hundreds column and 1 above the thousands. Add thousands: We get 4.

**29.**
```
   11
  5093
+ 3217
  8310
```
Add ones: We get 10 ones, or 1 ten + 0 ones. Write 0 in the ones column and 1 above the tens. Add tens: We get 11. Write 1 in the tens column and 1 above the hundreds. Add hundreds: We get 3 hundreds. Add thousands: We get 8 thousands.

**31.**
```
   11
  4825
+ 1783
  6608
```
Add ones: We get 8.  Add tens: We get 10 tens.  Write 0 in the tens column and 1 above the hundreds.  Add hundreds: We get 16 hundreds.  Write 6 in the hundreds column and 1 above the thousands. Add thousands: We get 6 thousands.

**33.**
```
   111
   9999
+  6785
 16,784
```
Add ones: We get 14 ones, or 1 ten + 4 ones. Write 4 in the ones column and 1 above the tens.  Add tens: We get 18 tens.  Write 8 in the tens column and 1 above the hundreds.  Add hundreds: We get 17 hundreds. Write 7 in the hundreds column and 1 above the thousands. Add thousands: We get 16 thousands.

**35.**
```
    111
  23,443
+ 10,989
  34,432
```
Add ones: We get 12 ones, or 1 ten + 2 ones. Write 2 in the ones column and 1 above the tens.  Add tens: We get 13 tens.  Write 3 in the tens column and 1 above the hundreds.  Add hundreds: We get 14 hundreds. Write 4 in the hundreds column and 1 above the thousands.  Add thousands: We get 4 thousands. Add ten thousands: We get 3 ten thousands.

**37.**
```
   1111
  77,543
+ 23,767
 101,310
```
Add ones: We get 10 ones, or 1 ten + 0 ones. Write 0 in the ones column and 1 above the tens. Add tens: We get 11 tens. Write 1 in the tens column and 1 above the hundreds.  Add hundreds: We get 13 hundreds.  Write 3 in the hundreds column and 1 above the thousands. Add thousands: We get 11 thousands. Write 1 in the thousands column and 1 above the ten thousands. Add ten thousands: We get 10 ten thousands.

**39.**
```
   1111
  99,999
+    102
 100,101
```
Add ones: We get 11 ones, or 1 ten + 1 one. Write 1 in the ones column and 1 above the tens. Add tens: We get 10 tens. Write 0 in the tens column and 1 above the hundreds.  Add hundreds: We get 11 hundreds.  Write 1 in the hundreds column and 1 above the thousands. Add thousands: We get 10 thousands. Write 0 in the thousands column and 1 above the ten thousands. Add ten thousands: We get 10 ten thousands.

**41.** Add from the top.

We first add 7 and 9, getting 16; then 16 and 4, getting 20; then 20 and 8, getting 28.

```
  7
  9  → 16
  4       4 → 20
+ 8            8 → 28
 28
```

Check by adding from the bottom.

We first add 8 and 4, getting 12; then 12 and 9, getting 21; then 21 and 7, getting 28.

```
  7              7  → 28
  9        9  → 21
  4  → 12
+ 8
─────
 28
```

**43.** Add from the top.

```
 8  → 14
 6     2  → 16
 2        3  → 19
 3           7  → 26
+7
────
 26
```

Check:

```
 8  → 26
 6    18
 2     6  → 12
 3  → 10
+7
────
 26
```

**45.** We look for pairs of numbers whose sums are 10, 20, 30, and so on.

```
  7 ────→  7
 18 ────→ 20
  3
 37 ────→ 40
+ 2
────      ──
 67       67
```

**47.** We look for pairs of numbers whose sums are 10, 20, 30, and so on.

```
 45 ────→ 70
 25
 36 ────→ 80
 44
+80 ────→ 80
────      ───
230       230
```

**49.**
```
   1
   2 3
   6 2
 + 4 5
 ──────
 1 3 0
```
Add ones: We get 10. Write 0 in the ones column and 1 above the tens. Add tens: We get 13 tens.

**51.**
```
     1
     4 5 1
       3 6
 +   8 6 2
 ─────────
   1 3 4 9
```
Add ones: We get 9. Add tens: We get 14 tens, or 1 hundred + 4 tens. Write 4 in the tens column and 1 above the hundreds. Add hundreds: We get 13 hundreds.

**53.**
```
   1       1
     2 6 0 3
   2 8, 2 1 4
 +   6 1 0 9
 ───────────
   3 6, 9 2 6
```
Add ones: We get 16 ones, or 1 ten + 6 ones. Write 6 in the ones column and 1 above the tens. Add tens: We get 2 tens. Add hundreds: We get 9 hundreds. Add thousands: We get 16 thousands, or 1 ten thousand + 6 thousands. Write 6 in the thousands column and 1 above the ten-thousands. Add ten thousands: We get 3 ten thousands.

**55.**
```
     1 1
   1 2, 0 7 0
     2 9 5 4
 +   3 4 0 0
 ───────────
   1 8, 4 2 4
```
Add ones: We get 4. Add tens: We get 12 tens, or 1 hundred + 2 tens. Write 2 in the tens column and 1 above the hundreds. Add hundreds: We get 14 hundreds, or 1 thousand + 4 hundreds. Write 4 in the hundreds column and 1 above the hundreds. Add thousands: We get 8 thousands. Add ten thousands: We get 1 ten thousand.

**57.**
```
   2 4
   3 2 7
   4 2 8
   5 6 9
   7 8 7
 + 2 0 9
 ───────
 2 3 2 0
```
Add ones: We get 40. Write 0 in the ones column and 4 above the tens. Add tens: We get 22 tens. Write 2 in the tens column and 2 above the hundreds. Add hundreds: We get 23 hundreds.

**59.**
```
   3 1 2
   4 8 3 5
     7 2 9
   9 2 0 4
   8 9 8 6
 + 7 9 3 1
 ─────────
 3 1, 6 8 5
```
Add ones: We get 25. Write 5 in the ones column and 2 above the tens. Add tens: We get 18 tens. Write 8 in the tens column and 1 above the hundreds. Add hundreds: We get 36 hundreds. Write 6 in the hundreds column and 3 above the thousands. Add thousands: We get 31 thousands.

**61.**
```
   1 1 1
   2 0 3 7
   4 9 2 3
   3 4 7 1
 + 1 2 4 8
 ─────────
 1 1, 6 7 9
```
Add ones: We get 19. Write 9 in the ones column and 1 above the tens. Add tens: We get 17 tens. Write 7 in the tens column and 1 above the hundreds. Add hundreds: We get 16 hundreds. Write 6 in the hundreds column and 1 above the thousands. Add thousands: We get 11 thousands.

**63.**
```
   1   1
   3 4 2 0
   8 7 1 9
   4 3 1 2
 + 6 2 0 3
 ─────────
 2 2, 6 5 4
```
Add ones: We get 14. Write 4 in the ones column and 1 above the tens. Add tens: We get 5 tens. Add hundreds: We get 16 hundreds. Write 6 in the hundreds column and 1 above the thousands. Add thousands: We get 22 thousands.

**65.**
```
   2 2 3 3 1 1
   5, 6 7 8, 9 8 7
   1, 4 0 9, 3 1 2
      8 9 8, 8 8 8
 + 4, 7 7 7, 9 1 0
 ─────────────────
 1 2, 7 6 5, 0 9 7
```

**67.** 7 thousands + 9 hundreds + 9 tens + 2 ones = 7992

**69.** 4 [8] 6, 2 0 5

The digit 8 means 8 ten thousands.

**71.** ◈

**73.** $5,987,943 + 328,959 + 49,738,765$

Using a calculator to carry out the addition, we find that the sum is 56,055,667.

**75.** One method is described in the answer section in the text. Another method is: $1 + 100 = 101, 2 + 99 = 101, \ldots, 50 + 51 = 101$. Then $50 \cdot 101 = 5050$.

## Exercise Set 1.3

**1.**
| Amount to begin with | | Amount spent | | Amount left |
|---|---|---|---|---|
| $1260 | − | $450 | = | ☐ |

**3.**
| Amount to begin with | | Amount poured out | | Amount left |
|---|---|---|---|---|
| 16 oz | − | 5 oz | = | ☐ |

**5.** $7 - 4 = 3$

↑

This number gets added (after 3).

$$7 = 3 + 4$$

(By the commutative law of addition, $7 = 4 + 3$ is also correct.)

**7.** $13 - 8 = 5$

↑

This number gets added (after 5).

$$13 = 5 + 8$$

(By the commutative law of addition, $13 = 8 + 5$ is also correct.)

**9.** $23 - 9 = 14$

↑

This number gets added (after 14).

$$23 = 14 + 9$$

(By the commutative law of addition, $23 = 9 + 14$ is also correct.)

**11.** $43 - 16 = 27$

↑

This number gets added (after 27).

$$43 = 27 + 16$$

(By the commutative law of addition, $43 = 16 + 27$ is also correct.)

**13.** $6 + 9 = 15$                $6 + 9 = 15$

↑                                          ↑

This addend gets subtracted from the sum.                This addend gets subtracted from the sum.

$6 = 15 - 9$                $9 = 15 - 6$

**15.** $8 + 7 = 15$                $8 + 7 = 15$

↑                                          ↑

This addend gets subtracted from the sum.                This addend gets subtracted from the sum.

$8 = 15 - 7$                $7 = 15 - 8$

**17.** $17 + 6 = 23$                $17 + 6 = 23$

↑                                          ↑

This addend gets subtracted from the sum.                This addend gets subtracted from the sum.

$17 = 23 - 6$                $6 = 23 - 17$

**19.** $23 + 9 = 32$                $23 + 9 = 32$

↑                                          ↑

This addend gets subtracted from the sum.                This addend gets subtracted from the sum.

$23 = 32 - 9$                $9 = 32 - 23$

**21.** We first write an addition sentence. Keep in mind that all numbers are in millions.

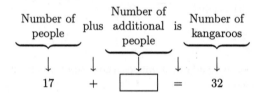

Now we write a related subtraction sentence.

$17 + \boxed{\phantom{00}} = 32$

$\boxed{\phantom{00}} = 32 - 17$   The addend 17 gets subtracted.

**23.** We first write an addition sentence.

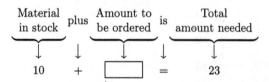

Now we write a related subtraction sentence.

$10 + \boxed{\phantom{00}} = 23$

$\boxed{\phantom{00}} = 23 - 10$   The addend 10 gets subtracted.

**25.**
```
   1 6
 -   4
 ─────
   1 2
```
Subtract ones, then subtract tens.

**27.**
```
   6 5
 - 2 1
 ─────
   4 4
```
Subtract ones, then subtract tens.

**29.**
```
   8 6 6
 - 3 3 3
 ───────
   5 3 3
```
Subtract ones, subtract tens, then subtract hundreds.

**31.**
```
    4 5 4 7
  − 3 4 2 1
  ─────────
    1 1 2 6
```
Subtract ones, subtract tens, subtract hundreds, then subtract thousands.

**33.**
```
      7 16
      8̸ 6̸
    − 4 7
    ───────
      3 9
```
We cannot subtract 7 ones from 6 ones. Borrow 1 ten to get 16 ones. Subtract ones, then subtract tens.

**35.**
```
        11
      5 7̸ 15
      6̸ 7̸ 5̸
    − 3 2 7
    ─────────
      2 9 8
```
We cannot subtract 7 ones from 5 ones. Borrow 1 ten to get 15 ones. Subtract ones. We cannot subtract 2 tens from 1 ten. Borrow 1 hundred to get 11 tens. Subtract tens, then subtract hundreds.

**37.**
```
      2 15
    8 3̸ 5̸
  − 6 0 9
  ─────────
    2 2 6
```
We cannot subtract 9 ones from 5 ones. Borrow 1 ten to get 15 ones. Subtract ones, subtract tens, then subtract hundreds.

**39.**
```
      7 11
    9 8̸ 1̸
  − 7 4 7
  ─────────
    2 3 4
```
We cannot subtract 7 ones from 1 one. Borrow 1 ten to get 11 ones. Subtract ones, subtract tens, then subtract hundreds.

**41.**
```
        6 16
    7 7 6̸ 9
  − 2 3 8 7
  ───────────
    5 3 8 2
```
Subtract ones. We cannot subtract 8 tens from 6 tens. Borrow 1 hundred to get 16 tens. Subtract tens, subtract hundreds, then subtract thousands.

**43.**
```
          17
      8 7 12
    3 9̸ 8̸ 2̸
  − 2 4 8 9
  ───────────
    1 4 9 3
```
We cannot subtract 9 ones from 2 ones. Borrow 1 ten to get 12 ones. Subtract ones. We cannot subtract 8 tens from 7 tens. Borrow 1 hundred to get 17 tens. Subtract tens, subtract hundreds, then subtract thousands.

**45.**
```
          13
      4 9 3̸ 16
    5̶-0̶-4̶ 6̸
  − 2 8 5 9
  ───────────
    2 1 8 7
```
We cannot subtract 9 ones from 6 ones. Borrow 1 ten to get 16 ones. Subtract ones. We cannot subtract 5 tens from 3 tens. We have 5 thousands or 50 hundreds. We borrow 1 hundred to get 13 tens. We have 49 hundreds. Subtract tens, subtract hundreds, then subtract thousands.

**47.**
```
    6 16 3 10
  7̸ 6̸ 4̸ 0̸
  − 3 8 0 9
  ───────────
    3 8 3 1
```
We cannot subtract 9 ones from 0 ones. Borrow 1 ten to get 10 ones. Subtract ones, then tens. We cannot subtract 8 hundreds from 6 hundreds. Borrow 1 thousand to get 16 hundreds. Subtract hundreds, then thousands.

**49.**
```
      11 15 13
      1̸ 3̸ 3̸ 17
    1 2̸, 6̸ 4̸ 7̸
  −    4 8 9 9
  ─────────────
       7 7 4 8
```

**51.**
```
          16 16
      5 6̸ 6̸ 11
    4 6̸, 7 7 1̸
  − 1 2, 9 7 7
  ─────────────
    3 3, 7 9 4
```

**53.**
```
      9 9 9 12
  1̶-0̶,-0̶-0̶ 2̸
  −    7 8 3 4
  ─────────────
       2 1 6 8
```
We have 1 ten thousand, or 1000 tens. We borrow 1 ten to get 10 ones. We then have 999 tens. Subtract ones, then tens, then hundreds, then thousands.

**55.**
```
    8 10 2 17
  9̸ 0̸, 2̸ 3̸ 7
  − 4 7, 2 0 9
  ─────────────
    4 3, 0 2 8
```

**57.**
```
    7 10
  8̸ 0̸
  − 2 4
  ───────
    5 6
```

**59.**
```
    8 10
  9̸ 0̸
  − 5 4
  ───────
    3 6
```

**61.**
```
      13
    3̸ 10
  1̸ 4̸ 0̸
  −   5 6
  ───────
      8 4
```

**63.**
```
      8 10
    6 9̸ 0̸
  − 2 3 6
  ───────
    4 5 4
```

**65.**
```
      8 10
    9̸ 0̸ 3
  − 1 3 2
  ───────
    7 7 1
```

**67.**
```
      2 9 10
    2̶-3̶-0̶ 0̸
  −    1 0 9
  ───────────
    2 1 9 1
```
We have 3 hundreds or 30 tens. We borrow 1 ten to get 10 ones. We then have 29 tens. Subtract ones, then tens, then hundreds, then thousands.

**69.**
$$\begin{array}{r} \overset{7\ \ 9\ \ 18}{6\,\cancel{8}\,\cancel{0}\,\cancel{8}} \\ -\ 3\ 0\ 5\ 9 \\ \hline 3\ 7\ 4\ 9 \end{array}$$
We have 8 hundreds or 80 tens. We borrow 1 ten to get 18 ones. We then have 79 tens. Subtract ones, then tens, then hundreds, then thousands.

**71.**
$$\begin{array}{r} \overset{8\ \ 12}{8\,0\,\cancel{9}\,\cancel{2}} \\ -\ 1\ 0\ 7\ 3 \\ \hline 7\ 0\ 1\ 9 \end{array}$$
Borrow 1 ten to get 12 ones. Subtract ones, then tens, then hundreds, then thousands.

**73.**
$$\begin{array}{r} \overset{\ \ 13}{\ \ 7\ \overset{3}{\cancel{8}}\ \cancel{\overset{13}{3}}} \\ 5\,\cancel{8}\,\cancel{4}\,\cancel{3} \\ -\ \ \ \ 9\ 8 \\ \hline 5\ 7\ 4\ 5 \end{array}$$

**75.**
$$\begin{array}{r} \overset{10\ \ 16}{\ \ \ \ 9\ \cancel{0}\ \cancel{6}\ 13} \\ \cancel{1\,0}\,1,\,7\,\cancel{3}\,4 \\ -\ \ \ \ \ \ 5\ 7\ 6\ 0 \\ \hline 9\ 5,\ 9\ 7\ 4 \end{array}$$

**77.**
$$\begin{array}{r} \overset{9\ \ 9\ \ 9\ \ 18}{\cancel{1\,0,\,0\,0}\,\cancel{8}} \\ -\ \ \ \ \ \ \ \ 1\ 9 \\ \hline 9\ \ 9\ 8\ 9 \end{array}$$
We have 1 ten thousand, or 1000 tens. We borrow 1 ten to get 10 ones. We then have 999 tens. Subtract ones, then tens, then hundreds, then thousands.

**79.**
$$\begin{array}{r} \overset{8\ \ 9\ \ 17}{8\,3,\,\cancel{9}\,\cancel{0}\,\cancel{7}} \\ -\ \ \ \ \ \ 8\ 9 \\ \hline 8\ 3,\ 8\ 1\ 8 \end{array}$$

**81.**
$$\begin{array}{r} \overset{6\ \ 9\ \ 9\ \ 10}{\cancel{7\,0\,0\,0}} \\ -\ 2\ 7\ 9\ 4 \\ \hline 4\ 2\ 0\ 6 \end{array}$$
We have 7 thousands or 700 tens. We borrow 1 ten to get 10 ones. We then have 699 tens. Subtract ones, then tens, then hundreds, then thousands.

**83.**
$$\begin{array}{r} \overset{7\ \ 9\ \ 9\ \ 10}{4\,\cancel{8,\,0\,0\,0}} \\ -\ 3\ 7,\ 6\ 9\ 5 \\ \hline 1\ 0,\ 3\ 0\ 5 \end{array}$$
We have 8 thousands or 800 tens. We borrow 1 ten to get 10 ones. We then have 799 tens. Subtract ones, then tens, then hundreds, then thousands, then ten thousands.

**85.** 6, 3 $\boxed{7}$ 5, 6 0 2

The digit 7 means 7 ten thousands.

**87.** 2 ten thousands + 9 thousands + 7 hundreds + 8 ones = 29,708

**89.** ◈

**91.** $3,928,124 - 1,098,947$

Using a calculator to carry out the subtraction, we find that the difference is 2,829,177.

**93.**
$$\begin{array}{r} 9,\ \_\ 4\ 8,\ 6\ 2\ 1 \\ -\ 2,\ 0\ 9\ 7,\ \_\ 8\ 1 \\ \hline 7,\ 2\ 5\ 1,\ 1\ 4\ 0 \end{array}$$

---

To subtract tens, we borrow 1 hundred to get 12 tens.
$$\begin{array}{r} \overset{\ \ \ \ \ \ \ \ 5\ \ 12}{9,\ \_\ 4\ 8,\ \cancel{6}\ \cancel{2}\ 1} \\ -\ 2,\ 0\ 9\ 7,\ \_\ 8\ 1 \\ \hline 7,\ 2\ 5\ 1,\ 1\ 4\ 0 \end{array}$$

In order to have 1 hundred in the difference, the missing digit in the subtrahend must be 4 $(5 - 4 = 1)$.
$$\begin{array}{r} \overset{\ \ \ \ \ \ \ \ 5\ \ 12}{9,\ \_\ 4\ 8,\ \cancel{6}\ \cancel{2}\ 1} \\ -\ 2,\ 0\ 9\ 7,\ 4\ 8\ 1 \\ \hline 7,\ 2\ 5\ 1,\ 1\ 4\ 0 \end{array}$$

In order to subtract ten thousands, we must borrow 1 hundred thousand to get 14 ten thousands. The number of hundred thousands left must be 2 since the hundred thousands place in the difference is 2 $(2 - 0 = 2)$. Thus, the missing digit in the minuend must be $2 + 1$, or 3.
$$\begin{array}{r} \overset{\ \ \ \ 2\ \ 14\ \ \ \ \ 5\ \ 12}{9,\ \cancel{3}\ \cancel{4}\ 8,\ \cancel{6}\ \cancel{2}\ 1} \\ -\ 2,\ 0\ 9\ 7,\ 4\ 8\ 1 \\ \hline 7,\ 2\ 5\ 1,\ 1\ 4\ 0 \end{array}$$

---

## Exercise Set 1.4

**1.** Round 48 to the nearest ten.

4 $\boxed{8}$
↑

The digit 4 is in the tens place. Consider the next digit to the right. Since the digit, 8, is 5 or higher, round 4 tens up to 5 tens. Then change the digit to the right of the tens digit to zero.

The answer is 50.

**3.** Round 67 to the nearest ten.

6 $\boxed{7}$
↑

The digit 6 is in the tens place. Consider the next digit to the right. Since the digit, 7, is 5 or higher, round 6 tens up to 7 tens. Then change the digit to the right of the tens digit to zero.

The answer is 70.

**5.** Round 731 to the nearest ten.

7 3 $\boxed{1}$
↑

The digit 3 is in the tens place. Consider the next digit to the right. Since the digit, 1, is 4 or lower, round down, meaning that 3 tens stays as 3 tens. Then change the digit to the right of the tens digit to zero.

The answer is 730.

**7.** Round 895 to the nearest ten.

8 9 $\boxed{5}$
↑

The digit 9 is in the tens place. Consider the next digit to the right. Since the digit, 5, is 5 or higher, we round up. The 89 tens become 90 tens. Then change the digit to the right of the tens digit to zero.

The answer is 900.

**9.** Round 146 to the nearest hundred.

1 4̄ 6
↑

The digit 1 is in the hundreds place. Consider the next digit to the right. Since the digit, 4, is 4 or lower, round down, meaning that 1 hundred stays as 1 hundred. Then change all digits to the right of the hundreds digit to zeros.

The answer is 100.

**11.** Round 957 to the nearest hundred.

9 5̄ 7
↑

The digit 9 is in the hundreds place. Consider the next digit to the right. Since the digit, 5, is 5 or higher, round up. The 9 hundreds become 10 hundreds. Then change all digits to the right of the hundreds digit to zeros.

The answer is 1000.

**13.** Round 9079 to the nearest hundred.

9 0 7̄ 9
↑

The digit 0 is in the hundreds place. Consider the next digit to the right. Since the digit, 7, is 5 or higher, round 0 hundreds up to 1 hundred. Then change all digits to the right of the hundreds digit to zeros.

The answer is 9100.

**15.** Round 32,850 to the nearest hundred.

3 2, 8 5̄ 0
↑

The digit 8 is in the hundreds place. Consider the next digit to the right. Since the digit, 5, is 5 or higher, round 8 hundreds up to 9 hundreds. Then change all digits to the right of the hundreds digit to zero.

The answer is 32,900.

**17.** Round 5876 to the nearest thousand.

5 8̄ 7 6
↑

The digit 5 is in the thousands place. Consider the next digit to the right. Since the digit, 8, is 5 or higher, round 5 thousands up to 6 thousands. Then change all digits to the right of the thousands digit to zeros.

The answer is 6000.

**19.** Round 7500 to the nearest thousand.

7 5̄ 0 0
↑

The digit 7 is in the thousands place. Consider the next digit to the right. Since the digit, 5, is 5 or higher, round 7 thousands up to 8 thousands. Then change all the digits to the right of the thousands digit to zeros.

The answer is 8000.

**21.** Round 45,340 to the nearest thousand.

4 5, 3̄ 4 0
↑

The digit 5 is in the thousands place. Consider the next digit to the right. Since the digit, 3, is 4 or lower, round down, meaning that 5 thousands stays as 5 thousands. Then change all the digits to the right of the thousands digit to zeros.

The answer is 45,000.

**23.** Round 373,405 to the nearest thousand.

3 7 3, 4̄ 0 5
↑

The digit 3 is in the thousands place. Consider the next digit to the right. Since the digit, 4, is 4 or lower, round down, meaning that 3 thousands stays as 3 thousands. Then change all the digits to the right of the thousands digit to zeros.

The answer is 373,000.

**25.**

|        | Rounded to the nearest ten |
|--------|----------------------------|
| 7 8    | 8 0                        |
| + 9 7  | + 1 0 0                    |
|        | 1 8 0 ← Estimated answer   |

**27.**

|         | Rounded to the nearest ten |
|---------|----------------------------|
| 8 0 7 4 | 8 0 7 0                    |
| − 2 3 4 7 | − 2 3 5 0                |
|         | 5 7 2 0 ← Estimated answer |

**29.**

|        | Rounded to the nearest ten |
|--------|----------------------------|
| 4 5    | 5 0                        |
| 7 7    | 8 0                        |
| 2 5    | 3 0                        |
| + 5 6  | + 6 0                      |
| 3 4 3  | 2 2 0 ← Estimated answer   |

The sum 343 seems to be incorrect since 220 is not close to 343.

**31.**

|         | Rounded to the nearest ten |
|---------|----------------------------|
| 6 2 2   | 6 2 0                      |
| 7 8     | 8 0                        |
| 8 1     | 8 0                        |
| + 1 1 1 | + 1 1 0                    |
| 9 3 2   | 8 9 0 ← Estimated answer   |

The sum 932 seems to be incorrect since 890 is not close to 932.

**33.**

|          | Rounded to the nearest hundred |
|----------|--------------------------------|
| 7 3 4 8  | 7 3 0 0                        |
| + 9 2 4 7 | + 9 2 0 0                     |
|          | 1 6, 5 0 0 ← Estimated answer  |

**35.**

|  | Rounded to the nearest hundred |
|---|---|
| $\begin{array}{r} 6\,8\,5\,2 \\ -\,1\,7\,4\,8 \\ \hline \end{array}$ | $\begin{array}{r} 6\,9\,0\,0 \\ -\,1\,7\,0\,0 \\ \hline 5\,2\,0\,0 \end{array}$ ← Estimated answer |

**37.**

|  | Rounded to the nearest hundred |
|---|---|
| $\begin{array}{r} 2\,1\,6 \\ 8\,4 \\ 7\,4\,5 \\ +\,5\,9\,5 \\ \hline 1\,6\,4\,0 \end{array}$ | $\begin{array}{r} 2\,0\,0 \\ 1\,0\,0 \\ 7\,0\,0 \\ +\,6\,0\,0 \\ \hline 1\,6\,0\,0 \end{array}$ ← Estimated answer |

The sum 1640 seems to be correct since 1600 is close to 1640.

**39.**

|  | Rounded to the nearest hundred |
|---|---|
| $\begin{array}{r} 7\,5\,0 \\ 4\,2\,8 \\ 6\,3 \\ +\,2\,0\,5 \\ \hline 1\,4\,4\,6 \end{array}$ | $\begin{array}{r} 8\,0\,0 \\ 4\,0\,0 \\ 1\,0\,0 \\ +\,2\,0\,0 \\ \hline 1\,5\,0\,0 \end{array}$ ← Estimated answer |

The sum 1446 seems to be correct since 1500 is close to 1446.

**41.**

|  | Rounded to the nearest thousand |
|---|---|
| $\begin{array}{r} 9\,6\,4\,3 \\ 4\,8\,2\,1 \\ 8\,9\,4\,3 \\ +\,7\,0\,0\,4 \\ \hline \end{array}$ | $\begin{array}{r} 1\,0,0\,0\,0 \\ 5\,0\,0\,0 \\ 9\,0\,0\,0 \\ +\,7\,0\,0\,0 \\ \hline 3\,1,0\,0\,0 \end{array}$ ← Estimated answer |

**43.**

|  | Rounded to the nearest thousand |
|---|---|
| $\begin{array}{r} 9\,2,1\,4\,9 \\ -\,2\,2,5\,5\,5 \\ \hline \end{array}$ | $\begin{array}{r} 9\,2,0\,0\,0 \\ -\,2\,3,0\,0\,0 \\ \hline 6\,9,0\,0\,0 \end{array}$ ← Estimated answer |

**45.**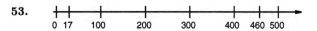

Since 0 is to the left of 17, we write $0 < 17$.

**47.**

Since 34 is to the right of 12, we write $34 > 12$.

**49.**

Since 1000 is to the left of 1001, we write $1000 < 1001$.

**51.**

Since 133 is to the right of 132, we write $133 > 132$.

**53.**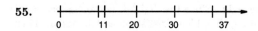

Since 460 is to the right of 17, we write $460 > 17$.

**55.**

Since 37 is to the right of 11, we write $37 > 11$.

**57.** $\begin{array}{r} \overset{1\ 1\ 1\ 1}{6\,7,7\,8\,9} \\ +\,1\,8,9\,6\,5 \\ \hline 8\,6,7\,5\,4 \end{array}$  Add ones. We get 14. Write 4 in the ones column and 1 above the tens. Add tens: We get 15 tens. Write 5 in the tens column and 1 above the hundreds. Add hundreds: We get 17 hundreds. Write 7 in the hundreds column and 1 above the thousands. Add thousands: We get 16 thousands. Write 6 in the thousands column and 1 above the ten thousands. Add ten thousands: We get 8 ten thousands.

**59.** $\begin{array}{r} \overset{\phantom{5}16}{\underset{\ }{5\ \not{6}\ 17}} \\ \not{6}\,\not{7},\not{7}\,8\,9 \\ -\,1\,8,9\,6\,5 \\ \hline 4\,8,8\,2\,4 \end{array}$  Subtract ones: We get 4. Subtract tens: We get 2. We cannot subtract 9 hundreds from 7 hundreds. We borrow 1 thousand to get 17 hundreds. Subtract hundreds. We cannot subtract 8 thousands from 6 thousands. We borrow 1 ten thousand to get 16 thousands. Subtract thousands, then ten thousands.

**61.**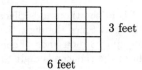

**63.** Using a calculator, we find that the sum is 30,411.

**65.** Using a calculator, we find that the difference is 69,594.

---

## Exercise Set 1.5

**1.** Think of a rectangular array consisting of 21 rows with 21 objects in each row.

$21 \cdot 21 = 441$

**3.** Repeated addition fits best in this case.

$$\underbrace{\boxed{12\ \text{oz}} + \boxed{12\ \text{oz}} + \boxed{12\ \text{oz}} + \cdots + \boxed{12\ \text{oz}}}_{8\ \text{addends}}$$

$8 \cdot 12\ \text{oz} = 96\ \text{oz}$

**5.** If we think of filling the rectangle with square feet, we have a rectangular array.

$A = l \times w = 6 \times 3 = 18$ square feet

**7.** $A = l \times w = 11\ \text{yd} \times 11\ \text{yd} = 121$ square yards

**9.** $A = l \times w = 48\ \text{mm} \times 3\ \text{mm} = 144$ square millimeters

**11.**
```
      8 7
  ×  1 0
    8 7 0
```
Multiplying by 1 ten (We write 0 and then multiply 87 by 1.)

**13.**
```
      2 3 4 0
  ×   1 0 0 0
  2, 3 4 0, 0 0 0
```
Multiplying by 1 thousand (We write 000 and then multiply 2340 by 1.)

**15.**
```
      4
      6 5
  ×     8
    5 2 0
```
Multiplying by 8

**17.**
```
      2
      9 4
  ×     6
    5 6 4
```
Multiplying by 6

**19.**
```
      6 5 2
  ×   1 0 0
  6 5, 2 0 0
```
Multiplying by 1 hundred (We write 00 and then multiply 652 by 1.)

**21.**
```
        4 3 7 1
  ×     1 0 0 0
  4, 3 7 1, 0 0 0
```
Multiplying by 1 thousand (We write 000 and then multiply 4371 by 1.)

**23.**
```
      2
      5 0 9
  ×       3
    1 5 2 7
```
Multiplying by 3

**25.**
```
    1 2 6
    9 2 2 9
  ×       7
  6 4, 6 0 3
```
Multiplying by 7

**27.**
```
      5 3
  ×   9 0
    4 7 7 0
```
Multiplying by 9 tens (We write 0 and then multiply 53 by 9.)

**29.**
```
      2
      3
      8 5
  ×   4 7
    5 9 5      Multiplying 85 by 7
  3 4 0 0      Multiplying 85 by 40
  3 9 9 5      Adding
```

**31.**
```
      2
      6 4 0
  ×     7 2
    1 2 8 0      Multiplying 640 by 2
  4 4 8 0 0      Multiplying 640 by 70
  4 6, 0 8 0      Adding
```

**33.**
```
    1 1
    1 1
    4 4 4
  ×     3 3
    1 3 3 2        Multiplying 444 by 3
  1 3 3 2 0        Multiplying 444 by 30
  1 4, 6 5 2        Adding
```

**35.**
```
        3
        7
      5 0 9
  ×   4 0 8
    4 0 7 2        Multiplying 509 by 8
  2 0 3 6 0 0      Multiplying 509 by 4 hundreds (We
  2 0 7, 6 7 2     write 00 and then multiply 509 by 4.)
```

**37.**
```
      4 2
      1
      3 1
      8 5 3
  ×   9 3 6
    5 1 1 8        Multiplying 853 by 6
  2 5 5 9 0        Multiplying 853 by 30
  7 6 7 7 0 0      Multiplying 853 by 900
  7 9 8, 4 0 8     Adding
```

**39.**
```
        2 2
        3 3
        4 8 9
  ×     3 4 0        Multiplying 489 by 4 tens (We write 0
    1 9 5 6 0  ←     and then multiply 489 by 4.)
  1 4 6 7 0 0  ←     Multiplying 489 by 3 hundreds (We
  1 6 6, 2 6 0       write 00 and then multiply 489 by 3.)
```

**41.**
```
          1 1
          2 4 4
          3 7 7
          1 3 3
          4 3 7 8
  ×       2 6 9 4
      1 7 5 1 2        Multiplying 4378 by 4
      3 9 4 0 2 0      Multiplying 4378 by 90
    2 6 2 6 8 0 0      Multiplying 4378 by 600
    8 7 5 6 0 0 0      Multiplying 4378 by 2000
  1 1, 7 9 4, 3 3 2    Adding
```

**43.**
```
          1   2
              1
              1
          1 1 3
          6 4 2 8
  ×       3 2 2 4
      2 5 7 1 2        Multiplying 6428 by 4
    1 2 8 5 6 0        Multiplying 6428 by 20
    1 2 8 5 6 0 0      Multiplying 6428 by 200
  1 9 2 8 4 0 0 0      Multiplying 6428 by 3000
  2 0, 7 2 3, 8 7 2    Adding
```

**45.**
```
        1 3
        3 4 8 2
  ×     1 0 4
    1 3 9 2 8        Multiplying 3482 by 4
  3 4 8 2 0 0        Multiplying 3482 by 1 hundred (We
  3 6 2, 1 2 8       write 00 and then multiply 3482 by 1.)
```

**47.**

$$\begin{array}{r} \overset{2}{\phantom{0}} \\ \overset{4}{\phantom{0}} \\ 5\,0\,0\,6 \\ \times\ 4\,0\,0\,8 \\ \hline 4\,0\,0\,4\,8 \\ 2\,0\,0\,2\,4\,0\,0\,0 \\ \hline 2\,0,0\,6\,4,0\,4\,8 \end{array}$$

Multiplying 5006 by 8
Multiplying 5006 by 4 thousands
(We write 000 and then multiply 5006 by 4.)

**49.**

$$\begin{array}{r} \overset{2}{\phantom{0}}\ \overset{3}{\phantom{0}} \\ \overset{3}{\phantom{0}}\ \overset{4}{\phantom{0}} \\ 5\,6\,0\,8 \\ \times\ 4\,5\,0\,0 \\ \hline 2\,8\,0\,4\,0\,0\,0 \\ 2\,2\,4\,3\,2\,0\,0\,0 \\ \hline 2\,5,2\,3\,6,0\,0\,0 \end{array}$$

Multiplying 5608 by 5 hundreds  (We write 00 and then multiply 5608 by 5.)
Multiplying 5608 by 4000
Adding

**51.**

$$\begin{array}{r} \overset{2}{\phantom{0}}\ \overset{1}{\phantom{0}} \\ \overset{3}{\phantom{0}}\ \overset{2}{\phantom{0}} \\ \overset{3}{\phantom{0}}\ \overset{3}{\phantom{0}} \\ 8\,7\,6 \\ \times\ 3\,4\,5 \\ \hline 4\,3\,8\,0 \\ 3\,5\,0\,4\,0 \\ 2\,6\,2\,8\,0\,0 \\ \hline 3\,0\,2,2\,2\,0 \end{array}$$

Multiplying 876 by 5
Multiplying 876 by 40
Multiplying 876 by 300
Adding

**53.**

$$\begin{array}{r} \overset{5}{\phantom{0}}\ \overset{5}{\phantom{0}}\ \overset{5}{\phantom{0}} \\ \overset{1}{\phantom{0}}\ \overset{1}{\phantom{0}}\ \overset{1}{\phantom{0}} \\ \overset{1}{\phantom{0}}\ \overset{1}{\phantom{0}}\ \overset{1}{\phantom{0}} \\ \overset{3}{\phantom{0}}\ \overset{3}{\phantom{0}}\ \overset{3}{\phantom{0}} \\ 7\,8\,8\,9 \\ \times\ 6\,2\,2\,4 \\ \hline 3\,1\,5\,5\,6 \\ 1\,5\,7\,7\,8\,0 \\ 1\,5\,7\,7\,8\,0\,0 \\ 4\,7\,3\,3\,4\,0\,0\,0 \\ \hline 4\,9,1\,0\,1,1\,3\,6 \end{array}$$

Multiplying 7889 by 4
Multiplying 7889 by 20
Multiplying 7889 by 200
Multiplying 7889 by 6000
Adding

**55.**

$$\begin{array}{r} \overset{2}{\phantom{0}}\ \overset{2}{\phantom{0}} \\ \overset{2}{\phantom{0}}\ \overset{2}{\phantom{0}} \\ 5\,5\,5 \\ \times\ \ 5\,5 \\ \hline 2\,7\,7\,5 \\ 2\,7\,7\,5\,0 \\ \hline 3\,0,5\,2\,5 \end{array}$$

Multiplying 555 by 5
Multiplying 555 by 50
Adding

**57.**

$$\begin{array}{r} \overset{1}{\phantom{0}}\ \overset{1}{\phantom{0}} \\ \overset{2}{\phantom{0}}\ \overset{2}{\phantom{0}} \\ 7\,3\,4 \\ \times\ 4\,0\,7 \\ \hline 5\,1\,3\,8 \\ 2\,9\,3\,6\,0\,0 \\ \hline 2\,9\,8,7\,3\,8 \end{array}$$

Multiplying 734 by 7
Multiplying 734 by 4 hundreds  (We write 00 and then multiply 734 by 4.)

**59.**

| | Rounded to the nearest ten |
|---|---|
| $\begin{array}{r}4\,5\\ \times\ 6\,7\\ \hline\end{array}$ | $\begin{array}{r}5\,0\\ \times\ 7\,0\\ \hline 3\,5\,0\,0\end{array}$ ← Estimated answer |

**61.**

| | Rounded to the nearest ten |
|---|---|
| $\begin{array}{r}3\,4\\ \times\ 2\,9\\ \hline\end{array}$ | $\begin{array}{r}3\,0\\ \times\ 3\,0\\ \hline 9\,0\,0\end{array}$ ← Estimated answer |

**63.**

| | Rounded to the nearest hundred |
|---|---|
| $\begin{array}{r}8\,7\,6\\ \times\ 3\,4\,5\\ \hline\end{array}$ | $\begin{array}{r}9\,0\,0\\ \times\ 3\,0\,0\\ \hline 2\,7\,0,0\,0\,0\end{array}$ ← Estimated answer |

**65.**

| | Rounded to the nearest hundred |
|---|---|
| $\begin{array}{r}4\,3\,2\\ \times\ 1\,9\,9\\ \hline\end{array}$ | $\begin{array}{r}4\,0\,0\\ \times\ 2\,0\,0\\ \hline 8\,0,0\,0\,0\end{array}$ ← Estimated answer |

**67.**

| | Rounded to the nearest thousand |
|---|---|
| $\begin{array}{r}5\,6\,0\,8\\ \times\ 4\,5\,7\,6\\ \hline\end{array}$ | $\begin{array}{r}6\,0\,0\,0\\ \times\ 5\,0\,0\,0\\ \hline 3\,0,0\,0\,0,0\,0\,0\end{array}$ ← Estimated answer |

**69.**

| | Rounded to the nearest thousand |
|---|---|
| $\begin{array}{r}7\,8\,8\,8\\ \times\ 6\,2\,2\,4\\ \hline\end{array}$ | $\begin{array}{r}8\,0\,0\,0\\ \times\ 6\,0\,0\,0\\ \hline 4\,8,0\,0\,0,0\,0\,0\end{array}$ ← Estimated answer |

**71.**

$$\begin{array}{r} \overset{1}{\phantom{000}} \\ 2\,0 \\ 8\,5\,0 \\ +\ 3\,5\,0\,0 \\ \hline 4\,3\,7\,0 \end{array}$$

Add ones: We get 0. Add tens: We get 7 tens. Add hundreds: We get 13 hundreds. Write 3 in the hundreds column and 1 above the thousands. Add thousands: We get 4 thousands.

**73.** Round 2 3 4 $\boxed{5}$ to the nearest ten.

The digit 4 is in the tens place. Since the next digit to the right, 5, is 5 or higher, round 4 tens up to 5 tens. Then change the digit to the right of the tens digit to zero.

The answer is 2350.

Round 2 3 $\boxed{4}$ 5 to the nearest hundred.

The digit 3 is in the hundreds place. Since the next digit to the right, 4, is 4 or lower, round down, meaning 3 hundreds stays as 3 hundreds. Then change all digits to the right of the hundreds digit to zeros.

The answer is 2300.

Round 2 $\boxed{3}$ 4 5 to the nearest thousand.

The digit 2 is in the thousands place. Since the next digit to the right, 3, is 4 or lower, round down, meaning 2 thousands stays as 2 thousands. Then change all digits to the right of the thousands digit to zeros.

The answer is 2000.

**75.**

## Exercise Set 1.6

**1.** Think of an array with 4 rows. The number of pounds in each row will go to a mule.

How many in each row?

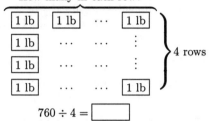

$$760 \div 4 = \boxed{\phantom{000}}$$

**3.** Think of an array with 5 rows. The number of mL in each row will go in a beaker.

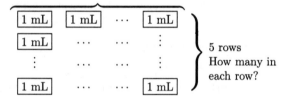

$$455 \div 5 = \boxed{\phantom{000}}$$

**5.** $18 \div 3 = 6$  The 3 moves to the right. A related multiplication sentence is $18 = 6 \cdot 3$. (By the commutative law of multiplication, there is also another multiplication sentence: $18 = 3 \cdot 6$.)

**7.** $22 \div 22 = 1$  The 22 on the right of the $\div$ symbol moves to the right. A related multiplication sentence is $22 = 1 \cdot 22$. (By the commutative law of multiplication, there is also another multiplication sentence: $22 = 22 \cdot 1$.)

**9.** $54 \div 6 = 9$  The 6 moves to the right. A related multiplication sentence is $54 = 9 \cdot 6$. (By the commutative law of multiplication, there is also another multiplication sentence: $54 = 6 \cdot 9$.)

**11.** $37 \div 1 = 37$  The 1 moves to the right. A related multiplication sentence is $37 = 37 \cdot 1$. (By the commutative law of multiplication, there is also another multiplication sentence: $37 = 1 \cdot 37$.)

**13.** $9 \times 5 = 45$

Move a factor to the other side and then write a division.

$$9 \times 5 = 45 \qquad\qquad 9 \times 5 = 45$$

$$9 = 45 \div 5 \qquad\qquad 5 = 45 \div 9$$

**15.** Two related division sentences for $37 \cdot 1 = 37$ are:

$$37 = 37 \div 1 \qquad ( \; 37 \cdot 1 = 37 \; )$$

and

$$1 = 37 \div 37 \qquad ( \; 37 \cdot 1 = 37 \; )$$

**17.** $8 \times 8 = 64$

Since the factors are both 8, moving either one to the other side gives the related division sentence $8 = 64 \div 8$.

**19.** Two related division sentences for $11 \cdot 6 = 66$ are:

$$11 = 66 \div 6 \qquad ( \; 11 \cdot 6 = 66 \; )$$

and

$$6 = 66 \div 11 \qquad ( \; 11 \cdot 6 = 66 \; )$$

**21.**
$$\begin{array}{r} 5\,5 \\ 5\overline{)2\,7\,7} \\ 2\,5\,0 \\ \hline 2\,7 \\ 2\,5 \\ \hline 2 \end{array}$$
Think: 2 hundreds $\div$ 5. There are no hundreds in the quotient.
Think: 27 tens $\div$ 5. Estimate 5 tens.
Think: 27 ones $\div$ 5. Estimate 5 ones.

The answer is 55 R 2.

**23.**
$$\begin{array}{r} 1\,0\,8 \\ 8\overline{)8\,6\,4} \\ 8\,0\,0 \\ \hline 6\,4 \\ 6\,4 \\ \hline 0 \end{array}$$
Think: 8 hundreds $\div$ 8. Estimate 1 hundred.
Think: 6 tens $\div$ 8. There are no tens in the quotient (other than the tens in 100). Write a 0 to show this.
Think: 64 ones $\div$ 8. Estimate 8 ones.

The answer is 108.

**25.**
$$\begin{array}{r} 3\,0\,7 \\ 4\overline{)1\,2\,2\,8} \\ 1\,2\,0\,0 \\ \hline 2\,8 \\ 2\,8 \\ \hline 0 \end{array}$$
Think: 12 hundreds $\div$ 4. Estimate 3 hundreds.
Think: 2 tens $\div$ 4. There are no tens in the quotient (other than the tens in 300). Write a 0 to show this.
Think: 28 ones $\div$ 4. Estimate 7 ones.

The answer is 307.

**27.**
$$\begin{array}{r} 7\,5\,3 \\ 6\overline{)4\,5\,2\,1} \\ 4\,2\,0\,0 \\ \hline 3\,2\,1 \\ 3\,0\,0 \\ \hline 2\,1 \\ 1\,8 \\ \hline 3 \end{array}$$
Think: 45 hundreds $\div$ 6. Estimate 7 hundreds.
Think: 32 tens $\div$ 6. Estimate 5 tens.
Think: 21 ones $\div$ 6. Estimate 3 ones.

The answer is 753 R 3.

**29.**
$$\begin{array}{r} 7\,4 \\ 4\overline{)2\,9\,7} \\ 2\,8\,0 \\ \hline 1\,7 \\ 1\,6 \\ \hline 1 \end{array}$$
Think: 29 tens $\div$ 4. Estimate 7 tens.
Think: 17 ones $\div$ 4. Estimate 4 ones.

The answer is 74 R 1.

**31.**
$$\begin{array}{r} 9\,2 \\ 8\overline{)7\,3\,8} \\ 7\,2\,0 \\ \hline 1\,8 \\ 1\,6 \\ \hline 2 \end{array}$$
Think: 73 tens $\div$ 8. Estimate 9 tens.
Think: 18 ones $\div$ 8. Estimate 2 ones.

The answer is 92 R 2.

**33.**
```
      1703
  5 ) 8515
    5000
    ─────
    3515
    3500
    ─────
      15
      15
    ─────
       0
```
Think: 8 thousands ÷ 5. Estimate 1 thousand.

Think: 35 hundreds ÷ 5. Estimate 7 hundreds.

Think: 1 ten ÷ 5. There are no tens in the quotient (other than the tens in 1700). Write a 0 to show this.

Think: 15 ones ÷ 5. Estimate 3 ones.

The answer is 1703.

**35.**
```
       987
  9 ) 8888
    8100
    ─────
     788
     720
    ─────
      68
      63
    ─────
       5
```
Think: 88 hundreds ÷ 9. Estimate 9 hundreds.

Think: 78 tens ÷ 9. Estimate 8 tens.

Think: 68 ones ÷ 9. Estimate 7 ones.

The answer is 987 R 5.

**37.**
```
        12,700
  10 ) 127,000
     100,000
     ───────
      27,000
      20,000
     ───────
       7000
       7000
     ───────
          0
```
Think: 12 ten thousands ÷ 10. Estimate 1 ten thousand.

Think: 27 thousands ÷ 10. Estimate 2 thousands.

Think: 70 hundreds ÷ 10. Estimate 7 hundreds.

Since the difference is 0, there are no tens or ones in the quotient (other than the tens and ones in 12,700). We write zeros to show this.

The answer is 12,700.

**39.**
```
           127
  1000 ) 127,000
       100,000
       ───────
        27,000
        20,000
       ───────
         7000
         7000
       ───────
            0
```
Think: 1270 hundreds ÷ 1000. Estimate 1 hundred.

Think: 2700 tens ÷ 1000. Estimate 2 tens.

Think: 7000 ones ÷ 1000. Estimate 7 ones.

The answer is 127.

**41.**
```
        52
  70 ) 3692
     3500
     ────
      192
      140
     ────
       52
```
Think: 369 tens ÷ 70. Estimate 5 tens.

Think: 192 ones ÷ 70. Estimate 2 ones.

The answer is 52 R 52.

**43.**
```
        29
  30 ) 875
     600
     ───
     275
     270
     ───
       5
```
Think: 87 tens ÷ 30. Estimate 2 tens.

Think: 275 ones ÷ 30. Estimate 9 ones.

The answer is 29 R 5.

**45.**
```
        40
  21 ) 852
     840
     ───
      12
```
Round 21 to 20.

Think: 85 tens ÷ 20. Estimate 4 tens.

Think: 12 ones ÷ 20. There are no ones in the quotient (other than the ones in 40). Write a 0 to show this.

The answer is 40 R 12.

**47.**
```
         8
  85 ) 7672
     6800
     ─────
      [87]2
```
Round 85 to 90.

Think: 767 tens ÷ 90. Estimate 8 tens.

Since 87 is larger than the divisor, the estimate is too low.

```
        90
  85 ) 7672
     7650
     ─────
       22
```
Think: 767 tens ÷ 90. Estimate 9 tens.

Think: 22 ones ÷ 90. There are no ones in the quotient (other than the ones in 90). Write a 0 to show this.

The answer is 90 R 22.

**49.**
```
          3
  111 ) 3219
      3330
      ────
```
Round 111 to 100.

Think: 321 tens ÷ 100. Estimate 3 tens.

Since we cannot subtract 3330 from 3219, the estimate is too high.

```
         29
  111 ) 3219
      2220
      ────
       999
       999
      ────
         0
```
Think: 321 tens ÷ 100. Estimate 2 tens.

Think: 999 ones ÷ 100. Estimate 9 ones.

The answer is 29.

**51.**
```
       105
  8 ) 843
    800
    ───
     43
     40
    ───
      3
```
Think: 8 hundreds ÷ 8. Estimate 1 hundred.

Think: 4 tens ÷ 8. There are no tens in the quotient (other than the tens in 100). Write a 0 to show this.

Think: 43 ones ÷ 8. Estimate 5 ones.

The answer is 105 R 3.

**53.**
```
      1609
  5 ) 8047
    5000
    ─────
    3047
    3000
    ─────
      47
      45
    ─────
       2
```
Think: 8 thousands ÷ 5. Estimate 1 thousand.

Think: 30 hundreds ÷ 5. Estimate 6 hundreds.

Think: 4 tens ÷ 5. There are no tens in the quotient (other than the tens in 1600). Write a 0 to show this.

Think: 47 ones ÷ 5. Estimate 9 ones.

The answer is 1609 R 2.

**55.**
```
    1 0 0 7
5 | 5 0 3 6
    5 0 0 0
    ─────
        3 6
        3 5
    ─────
          1
```
Think: 5 thousands ÷ 5. Estimate 1 thousand.

Think: 0 hundreds ÷ 5. There are no hundreds in the quotient (other than the hundreds in 1000). Write a 0 to show this.

Think: 3 tens ÷ 5. There are no tens in the quotient (other than the tens in 1000). Write a 0 to show this.

Think: 36 ones ÷ 5. Estimate 7 ones.

The answer is 1007 R 1.

**57.**
```
      2 2
4 6 | 1 0 5 8
      9 2 0
    ─────────
      1 3 8
        9 2
    ─────────
        4 6
```
Round 46 to 50.

Think: 105 tens ÷ 50. Estimate 2 tens.

Think: 138 ones ÷ 50. Estimate 2 ones.

Since 46 is not smaller than the divisor, 46, the estimate is too low.

```
      2 3
4 6 | 1 0 5 8
      9 2 0
    ─────────
      1 3 8
      1 3 8
    ─────────
          0
```
Think: 138 ones ÷ 50. Estimate 3 ones.

The answer is 23.

**59.**
```
      1 0 7
3 2 | 3 4 2 5
      3 2 0 0
    ─────────
        2 2 5
        2 2 4
    ─────────
            1
```
Round 32 to 30.

Think: 34 hundreds ÷ 30. Estimate 1 hundred.

Think: 22 tens ÷ 30. There are no tens in the quotient (other than the tens in 100). Write 0 to show this.

Think: 225 ones ÷ 30. Estimate 7 ones.

The answer is 107 R 1.

**61.**
```
        4
2 4 | 8 8 8 0
      9 6 0 0
    ─────────
```
Round 24 to 20.

Think: 88 hundreds ÷ 20. Estimate 4 hundreds.

Since we cannot subtract 9600 from 8880, the estimate is too high.

```
        3 8
2 4 | 8 8 8 0
      7 2 0 0
    ─────────
      1 6 8 0
      1 9 2 0
```
Think: 88 hundreds ÷ 20. Estimate 3 hundreds.

Think: 168 tens ÷ 20. Estimate 8 tens.

Since we cannot subtract 1920 from 1680, the estimate is too high.

```
        3 7 0
2 4 | 8 8 8 0
      7 2 0 0
    ─────────
      1 6 8 0
      1 6 8 0
    ─────────
            0
```
Think: 168 tens ÷ 20. Estimate 7 tens.

Think: 0 ones ÷ 20. There are no ones in the quotient (other than the ones in 370). Write a 0 to show this.

The answer is 370.

**63.**
```
              5
2 8 | 1 7, 0 6 7
      1 4 0 0 0
    ───────────
        3 0  6 7
```
Round 28 to 30. Think: 170 hundreds ÷ 30. Estimate 5 hundreds.

Since 30 is larger than the divisor, 28, the estimate is too low.

```
            6 0 8
2 8 | 1 7, 0 6 7
      1 6 8 0 0
    ───────────
          2 6 7
          2 2 4
    ───────────
          4 3
```
Think: 170 hundreds ÷ 30. Estimate 6 hundreds.

Think: 26 tens ÷ 30. There are no tens in the quotient (other than the tens in 600.) Write a zero to show this.

Think: 267 ones ÷ 30. Estimate 8 ones.

Since 43 is larger than the divisor, 28, the estimate is too low.

```
            6 0 9
2 8 | 1 7, 0 6 7
      1 6 8 0 0
    ───────────
          2 6 7
          2 5 2
    ───────────
            1 5
```
Think: 267 ones ÷ 30. Estimate 9 ones.

The answer is 609 R 15.

**65.**
```
            3 0 4
8 0 | 2 4, 3 2 0
      2 4 0 0 0
    ───────────
            3 2 0
            3 2 0
    ───────────
                0
```
Think: 243 hundreds ÷ 80. Estimate 3 hundreds.

Think: 32 tens ÷ 80. There are no tens in the quotient (other than the tens in 300). Write a 0 to show this.

Think: 320 ones ÷ 80. Estimate 4 ones.

The answer is 304.

**67.**
```
                3 5 0 8
2 8 5 | 9 9 9, 9 9 9
        8 5 5 0 0 0
      ─────────────
        1 4 4 9 9 9
        1 4 2 5 0 0
      ─────────────
            2 4 9 9
            2 2 8 0
      ─────────────
              2 1 9
```

The answer is 3508 R 219.

**69.**

$$
\begin{array}{r}
8\,0\,7\,0 \\
456\overline{\smash{\big)}3,6\,7\,9,9\,2\,0} \\
\underline{3\,6\,4\,8\,0\,0\,0} \\
3\,1\,9\,2\,0 \\
\underline{3\,1\,9\,2\,0} \\
0
\end{array}
$$

The answer is 8070.

**71.** $7882 = 7$ thousands $+ 8$ hundreds $+ 8$ tens $+ 2$ ones

**73.** $21 - 16 = 5$

This number gets added (after 5).

$$21 = 5 + 16$$

(By the commutative law of addition, $21 = 16 + 5$ is also correct.)

**75.** $47 + 9 = 56$

This addend gets subtracted from the sum.

$$47 = 56 - 9$$

$47 + 9 = 56$

This addend gets subtracted from the sum.

$$9 = 56 - 47$$

**77.** ◈

**79.** We divide 1231 by 42:

$$
\begin{array}{r}
2\,9 \\
42\overline{\smash{\big)}1\,2\,3\,1} \\
\underline{8\,4\,0} \\
3\,9\,1 \\
\underline{3\,7\,8} \\
1\,3
\end{array}
$$

The answer is 29 R 13. Since 13 students will be left after 29 buses are filled, then 30 buses are needed.

---

## Exercise Set 1.7

**1.** $x + 0 = 14$

We replace $x$ by different numbers until we get a true equation. If we replace $x$ by 14, we get a true equation: $14 + 0 = 14$. No other replacement makes the equation true, so the solution is 14.

**3.** $y \cdot 17 = 0$

We replace $y$ by different numbers until we get a true equation. If we replace $y$ by 0, we get a true equation: $0 \cdot 17 = 0$. No other replacement makes the equation true, so the solution is 0.

**5.**
$$
\begin{aligned}
13 + x &= 42 \\
13 + x - 13 &= 42 - 13 \quad \text{Subtracting 13 on both sides} \\
0 + x &= 29 \qquad\quad \text{13 plus } x \text{ minus 13 is } 0 + x. \\
x &= 29
\end{aligned}
$$
The solution is 29.

**7.**
$$
\begin{aligned}
12 &= 12 + m \\
12 - 12 &= 12 + m - 12 \quad \text{Subtracting 12 on both sides} \\
0 &= 0 + m \qquad\quad \text{12 plus } m \text{ minus 12 is } 0 + m. \\
0 &= m
\end{aligned}
$$
The solution is 0.

**9.** $3 \cdot x = 24$

$$
\begin{aligned}
\frac{3 \cdot x}{3} &= \frac{24}{3} \quad \text{Dividing by 3 on both sides} \\
x &= 8 \qquad \text{3 times } x \text{ divided by 3 is } x.
\end{aligned}
$$
The solution is 8.

**11.** $112 = n \cdot 8$

$$
\begin{aligned}
\frac{112}{8} &= \frac{n \cdot 8}{8} \quad \text{Dividing by 8 on both sides} \\
14 &= n
\end{aligned}
$$
The solution is 14.

**13.** $45 \times 23 = x$

To solve the equation we carry out the calculation.

$$
\begin{array}{r}
4\,5 \\
\times\,2\,3 \\
\hline
1\,3\,5 \\
9\,0\,0 \\
\hline
1\,0\,3\,5
\end{array}
$$
The solution is 1035.

**15.** $t = 125 \div 5$

To solve the equation we carry out the calculation.

$$
\begin{array}{r}
2\,5 \\
5\overline{\smash{\big)}1\,2\,5} \\
\underline{1\,0\,0} \\
2\,5 \\
\underline{2\,5} \\
0
\end{array}
$$
The solution is 25.

**17.** $p = 908 - 458$

To solve the equation we carry out the calculation.

$$
\begin{array}{r}
9\,0\,8 \\
-\,4\,5\,8 \\
\hline
4\,5\,0
\end{array}
$$
The solution is 450.

**19.** $x = 12{,}345 + 78{,}555$

To solve the equation we carry out the calculation.

$$
\begin{array}{r}
1\,2,3\,4\,5 \\
+\,7\,8,5\,5\,5 \\
\hline
9\,0,9\,0\,0
\end{array}
$$
The solution is 90,900.

**21.** $3 \cdot m = 96$

$$
\begin{aligned}
\frac{3 \cdot m}{3} &= \frac{96}{3} \quad \text{Dividing by 3 on both sides} \\
m &= 32
\end{aligned}
$$
The solution is 32.

**23.** $715 = 5 \cdot z$

$$
\begin{aligned}
\frac{715}{5} &= \frac{5 \cdot z}{5} \quad \text{Dividing by 5 on both sides} \\
143 &= z
\end{aligned}
$$
The solution is 143.

**25.**
$$10 + x = 89$$
$$10 + x - 10 = 89 - 10$$
$$x = 79$$
The solution is 79.

**27.**
$$61 = 16 + y$$
$$61 - 16 = 16 + y - 16$$
$$45 = y$$
The solution is 45.

**29.**
$$6 \cdot p = 1944$$
$$\frac{6 \cdot p}{6} = \frac{1944}{6}$$
$$p = 324$$
The solution is 324.

**31.**
$$5 \cdot x = 3715$$
$$\frac{5 \cdot x}{5} = \frac{3715}{5}$$
$$x = 743$$
The solution is 743.

**33.**
$$47 + n = 84$$
$$47 + n - 47 = 84 - 47$$
$$n = 37$$
The solution is 37.

**35.**
$$x + 78 = 144$$
$$x + 78 - 78 = 144 - 78$$
$$x = 66$$
The solution is 66.

**37.**
$$165 = 11 \cdot n$$
$$\frac{165}{11} = \frac{11 \cdot n}{11}$$
$$15 = n$$
The solution is 15.

**39.**
$$624 = t \cdot 13$$
$$\frac{624}{13} = \frac{t \cdot 13}{13}$$
$$48 = t$$
The solution is 48.

**41.**
$$x + 214 = 389$$
$$x + 214 - 214 = 389 - 214$$
$$x = 175$$
The solution is 175.

**43.**
$$567 + x = 902$$
$$567 + x - 567 = 902 - 567$$
$$x = 335$$
The solution is 335.

**45.**
$$18 \cdot x = 1872$$
$$\frac{18 \cdot x}{18} = \frac{1872}{18}$$
$$x = 104$$
The solution is 104.

**47.**
$$40 \cdot x = 1800$$
$$\frac{40 \cdot x}{40} = \frac{1800}{40}$$
$$x = 45$$
The solution is 45.

**49.**
$$2344 + y = 6400$$
$$2344 + y - 2344 = 6400 - 2344$$
$$y = 4056$$
The solution is 4056.

**51.**
$$8322 + 9281 = x$$
$$17,603 = x \quad \text{Doing the addition}$$
The solution is 17,603.

**53.**
$$234 \cdot 78 = y$$
$$18,252 = y \quad \text{Doing the multiplication}$$
The solution is 18,252.

**55.**
$$58 \cdot m = 11,890$$
$$\frac{58 \cdot m}{58} = \frac{11,890}{58}$$
$$m = 205$$
The solution is 205.

**57.** $7 + 8 = 15$       $7 + 8 = 15$
   ↑                ↑

This number gets      This number gets
subtracted from the    subtracted from the
sum.          ↓     sum.          ↓
      $7 = 15 - 8$         $8 = 15 - 7$

**59.** Since 123 is to the left of 789 on a number line, we write $123 < 789$.

**61.**

```
        1 4 2
    9 ⟌ 1 2 8 3      Think: 12 hundreds ÷ 9.  Estimate 1
        9 0 0        hundred.
        ─────
        3 8 3        Think: 38 tens ÷ 9.  Estimate 4 tens.
        3 6 0
        ─────
          2 3        Think: 23 ones ÷ 9.  Estimate 2 ones.
          1 8
        ─────
            5
```

The answer is 142 R 5.

**63.** ◈

**65.**
$$23,465 \cdot x = 8,142,355$$
$$\frac{23,465 \cdot x}{23,465} = \frac{8,142,355}{23,465}$$
$$x = 347 \quad \text{Using a calculator to divide}$$
The solution is 347.

## Exercise Set 1.8

**1. *Familiarize*.** We visualize the situation. Let $s$ = the total sales during the given period. We are combining amounts, so we use addition.

$$\boxed{\$3572} + \boxed{\$2718} + \boxed{\$2809} + \boxed{\$3177} = s$$

in      in      in      in    Total
January   February   March   April   sales

***Translate*.** We translate to an equation.

$$3572 + 2718 + 2809 + 3177 = s$$

***Solve*.** We carry out the addition.

$$\begin{array}{r} {\scriptstyle 2\ 1\ 2} \\ 3\ 5\ 7\ 2 \\ 2\ 7\ 1\ 8 \\ 2\ 8\ 0\ 9 \\ +\ 3\ 1\ 7\ 7 \\ \hline 1\ 2,2\ 7\ 6 \end{array}$$

Thus, $12,276 = s$, or $s = 12,276$.

***Check*.** We can repeat the calculation. We can also estimate by rounding, say to the nearest thousand.

$$3572 + 2718 + 2809 + 3177 \approx 4000 + 3000 + 3000 + 3000 =$$
$$13,000 \approx 12,276.$$

Since the estimated answer is close to the calculated answer, our result is probably correct.

***State*.** The total sales from January through April were $12,276.

**3. *Familiarize*.** We visualize the situation. Let $t$ = the total sales for 1993 and 1994, in millions of dollars. We are combining amounts, so we use addition.

$$\boxed{\$3534} + \boxed{\$3470} = t$$

in 1993    in 1994    Total sales

***Translate*.** We translate to an equation.

$$3534 + 3470 = t$$

***Solve*.** We carry out the addition.

$$\begin{array}{r} {\scriptstyle 1\ 1} \\ 3\ 5\ 3\ 4 \\ +\ 3\ 4\ 7\ 0 \\ \hline 7\ 0\ 0\ 4 \end{array}$$

Thus, $7004 = t$, or $t = 7004$.

***Check*.** We can repeat the calculation. We can also estimate by rounding, say to the nearest hundred.

$$3534 + 3470 \approx 3500 + 3500 = 7000 \approx 7004$$

Since the estimated answer is close to the calculated answer, our result is probably correct.

***State*.** Total sales for 1993 and 1994 were $7004 million, or $7,004,000,000.

**5. *Familiarize*.** We visualize the situation. Let $s$ = the amount by which sales in 1994 exceeded sales in 1993, in millions of dollars.

| 1994 sales | Excess |
|------------|--------|
| $3470 | $s$ |
| 1993 sales | |
| $3534 | |

***Translate*.** This is a "how much more" situation. We translate to an equation.

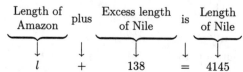

1994 sales   plus   How much more in sales   is   1993 sales

$$3470 \quad + \quad s \quad = \quad 3534$$

***Solve*.** We solve the equation

$$3470 + s = 3534$$
$$3470 + s - 3470 = 3534 - 3470 \quad \text{Subtracting 3470 on}$$
$$\qquad\qquad\qquad\qquad\qquad\qquad \text{both sides}$$
$$s = 64$$

***Check*.** We can check by adding the difference, 64, to the subtrahend, 3470: $3470 + 64 = 3534$. We get the original minuend, 3534, so our answer checks.

***State*.** Sales in 1993 were $64 million, or $64,000,000, more than sales in 1994.

**7. *Familiarize*.** We visualize the situation. Let $l$ = the length of the Amazon River, in miles.

| Length of Amazon | Excess length of Nile |
|------------------|------------------------|
| $l$ | 138 miles |
| Length of Nile | |
| 4145 miles | |

***Translate*.** This is a "how much more" situation. We translate to an equation.

Length of Amazon   plus   Excess length of Nile   is   Length of Nile

$$l \quad + \quad 138 \quad = \quad 4145$$

***Solve*.** We solve the equation

$$l + 138 = 4145$$
$$l + 138 - 138 = 4145 - 138 \quad \text{Subtracting 138 on}$$
$$\qquad\qquad\qquad\qquad\qquad \text{both sides}$$
$$l = 4007$$

***Check*.** We can check by adding the difference, 4007, to the subtrahend, 138: $138 + 4007 = 4145$. Our answer checks.

We could also have estimated: $138 + 4000 \approx 140 + 4000 = 4140 \approx 4145$. Since the estimate is close to the calculated answer, our result is probably correct.

***State*.** The length of the Amazon is 4007 miles.

**9. *Familiarize*.** We first make a drawing.

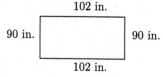

102 in.

90 in.             90 in.

102 in.

Let $P$ = the perimeter of the sheet, in inches.

*Translate*. The perimeter is the sum of the lengths of the sides of the sheet. We translate to an equation.

$$P = 102 + 90 + 102 + 90$$

*Solve*. We carry out the addition.

$$P = 102 + 90 + 102 + 90 = 384$$

*Check*. We can repeat the calculation. We can also estimate by rounding, say to the nearest hundred.

$$102 + 90 + 102 + 90 \approx 100 + 100 + 100 + 100 = 400 \approx 384$$

Our answer checks.

*State*. The perimeter of the sheet is 384 in.

**11.** *Familiarize*. The drawing in the text visualizes the situation. Let $p$ = the number of sheets in 9 reams of paper. Repeated addition works well here.

*Translate*. We translate to an equation.

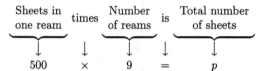

*Solve*. We carry out the multiplication.

$$\begin{array}{r} 5\,0\,0 \\ \times \quad 9 \\ \hline 4\,5\,0\,0 \end{array}$$

Thus, $4500 = p$, or $p = 4500$.

*Check*. We can repeat our calculation. The answer checks.

*State*. There are 4500 sheets in 9 reams of paper.

**13.** *Familiarize*. We visualize the situation. Let $n$ = the number by which the number of Elvis impersonators in 1995 exceeded the number in 1977.

| Number in 1977 | |
|:---:|:---:|
| 48 | $n$ |
| Number in 1995 | |
| 7328 | |

*Translate*. This is a "how much more" situation. We translate to an equation.

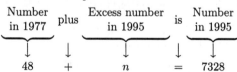

*Solve*. We solve the equation

$$48 + n = 7328$$
$$48 + n - 48 = 7328 - 48 \quad \text{Subtracting 48}$$
$$\qquad\qquad\qquad\qquad \text{on both sides}$$
$$n = 7280$$

*Check*. We can check by adding the difference, 7280, to the subtrahend, 48: $7280 + 48 = 7328$. Our answer checks.

We could also have estimated: $7328 - 48 \approx 7300 - 50 = 7250 \approx 7280$. Since the estimate is close to the calculated answer, our result is probably correct.

*State*. In 1995 there were 7280 more Elvis impersonators than in 1977.

**15.** *Familiarize*. We first draw a picture. We let $x$ = the amount of each payment.

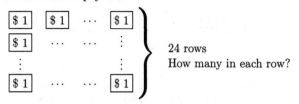

*Translate*. We translate to an equation.

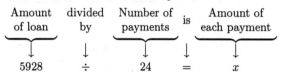

*Solve*. We carry out the division.

$$\begin{array}{r} 2\,4\,7 \\ 2\,4\,\overline{)\,5\,9\,2\,8} \\ 4\,8\,0\,0 \\ \hline 1\,1\,2\,8 \\ 9\,6\,0 \\ \hline 1\,6\,8 \\ 1\,6\,8 \\ \hline 0 \end{array}$$

Thus, $247 = x$, or $x = 247$.

*Check*. We can check by multiplying 247 by 24: $24 \cdot 247 = 5928$. The answer checks.

*State*. Each payment is $247.

**17.** *Familiarize*. We first draw a picture. Let $w$ = the number of full weeks the episodes can run.

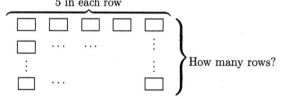

*Translate*. We translate to an equation.

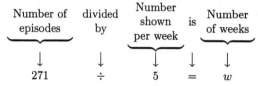

*Solve*. We carry out the division.

$$\begin{array}{r} 5\,4 \\ 5\,\overline{)\,2\,7\,1} \\ 2\,5\,0 \\ \hline 2\,1 \\ 2\,0 \\ \hline 1 \end{array}$$

*Check*. We can check by multiplying the number of weeks by 5 and adding the remainder, 1:

$$5 \cdot 54 = 270, \qquad 270 + 1 = 271$$

*State*. 54 full weeks will pass before the station must start over. There will be 1 episode left over.

**19. Familiarize.** We first draw a picture. Let $h$ = the number of hours in a week. Repeated addition works well here.

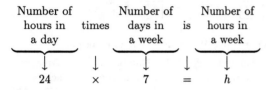

$$\underbrace{\boxed{24 \text{ hours}} + \boxed{24 \text{ hours}} + \cdots + \boxed{24 \text{ hours}}}_{7 \text{ addends}}$$

**Translate.** We translate to an equation.

$$\underbrace{\substack{\text{Number of} \\ \text{hours in} \\ \text{a day}}}_{24} \quad \substack{\text{times} \\ \downarrow \\ \times} \quad \underbrace{\substack{\text{Number of} \\ \text{days in} \\ \text{a week}}}_{7} \quad \substack{\text{is} \\ \downarrow \\ =} \quad \underbrace{\substack{\text{Number of} \\ \text{hours in} \\ \text{a week}}}_{h}$$

**Solve.** We carry out the multiplication.

$$\begin{array}{r} 2\,4 \\ \times\ \ 7 \\ \hline 1\,6\,8 \end{array}$$

Thus, $168 = h$, or $h = 168$.

**Check.** We can repeat the calculation. We an also estimate:

$$24 \times 7 \approx 20 \times 10 = 200 \approx 168$$

Our answer checks.

**State.** There are 168 hours in a week.

**21. Familiarize.** Let $a$ = the amount left in the account. We will start with \$568, subtract the amount of each check, and add the amount of the deposit.

**Translate.** We translate to an equation.

$$a = 568 - 46 - 87 - 129 + 94$$

**Solve.** We carry out the calculations.

$$a = 568 - 46 - 87 - 129 + 94 = 400$$

**Check.** We can repeat the calculations. We can also estimate:

$$568 - 46 - 87 - 129 + 94 \approx 570 - 50 - 90 - 130 + 100 = 400$$

Our answer checks.

**State.** \$400 is left in the account.

**23. Familiarize.** We first draw a picture. Let $A$ = the area and $P$ = the perimeter of the court, in feet.

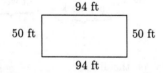

94 ft

50 ft            50 ft

94 ft

**Translate.** We write one equation to find the area and another to find the perimeter.

a) Using the formula for the area of a rectangle, we have

$$A = l \cdot w = 94 \cdot 50$$

b) Recall that the perimeter is the distance around the court.

$$P = 94 + 50 + 94 + 50$$

**Solve.** We carry out the calculations.

a)
$$\begin{array}{r} 5\,0 \\ \times\ 9\,4 \\ \hline 2\,0\,0 \\ 4\,5\,0\,0 \\ \hline 4\,7\,0\,0 \end{array}$$

Thus, $A = 4700$.

b) $P = 94 + 50 + 94 + 50 = 288$

**Check.** We can repeat the calculation. The answers check.

**State.** a) The area of the court is 4700 square feet.

b) The perimeter of the court is 288 ft.

**25. Familiarize.** We first draw a picture. We let $c$ = the number of cartons needed.

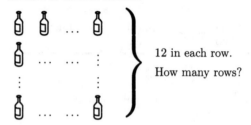

12 in each row.
How many rows?

**Translate.**

$$\underbrace{\substack{\text{Number} \\ \text{of bottles}}}_{528} \quad \substack{\text{divided} \\ \text{by} \\ \downarrow \\ \div} \quad \underbrace{\substack{\text{Number} \\ \text{per carton}}}_{12} \quad \substack{\text{is} \\ \downarrow \\ =} \quad \underbrace{\substack{\text{Number of} \\ \text{cartons.}}}_{c}$$

**Solve.** We carry out the division.

$$\begin{array}{r} 4\,4 \\ 12\overline{)5\,2\,8} \\ 4\,8\,0 \\ \hline 4\,8 \\ 4\,8 \\ \hline 0 \end{array}$$

Thus, $44 = c$, or $c = 44$.

**Check.** We can check by multiplying: $12 \cdot 44 = 528$. The answer checks.

**State.** It will take 44 cartons to ship 528 bottles of catsup.

**27. Familiarize.** We first draw a picture. Let $c$ = the number of cartons that can be filled.

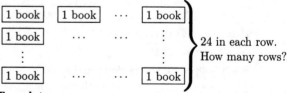

24 in each row.
How many rows?

**Translate.**

$$\underbrace{\substack{\text{Number} \\ \text{of books}}}_{1355} \quad \substack{\text{divided} \\ \text{by} \\ \downarrow \\ \div} \quad \underbrace{\substack{\text{Number} \\ \text{per carton}}}_{24} \quad \substack{\text{is} \\ \downarrow \\ =} \quad \underbrace{\substack{\text{Number of} \\ \text{full cartons.}}}_{c}$$

**Solve**. We carry out the division.

```
        5 6
  2 4 ⟌ 1 3 5 5
        1 2 0 0
        ───────
          1 5 5
          1 4 4
          ─────
            1 1
```

**Check**. We can check by multiplying the number of cartons by 24 and adding the remainder, 11:

$$24 \cdot 56 = 1344, \qquad 1344 + 11 = 1355$$

Our answer checks.

**State**. 56 cartons can be filled. There will be 11 books left over.

29. **Familiarize**. First we find the distance in reality between two cities that are 25 in. apart on the map. We make a drawing. Let $d$ = the distance between the cities, in miles. Repeated addition works well here.

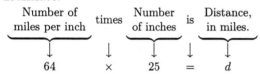

25 addends

**Translate**.

| Number of miles per inch | times | Number of inches | is | Distance, in miles. |
|:---:|:---:|:---:|:---:|:---:|
| ↓ | ↓ | ↓ | ↓ | ↓ |
| 64 | × | 25 | = | $d$ |

**Solve**. We carry out the multiplication.

```
      2 5
    × 6 4
    ─────
    1 0 0
  1 5 0 0
  ───────
  1 6 0 0
```

Thus, $1600 = d$, or $d = 1600$.

**Check**. We can repeat the calculation or estimate the product. Our answer checks.

**State**. Two cities that are 25 in. apart on the map are 1600 miles apart in reality.

Next we find distance on the map between two cities that, in reality, are 1728 mi apart.

**Familiarize**. We visualize the situation. Let $m$ = the distance between the cities on the map.

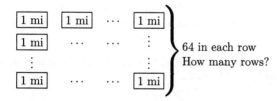

64 in each row
How many rows?

**Translate**.

| Number of miles | divided by | Number of miles per inch | is | Distance, in inches. |
|:---:|:---:|:---:|:---:|:---:|
| ↓ | ↓ | ↓ | ↓ | ↓ |
| 1728 | ÷ | 64 | = | $m$ |

**Solve**. We carry out the division.

```
        2 7
  6 4 ⟌ 1 7 2 8
        1 2 8 0
        ───────
          4 4 8
          4 4 8
          ─────
              0
```

Thus, $27 = m$, or $m = 27$.

**Check**. We can check by multiplying: $64 \cdot 27 = 1728$. Our answer checks.

**State**. The cities are 27 in. apart on the map.

31. **Familiarize**. We first make a drawing. Let $r$ = the number of rows.

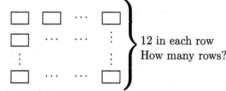

12 in each row
How many rows?

**Translate**.

| Number of holes | divided by | Number per row | is | Number of rows. |
|:---:|:---:|:---:|:---:|:---:|
| ↓ | ↓ | ↓ | ↓ | ↓ |
| 216 | ÷ | 12 | = | $r$ |

**Solve**. We carry out the division.

```
        1 8
  1 2 ⟌ 2 1 6
        1 2 0
        ─────
          9 6
          9 6
          ───
            0
```

Thus, $18 = r$, or $r = 18$.

**Check**. We can check by multiplying: $12 \cdot 18 = 216$. Our answer checks.

**State**. There are 18 rows.

33. **Familiarize**. This is a multistep problem.

We must find the total price of the 5 video games. Then we must find how many 10's there are in the total price. Let $p$ = the total price of the games.

To find the total price of the 5 video games we can use repeated addition.

$$\boxed{\$44} + \boxed{\$44} + \boxed{\$44} + \boxed{\$44} + \boxed{\$44}$$

5 addends

**Translate**.

| Price per game | times | Number of games | is | Total price of games |
|:---:|:---:|:---:|:---:|:---:|
| ↓ | ↓ | ↓ | ↓ | ↓ |
| 44 | · | 5 | = | $p$ |

**Solve**. First we carry out the multiplication.

$$44 \cdot 5 = p$$
$$220 = p$$

The total price of the 5 video games is $220. Repeated addition can be used again to find how many 10's there are in $220. We let $x$ = the number of $10 bills required.

| $220 | | | |
|---|---|---|---|
| $10 | $10 | $\cdots$ | $10 |

Translate to an equation and solve.

$$10 \cdot x = 220$$
$$\frac{10 \cdot x}{10} = \frac{220}{10}$$
$$x = 22$$

**Check**. We repeat the calculations. The answer checks.

**State**. It took 22 ten dollar bills.

**35. Familiarize**. This is a multistep problem.

We must find the total cost of the 4 shirts and the total cost of the 6 pairs of pants. The total cost of the clothing is the sum of these two totals.

Repeated addition works well in finding the total cost of the 4 shirts and the total cost of the 6 pairs of pants. We let $x$ = the total cost of the shirts and $y$ = the total cost of the pants.

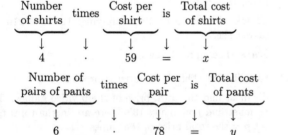

**Translate**. We translate to two equations.

| Number of shirts | times | Cost per shirt | is | Total cost of shirts |
|---|---|---|---|---|
| $\downarrow$ | $\downarrow$ | $\downarrow$ | $\downarrow$ | $\downarrow$ |
| 4 | $\cdot$ | 59 | = | $x$ |

| Number of pairs of pants | times | Cost per pair | is | Total cost of pants |
|---|---|---|---|---|
| $\downarrow$ | $\downarrow$ | $\downarrow$ | $\downarrow$ | $\downarrow$ |
| 6 | $\cdot$ | 78 | = | $y$ |

**Solve**. To solve these equations, we carry out the multiplications.

$$\begin{array}{r} 5\,9 \\ \times\quad 4 \\ \hline 2\,3\,6 \end{array} \quad \text{Thus, } x = \$236.$$

$$\begin{array}{r} 7\,8 \\ \times\quad 6 \\ \hline 4\,6\,8 \end{array} \quad \text{Thus } y = \$468.$$

We let $a$ = the total amount spent.

| Total cost of shirts | plus | Total cost of pants | is | Amount spent |
|---|---|---|---|---|
| $\downarrow$ | $\downarrow$ | $\downarrow$ | $\downarrow$ | $\downarrow$ |
| 236 | + | 468 | = | $a$ |

To solve the equation, carry out the addition.

$$\begin{array}{r} 2\,3\,6 \\ +\,4\,6\,8 \\ \hline 7\,0\,4 \end{array}$$

**Check**. We repeat the calculations. The answer checks.

**State**. The total cost of the clothing is $704.

**37. Familiarize**. This is a multistep problem.

We must find how many 100's there are in 3500. Then we must find that number times 15.

First we draw a picture

| One pound | | | |
|---|---|---|---|
| 3500 calories | | | |
| 100 cal | 100 cal | $\cdots$ | 100 cal |
| 15 min | 15 min | $\cdots$ | 15 min |

In Example 9 it was determined that there are 35 100's in 3500. We let $t$ = the time you have to do aerobic exercises to lose a pound.

**Translate**. We know that to do aerobic exercises for 15 min will burn 100 calories, so we need to do this 35 times to burn off one pound. We translate to an equation.

$$35 \times 15 = t$$

**Solve**. Carry out the multiplication.

$$35 \times 15 = t$$
$$525 = t$$

**Check**. Suppose you do aerobic exercises for 525 minutes. If we divide 525 by 15, we get 35, and 35 times 100 is 3500, the number of calories that must be burned off to lose one pound. The answer checks.

**State**. You must do aerobic exercises for 525 min, or 8 hr 45 min, to lose one pound.

**39. Familiarize**. This is a multistep problem. First we find the area of one side of one card. Then we double this to find the area of the front and back of one card. Finally, we find the area of the front and back sides of 100 cards.

**Translate**. We begin by using the formula for the area of a rectangle.

$$A = l \cdot w = 5 \cdot 3$$

**Solve**. First we carry out the multiplication.

$$A = 5 \cdot 3 = 15$$

This is the area of one side of one card. We multiply by 2 to find the area, $x$, of the front and back of one card:

$$x = 2 \cdot 15 = 30$$

Now let $t$ = the total writing area of the front and back sides of 100 cards.

$$t = 100 \cdot 30 = 3000$$

**Check**. We repeat the calculations. The answer checks.

**State**. The total writing area on the front and back sides of a package of 100 index cards is 3000 square inches.

**41.** Round 234,562 to the nearest hundred.

$$2\ 3\ 4,\ 5\ \boxed{6}\ 2$$
$$\uparrow$$

The digit 5 is in the hundreds place. Consider the next digit to the right. Since the digit, 6, is 5 or higher, round 5 hundreds up to 6 hundreds. Then change all digits to the right of the hundreds place to zeros.

The answer is 234,600.

**43.**

| | Rounded to the nearest thousand |
|---|---|
| 2 7 8 3 | 3 0 0 0 |
| 4 6 0 2 | 5 0 0 0 |
| 5 7 9 7 | 6 0 0 0 |
| + 8 1 1 1 | + 8 0 0 0 |
| | 2 2, 0 0 0 ← Estimated answer |

**45.**

| | Rounded to the nearest hundred |
|---|---|
| 7 8 7 | 8 0 0 |
| × 3 6 3 | × 4 0 0 |
| | 3 2 0, 0 0 0 ← Estimated answer |

**47.** ◈

**49.** *Familiarize*. This is a multistep problem. First we will find the differences in the distances traveled in 1 second. Then we will find the differences for 18 seconds. Let $d =$ the difference in the number of miles light would travel per second in a vacuum and in ice. Let $g =$ the difference in the number of miles light would travel per second in a vacuum and in glass.

*Translate*. Each is a "how much more" situation.

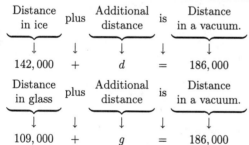

*Solve*. We begin by solving each equation.

$$142,000 + d = 186,000$$
$$142,000 + d - 142,000 = 186,000 - 142,000$$
$$d = 44,000$$

$$109,000 + g = 186,000$$
$$109,000 + g - 109,000 = 186,000 - 109,000$$
$$g = 77,000$$

Now to find the differences in the distances in 18 seconds, we multiply each solution by 18.

For ice: $18 \cdot 44,000 = 792,000$

For glass: $18 \cdot 77,000 = 1,386,000$

*Check*. We repeat the calculations. Our answers check.

*State*. In 18 seconds light travels 792,000 miles farther in ice and 1,386,000 miles farther in glass than in a vacuum.

---

## Exercise Set 1.9

**1.** Exponential notation for $3 \cdot 3 \cdot 3 \cdot 3$ is $3^4$.

**3.** Exponential notation for $5 \cdot 5$ is $5^2$.

**5.** Exponential notation for $7 \cdot 7 \cdot 7 \cdot 7 \cdot 7$ is $7^5$.

**7.** Exponential notation for $10 \cdot 10 \cdot 10$ is $10^3$.

**9.** $7^2 = 7 \cdot 7 = 49$

**11.** $9^3 = 9 \cdot 9 \cdot 9 = 729$

**13.** $12^4 = 12 \cdot 12 \cdot 12 \cdot 12 = 20,736$

**15.** $11^2 = 11 \cdot 11 = 121$

**17.** $12 + (6 + 4) = 12 + 10$    Doing the calculation inside the parentheses
$$= 22 \quad \text{Adding}$$

**19.** $52 - (40 - 8) = 52 - 32$    Doing the calculation inside the parentheses
$$= 20 \quad \text{Subtracting}$$

**21.** $1000 \div (100 \div 10)$
$$= 1000 \div 10 \quad \text{Doing the calculation inside the parentheses}$$
$$= 100 \quad \text{Dividing}$$

**23.** $(256 \div 64) \div 4 = 4 \div 4$    Doing the calculation inside the parentheses
$$= 1 \quad \text{Dividing}$$

**25.** $(2 + 5)^2 = 7^2$    Doing the calculation inside the parentheses
$$= 49 \quad \text{Evaluating the exponential expression}$$

**27.** $(11 - 8)^2 - (18 - 16)^2$
$$= 3^2 - 2^2 \quad \text{Doing the calculations inside the parentheses}$$
$$= 9 - 4 \quad \text{Evaluating the exponential expressions}$$
$$= 5 \quad \text{Subtracting}$$

**29.** $16 \cdot 24 + 50 = 384 + 50$    Doing all multiplications and divisions in order from left to right
$$= 434 \quad \text{Doing all additions and subtractions in order from left to right}$$

**31.** $83 - 7 \cdot 6 = 83 - 42$    Doing all multiplications and divisions in order from left to right
$$= 41 \quad \text{Doing all additions and subtractions in order from left to right}$$

**33.** $10 \cdot 10 - 3 \times 4$

$\qquad = 100 - 12$  Doing all multiplications and divisions in order from left to right

$\qquad = 88$  Doing all additions and subtractions in order from left to right

**35.** $4^3 \div 8 - 4$

$\qquad = 64 \div 8 - 4$  Evaluating the exponential expression

$\qquad = 8 - 4$  Doing all multiplications and divisions in order from left to right

$\qquad = 4$  Doing all additions and subtractions in order from left to right

**37.** $17 \cdot 20 - (17 + 20)$

$\qquad = 17 \cdot 20 - 37$  Carrying out the operation inside parentheses

$\qquad = 340 - 37$  Doing all multiplications and divisions in order from left to right

$\qquad = 303$  Doing all additions and subtractions in order from left to right

**39.** $6 \cdot 10 - 4 \cdot 10$

$\qquad = 60 - 40$  Doing all multiplications and divisions in order from left to right

$\qquad = 20$  Doing all additions and subtractions in order from left to right

**41.** $300 \div 5 + 10$

$\qquad = 60 + 10$  Doing all multiplications and divisions in order from left to right

$\qquad = 70$  Doing all additions and subtractions in order from left to right

**43.** $3 \cdot (2 + 8)^2 - 5 \cdot (4 - 3)^2$

$\qquad = 3 \cdot 10^2 - 5 \cdot 1^2$  Carrying out operations inside parentheses

$\qquad = 3 \cdot 100 - 5 \cdot 1$  Evaluating the exponential expressions

$\qquad = 300 - 5$  Doing all multiplications and divisions in order from left to right

$\qquad = 295$  Doing all additions and subtractions in order from left to right

**45.** $4^2 + 8^2 \div 2^2 = 16 + 64 \div 4$

$\qquad\qquad\qquad\quad = 16 + 16$

$\qquad\qquad\qquad\quad = 32$

**47.** $10^3 - 10 \cdot 6 - (4 + 5 \cdot 6) = 10^3 - 10 \cdot 6 - (4 + 30)$

$\qquad\qquad\qquad\qquad\qquad\quad = 10^3 - 10 \cdot 6 - 34$

$\qquad\qquad\qquad\qquad\qquad\quad = 1000 - 10 \cdot 6 - 34$

$\qquad\qquad\qquad\qquad\qquad\quad = 1000 - 60 - 34$

$\qquad\qquad\qquad\qquad\qquad\quad = 940 - 34$

$\qquad\qquad\qquad\qquad\qquad\quad = 906$

**49.** $6 \times 11 - (7 + 3) \div 5 - (6 - 4) = 6 \times 11 - 10 \div 5 - 2$

$\qquad\qquad\qquad\qquad\qquad\qquad\quad = 66 - 2 - 2$

$\qquad\qquad\qquad\qquad\qquad\qquad\quad = 64 - 2$

$\qquad\qquad\qquad\qquad\qquad\qquad\quad = 62$

**51.** $\quad 120 - 3^3 \cdot 4 \div (5 \cdot 6 - 6 \cdot 4)$

$\qquad = 120 - 3^3 \cdot 4 \div (30 - 24)$

$\qquad = 120 - 3^3 \cdot 4 \div 6$

$\qquad = 120 - 27 \cdot 4 \div 6$

$\qquad = 120 - 108 \div 6$

$\qquad = 120 - 18$

$\qquad = 102$

**53.** We add the numbers and then divide by the number of addends.

$$(\$64 + \$97 + \$121) \div 3 = \$282 \div 3$$
$$= \$94$$

**55.** $8 \times 13 + \{42 \div [18 - (6 + 5)]\}$

$\qquad = 8 \times 13 + \{42 \div [18 - 11]\}$

$\qquad = 8 \times 13 + \{42 \div 7\}$

$\qquad = 8 \times 13 + 6$

$\qquad = 104 + 6$

$\qquad = 110$

**57.** $[14 - (3 + 5) \div 2] - [18 \div (8 - 2)]$

$\qquad = [14 - 8 \div 2] - [18 \div 6]$

$\qquad = [14 - 4] - 3$

$\qquad = 10 - 3$

$\qquad = 7$

**59.** $(82 - 14) \times [(10 + 45 \div 5) - (6 \cdot 6 - 5 \cdot 5)]$

$\qquad = (82 - 14) \times [(10 + 9) - (36 - 25)]$

$\qquad = (82 - 14) \times [19 - 11]$

$\qquad = 68 \times 8$

$\qquad = 544$

**61.** $4 \times \{(200 - 50 \div 5) - [(35 \div 7) \cdot (35 \div 7) - 4 \times 3]\}$

$\qquad = 4 \times \{(200 - 10) - [5 \cdot 5 - 4 \times 3]\}$

$\qquad = 4 \times \{190 - [25 - 12]\}$

$\qquad = 4 \times \{190 - 13\}$

$\qquad = 4 \times 177$

$\qquad = 708$

**63.**
$$x + 341 = 793$$
$$x + 341 - 341 = 793 - 341$$
$$x = 452$$

The solution is 452.

**65.** *Familiarize.* We first make a drawing.

$$\boxed{\phantom{xxxxxxxxxxxx}}\ \text{270 mi}$$
$$\text{380 mi}$$

*Translate.* We use the formula for the area of a rectangle.

$$A = l \cdot w = 380 \cdot 270$$

*Solve.* We carry out the multiplication.

$$A = 380 \cdot 270 = 102,600$$

*Check.* We repeat the calculation. The answer checks.

*State.* The area is 102,600 square miles.

**67.**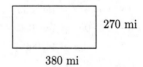

**69.** $15(23 - 4 \cdot 2)^3 \div (3 \cdot 25)$

$\quad = 15(23 - 8)^3 \div 75 \quad$ Multiplying inside parentheses

$\quad = 15 \cdot 15^3 \div 75 \quad$ Subtracting inside parentheses

$\quad = 15 \cdot 3375 \div 75 \quad$ Evaluating the exponential expression

$\quad = 50,625 \div 75 \quad$ Doing all multiplication and

$\quad = 675 \quad$ divisions in order from left to right

**71.** $1 + 5 \cdot 4 + 3 = 1 + 20 + 3$

$\qquad\qquad\quad = 24 \qquad$ Correct answer

To make the incorrect answer correct we add parentheses:

$\quad 1 + 5 \cdot (4 + 3) = 36$

**73.** $12 \div 4 + 2 \cdot 3 - 2 = 3 + 6 - 2$

$\qquad\qquad\qquad\quad = 7 \qquad$ Correct answer

To make the incorrect answer correct we add parentheses:

$\quad 12 \div (4 + 2) \cdot 3 - 2 = 4$

# Chapter 2

# Introduction to Integers and Algebraic Expressions

## Exercise Set 2.1

1. The integer $-2$ corresponds to a drop of 2 points.

3. The integer $-1286$ corresponds to 1286 ft below sea level; the integer 29,028 corresponds to 29,028 ft above sea level.

5. The integer 850 corresponds to an \$850 deposit; the integer $-432$ corresponds to a \$432 withdrawal.

7. Since 7 is to the right of 0, we have $7 > 0$.

9. Since $-9$ is to the left of 5, we have $-9 < 5$.

11. Since $-6$ is to the left of 6, we have $-6 < 6$.

13. Since $-8$ is to the left of $-5$, we have $-8 < -5$.

15. Since $-5$ is to the right of $-11$, we have $-5 > -11$.

17. Since $-6$ is to the left of $-5$, we have $-6 < -5$.

19. The distance from 23 to 0 is 23, so $|23| = 23$.

21. The distance from 0 to 0 is 0, so $|0| = 0$.

23. The distance from $-24$ to 0 is 24, so $|-24| = 24$.

25. The distance from 53 to 0 is 53, so $|53| = 53$.

27. This distance from $-8$ to 0 is 8, so $|-8| = 8$.

29. To find the opposite of $x$ when $x$ is $-8$, we reflect $-8$ to the other side of 0. We have $-(-8) = 8$. The opposite of $-8$ is 8.

31. To find the opposite of $x$ when $x$ is $-7$, we reflect $-7$ to the other side of 0. We have $-(-7) = 7$. The opposite of $-7$ is 7.

33. When we try to reflect 0 to the other side of 0, we go nowhere. The opposite of 0 is 0.

35. To find the opposite of $x$ when $x$ is $-19$, we reflect $-19$ to the other side of 0. We have $-(-19) = 19$.

37. To find the opposite of $x$ when $x$ is 42, we reflect 42 to the other side of 0. We have $-(42) = -42$.

39. The opposite of $-8$ is 8. $\quad -(-8) = 8$

41. The opposite of 7 is $-7$. $\quad -(7) = 7$

43. The opposite of $-29$ is 29. $\quad -(-29) = 29$

45. The opposite of $-22$ is 22. $\quad -(-22) = 22$

47. The opposite of 1 is $-1$. $\quad -(1) = -1$

49. We replace $x$ by 3. We wish to find $-(-3)$. Reflecting 3 to the other side of 0 gives us $-3$ and then reflecting back gives us 3. Thus, $-(-x) = 3$ when $x$ is 3.

51. We replace $x$ by $-8$. We wish to find $-(-(-8))$. Reflecting $-8$ to the other side of 0 gives us 8 and then reflecting back gives us $-8$. Thus, $-(-x) = -8$ when $x$ is $-8$.

53. We replace $x$ by 2. We wish to find $-(-2)$. Reflecting 2 to the other side of 0 gives us $-2$ and then reflecting back gives us 2. Thus, $-(-x) = 2$ when $x$ is 2.

55. We replace $x$ by 0. We wish to find $-(-0)$. When we try to reflect 0 "to the other side of 0" we go nowhere. The same thing happens when we try to reflect back. Thus, $-(-x) = 0$ when $x$ is 0.

57. We replace $x$ by $-34$. We wish to find $-(-(-34))$. Reflecting $-34$ to the other side of 0 gives us 34 and then reflecting back gives us $-34$. Thus, $-(-x) = -34$ when $x$ is $-34$.

59.
$$
\begin{array}{r}
{\scriptstyle 1\ 1} \\
3\ 2\ 7 \\
+\ 4\ 9\ 8 \\
\hline
8\ 2\ 5
\end{array}
$$

61.
$$
\begin{array}{r}
{\scriptstyle \phantom{0}2} \\
{\scriptstyle \phantom{0}3} \\
2\ \overset{}{0}\ 9 \\
\times\ \ \ 3\ 4 \\
\hline
8\ 3\ 6 \\
6\ 2\ 7\ 0 \\
\hline
7\ 1\ 0\ 6
\end{array}
$$
  Multiplying 209 by 4
  Multiplying 209 by 30
  Adding

63. $9^2 = 9 \cdot 9 = 81$

65. ◈

67. ◈

69. Answers may vary. On many scientific calculators we would multiply 327 and 83 and then take the opposite of the product.

$$\boxed{3}\ \boxed{2}\ \boxed{7}\ \boxed{\times}\ \boxed{8}\ \boxed{3}\ \boxed{=}\ \boxed{+/-}$$

71. $|-8| = 8$, so $-|-8| = -(8) = -8$.

73. $|7| = 7$, so $-|7| = -(7) = -7$.

75. The integers whose distance from 0 is less than 2 are $-1$, 0, and 1. These are the solutions.

77. First note that $2^{10} = 1024$, $|-6| = 6$, $|3| = 3$, $2^7 = 128$, $7^2 = 49$, and $10^2 = 100$. Listing the entire set of integers in order from least to greatest, we have $-100$, $-5$, 0, $|3|$, 4, $|-6|$, $7^2$, $10^2$, $2^7$, $2^{10}$.

## Exercise Set 2.2

**1.** Add: $-7 + 2$

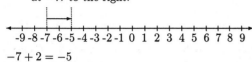

$-7 + 2 = -5$

**3.** Add: $-9 + 5$

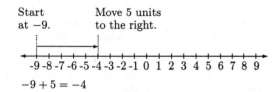

$-9 + 5 = -4$

**5.** Add: $-3 + 9$

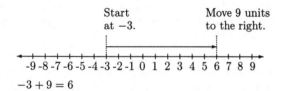

$-3 + 9 = 6$

**7.** $-7 + 7$

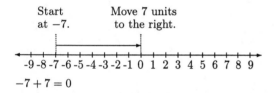

$-7 + 7 = 0$

**9.** $-3 + (-1)$

$-3 + (-1) = -4$

**11.** $-4 + (-11)$   Two negative integers

Add the absolute values: $4 + 11 = 15$
Make the answer negative: $-4 + (-11) = -15$

**13.** $-6 + (-5)$   Two negative integers

Add the absolute values: $6 + 5 = 11$
Make the answer negative: $-6 + (-5) = -11$

**15.** $9 + (-9) = 0$

For any integer $a$, $a + (-a) = 0$.

**17.** $-2 + 2 = 0$

For any integer $a$, $-a + a = 0$.

**19.** $0 + 8 = 8$

For any integer $a$, $0 + a = a$.

**21.** $0 + (-8) = -8$

For any integer $a$, $0 + a = a$.

**23.** $-25 + 0 = -25$

For any integer $a$, $a + 0 = a$.

**25.** $0 + (-27) = -27$

For any integer $a$, $0 + a = a$.

**27.** $17 + (-17) = 0$

For any integer $a$, $a + (-a) = 0$.

**29.** $-25 + 25 = 0$

For any integer $a$, $-a + a = 0$.

**31.** $8 + (-5)$   The absolute values are 8 and 5. The difference is 3. The positive number has the larger absolute value, so the answer is positive. $8 + (-5) = 3$

**33.** $-4 + (-5)$   Two negative integers

Add the absolute values: $4 + 5 = 9$
Make the answer negative: $-4 + (-5) = -9$

**35.** $0 + (-5) = -5$

For any integer $a$, $0 + a = a$.

**37.** $14 + (-5)$   The absolute values are 14 and 5. The difference is 9. The positive number has the larger absolute value, so the answer is positive. $14 + (-5) = 9$

**39.** $-11 + 8$   The absolute values are 11 and 8. The difference is 3. Since the negative number has the larger absolute value, the answer is negative. $-11 + 8 = -3$

**41.** $-19 + 19 = 0$

For any integer $a$, $-a + a = 0$.

**43.** $-17 + 7$   The absolute values are 17 and 7. The difference is 10. Since the negative number has the larger absolute value, the answer is negative. $-17 + 7 = -10$

**45.** $-17 + (-7)$   Two negative integers

Add the absolute values: $17 + 7 = 24$
Make the answer negative: $-17 + (-7) = -24$

**47.** $11 + (-16)$   The absolute values are 11 and 16. The difference is 5. Since the negative number has the larger absolute value, the answer is negative. $11 + (-16) = -5$

**49.** $-15 + (-6)$   Two negative integers

Add the absolute values: $15 + 6 = 21$
Make the answer negative: $-15 + (-6) = -21$

**51.** $11 + (-9)$   The absolute values are 11 and 9. The difference is 2. The positive number has the larger absolute value, so the answer is positive. $11 + (-9) = 2$

**53.** $-20 + (-6)$   Two negative integers

Add the absolute values: $20 + 6 = 26$
Make the answer negative: $-20 + (-6) = -26$

**55.** We will add from left to right.

$$-15 + (-7) + 1 = -22 + 1$$
$$= -21$$

**57.** We will add from left to right.
$$30 + (-10) + 5 = 20 + 5$$
$$= 25$$

**59.** We will add from left to right.
$$-23 + (-9) + 15 = -32 + 15$$
$$= -17$$

**61.** We will add from left to right.
$$40 + (-40) + 6 = 0 + 6$$
$$= 6$$

**63.** We will add from left to right.
$$85 + (-65) + (-12) = 20 + (-12)$$
$$= 8$$

**65.** We will add from left to right.
$$
\begin{aligned}
& -24 + (-37) + (-19) + (-45) + (-35) \\
=\ & -61 + (-19) + (-45) + (-35) \\
=\ & -80 + (-45) + (-35) \\
=\ & -125 + (-35) \\
=\ & -160
\end{aligned}
$$

**67.** $28 + (-44) + 17 + 31 + (-94)$

a) $28 + 17 + 31 = 76$   Adding the positive numbers

b) $-44 + (-94) = -138$   Adding the negative numbers

c) $76 + (-138) = -62$   Adding the results

**69.** 3 ten thousands + 9 thousands + 4 hundreds + 1 ten + 7 ones

**71.** a) Locate the digit in the thousands place.
$$3\,2,\ \boxed{8}\ 3\,1$$
$$\uparrow$$

b) Then consider the next digit to the right.

c) Since the digit is 5 or higher, round 2 thousands up to 3 thousands.

d) Change all digits to the right of thousands to zeros.

The answer is 33,000.

**73.**
$$
\begin{array}{r}
3\,2 \\
9\overline{)2\,8\,8} \\
2\,7\,0 \\
\hline
1\,8 \\
1\,8 \\
\hline
0
\end{array}
$$

The answer is 32.

**75.** ◈

**77.** ◈

**79.** $|-32| + (-|15|) = 32 + (-15) = 17$

**81.** Use a calculator.
$$497 + (-3028) = -2531$$

**83.** Think of starting at $-7$ on the number line and moving $x$ units right or left. Since we are starting at a negative number, moving to the left will take us to another negative number. If we move 7 units to the right, we move to 0. If we move more than 7 units to the right, we move to a positive number, and a move of less than 7 units to the right takes us to a negative number. Thus, for all numbers $x$ less than 7, $-7 + x$ is negative.

**85.** If $n = m$ and $n$ is negative, then $m$ is also negative and $-n$ and $-m$ are both positive. Thus, $-n + (-m)$, the sum of two positive numbers, is positive.

**87.** If $n$ is positive and $m$ is greater than $n$, then $m$ is also positive. Thus $n + m$, the sum of two positive numbers, is positive.

## Exercise Set 2.3

**1.** $4 - 7 = 4 + (-7) = -3$

**3.** $0 - 8 = 0 + (-8) = -8$

**5.** $-8 - (-4) = -8 + 4 = -4$

**7.** $-11 - (-11) = -11 + 11 = 0$

**9.** $12 - 17 = 12 + (-17) = -5$

**11.** $20 - 27 = 20 + (-27) = -7$

**13.** $-9 - (-5) = -9 + 5 = -4$

**15.** $-40 - (-40) = -40 + 40 = 0$

**17.** $7 - 7 = 7 + (-7) = 0$

**19.** $7 - (-7) = 7 + 7 = 14$

**21.** $8 - (-3) = 8 + 3 = 11$

**23.** $-6 - 8 = -6 + (-8) = -14$

**25.** $-4 - (-9) = -4 + 9 = 5$

**27.** $1 - 8 = 1 + (-8) = -7$

**29.** $-6 - (-5) = -6 + 5 = -1$

**31.** $8 - (-10) = 8 + 10 = 18$

**33.** $0 - 10 = 0 + (-10) = -10$

**35.** $-5 - (-2) = -5 + 2 = -3$

**37.** $-7 - 14 = -7 + (-14) = -21$

**39.** $0 - (-5) = 0 + 5 = 5$

**41.** $-8 - 0 = -8 + 0 = -8$

**43.** $7 - (-5) = 7 + 5 = 12$

**45.** $2 - 25 = 2 + (-25) = -23$

**47.** $-42 - 26 = -42 + (-26) = -68$

**49.** $-71 - 2 = -71 + (-2) = -73$

**51.** $24 - (-92) = 24 + 92 = 116$

**53.** $-50 - (-50) = -50 + 50 = 0$

**55.** $-30 - (-85) = -30 + 85 = 55$

**57.** $7 - (-5) + 4 - (-3) = 7 + 5 + 4 + 3 = 19$

**59.** $\phantom{=}-31 + (-28) - (-14) - 17$
$= -31 + (-28) + 14 + (-17)$
$= -31 + (-28) + (-17) + 14$  Using a commutative law
$= -76 + 14$  Adding the negative numbers
$= -62$

**61.** $-34 - 28 + (-33) - 44 = (-34) + (-28) + (-33) + (-44) = -139$

**63.** $\phantom{=}-93 - (-84) - 41 - (-56)$
$= -93 + 84 + (-41) + 56$
$= -93 + (-41) + 84 + 56$  Using a commutative law
$= -134 + 140$  Adding negatives and adding positives
$= 6$

**65.** $\phantom{=}-5 - (-30) + 30 + 40 - (-12)$
$= -5 + 30 + 30 + 40 + 12$
$= -5 + 112$  Adding the positive numbers
$= 107$

**67.** $132 - (-21) + 45 - (-21) = 132 + 21 + 45 + 21 = 219$

**69.** We subtract 19 from 12 to find the counter-reading:
$$12 - 19 = -7$$
The counter will read $-7$ min.

**71.** The integer $-120$ corresponds to a \$120 debt. We subtract $-120$ from 350:
$$350 - (-120) = 350 + 120 = 470$$
Jan needs to earn \$470 to raise her total assets to \$350.

**73.** To find the elevation that is 360 ft deeper than $-2860$ ft, we subtract the additional depth from the current depth:
$$-2860 - 360 = -2860 + (-360) = -3220$$
In 1998 the elevation of the world's deepest offshore oil well will be $-3220$ ft.

**75.** $4^3 = 4 \cdot 4 \cdot 4 = 64$

**77.** *Familiarize.* Let $n =$ the number of 12-oz cans that can be filled. We think of an array consisting of 96 oz with 12 oz in each row.

The number $n$ corresponds to the number of rows in the array.

*Translate and Solve.* We translate to an equation and solve it.

$$96 \div 12 = n \qquad \begin{array}{r} 8 \\ 12\overline{)96} \\ \underline{96} \\ 0 \end{array}$$

*Check.* We multiply the number of cans by 12: $8 \cdot 12 = 96$. The result checks.

*State.* Eight 12-oz cans can be filled.

**79.** $\phantom{=}5 + 4^2 + 2 \cdot 7$
$= 5 + 16 + 2 \cdot 7$
$= 5 + 16 + 14$
$= 21 + 14$
$= 35$

**81.** ◈

**83.** ◈

**85.** Use a calculator.
$$23,011 - (-60,432) = 83,443$$

**87.** False; $0 - 3 \neq 3$.

**89.** True

**91.** False; $3 - 3 = 0$, but $3 \neq -3$.

**93.** a)  We add the values of the cards:
$-1 + (-1) + 1 + 1 + 1 + (-1) + (-1) + 0 + (-1) + (-1) + 1 = -2$

b)  Since the final count on the sequence of cards is negative, the player has a winning edge.

## Exercise Set 2.4

**1.** $-3 \cdot 7 = -21$

**3.** $-9 \cdot 2 = -18$

**5.** $8 \cdot (-6) = -48$

**7.** $-10 \cdot 3 = -30$

**9.** $-2 \cdot (-5) = 10$

**11.** $-9 \cdot (-2) = 18$

**13.** $-7 \cdot (-6) = 42$

**15.** $-10(-3) = 30$

**17.** $12(-10) = -120$

**19.** $-6(-50) = 300$

**21.** $(-72)(-1) = 72$

**23.** $(-20)17 = -340$

**25.** $-23 \cdot 0 = 0$

**27.** $0(-14) = 0$

**29.** $\phantom{=}3 \cdot (-8) \cdot 4$
$= -24 \cdot 4$  Multiplying the first two numbers
$= -96$

**31.** $\phantom{=}7(-4)(-3)5$
$= 7 \cdot 12 \cdot 5$  Multiplying the negative numbers
$= 84 \cdot 5$
$= 420$

**33.**  $\qquad -2(-5)(-7)$
$\qquad = 10 \cdot (-7) \qquad$ Multiplying the first two numbers
$\qquad = -70$

**35.**  $\qquad (-5)(-2)(-3)(-1)$
$\qquad = 10 \cdot 3 \qquad\qquad$ Multiplying the first two numbers and the last two numbers
$\qquad = 30$

**37.**  $\qquad (-15)(-29)0 \cdot 8$
$\qquad = 435 \cdot 0 \qquad$ Multiplying the first two numbers and the last two numbers
$\qquad = 0$

(We might have noted at the outset that the product would be 0 since one of the numbers in the product is 0.)

**39.**  $\qquad (-7)(-1)(7)(-6)$
$\qquad = 7(-42) \qquad$ Multiplying the first two numbers and the last two numbers
$\qquad = -294$

**41.**  $(-5)^2 = (-5)(-5) = 25$

**43.**  $(-5)^3 = (-5)(-5)(-5)$
$\qquad\qquad = 25(-5)$
$\qquad\qquad = -125$

**45.**  $(-10)^4 = (-10)(-10)(-10)(-10)$
$\qquad\qquad\quad = 100 \cdot 100$
$\qquad\qquad\quad = 10,000$

**47.**  $-2^4 = -1 \cdot 2^4$
$\qquad\quad = -1 \cdot 2 \cdot 2 \cdot 2 \cdot 2$
$\qquad\quad = -1 \cdot 4 \cdot 4$
$\qquad\quad = -1 \cdot 16$
$\qquad\quad = -16$

**49.**  $(-3)^5 = (-3)(-3)(-3)(-3)(-3)$
$\qquad\qquad = 9 \cdot 9 \cdot (-3)$
$\qquad\qquad = 81(-3)$
$\qquad\qquad = -243$

**51.**  $\qquad (-1)^{12}$
$\qquad = (-1) \cdot (-1) \cdot (-1) \cdot (-1) \cdot (-1) \cdot (-1) \cdot (-1) \cdot (-1) \cdot$
$\qquad\quad (-1) \cdot (-1) \cdot (-1) \cdot (-1)$
$\qquad = 1 \cdot 1 \cdot 1 \cdot 1 \cdot 1 \cdot 1$
$\qquad = 1 \cdot 1 \cdot 1$
$\qquad = 1 \cdot 1$
$\qquad = 1$

**53.**  $\qquad -3^6$
$\qquad = -1 \cdot 3^6$
$\qquad = -1 \cdot 3 \cdot 3 \cdot 3 \cdot 3 \cdot 3 \cdot 3$
$\qquad = -1 \cdot 9 \cdot 9 \cdot 9$
$\qquad = -9 \cdot 81$
$\qquad = -729$

**55.**  $-5^3 = -1 \cdot 5^3$
$\qquad\quad = -1 \cdot 5 \cdot 5 \cdot 5$
$\qquad\quad = -5 \cdot 25$
$\qquad\quad = -125$

**57.**  $-7^4$ is read "the opposite of seven to the fourth power."

**59.**  $(-9)^6$ is read "negative nine to the sixth power."

**61.**  a) Locate the digit in the hundreds place.

$\qquad\qquad 5\ 3\ 2,\ 4\ \boxed{5}\ 1$
$\qquad\qquad\qquad\quad \uparrow$

b) Then consider the next digit to the right.

c) Since that digit is 5 or higher, round 4 hundreds up to 5 hundreds.

d) Change all digits to the right of hundreds to zeros.

The answer is 532,500.

**63.**
$$
\begin{array}{r}
8\ 0 \\
3\ 6\ \overline{\smash{)}\ 2\ 8\ 8\ 0} \\
\underline{2\ 8\ 8\ 0} \\
0 \\
\underline{0} \\
0
\end{array}
$$

The answer is 80.

**65.**  *Familiarize*. We first make a drawing.

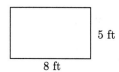

5 ft

8 ft

Let $A =$ the area.

*Translate*. Using the formula for area, we have $A = l \cdot w = 8 \cdot 5$.

*Solve*. We carry out the multiplication.

$\qquad A = 8 \cdot 5 = 40$

*Check*. We repeat our calculation.

*State*. The area of the rug is 40 ft².

**67.**  ◈

**69.**  ◈

**71.**  $\qquad (-2)^3 \cdot [(-1)^{29}]^{46}$
$\qquad = (-2)^3 \cdot [-1]^{46}$
$\qquad = -8 \cdot 1$
$\qquad = -8$

**73.**  $\qquad -5^2(-1)^{29}$
$\qquad = -1 \cdot 5^2(-1)^{29}$
$\qquad = -1 \cdot 5^2 \cdot (-1)$
$\qquad = -1 \cdot 25 \cdot (-1)$
$\qquad = -25 \cdot (-1)$
$\qquad = 25$

**75.**  $\qquad |-12(-3)^2 - 5^3 - 6^2 - (-5)^2|$
$\qquad = |-12 \cdot 9 - 125 - 36 - 25|$
$\qquad = |-108 - 125 - 36 - 25|$
$\qquad = |-294|$
$\qquad = 294$

**77.** Use a calculator.

$(-17)^4(129 - 133)^5 = -85,525,504$

**79.** The integer $-95$ corresponds to the elevation of 95 m below the surface. At a rate of 7 meters per minute, in 9 min the diver rises $7 \cdot 9$ m. We have:

$$-95 + 7 \cdot 9 = -95 + 63$$
$$= -32$$

The diver's new elevation is 32 m below the surface, or $-32$ m.

**81.** (a) If $-mn$ is positive, then $mn$ is negative so $m$ and $n$ must have different signs.

   (b) If $-mn$ is zero, then at least one of $m$ and $n$ must be zero.

   (c) If $-mn$ is negative, then $mn$ is positive so $m$ and $n$ must have the same sign.

## Exercise Set 2.5

**1.** $28 \div (-4) = -7$    Check: $-7(-4) = 28$

**3.** $\dfrac{28}{-2} = -14$    Check: $-14(-2) = 28$

**5.** $\dfrac{18}{-2} = -9$    Check: $-9(-2) = 18$

**7.** $\dfrac{-48}{-12} = 4$    Check: $4(-12) = -48$

**9.** $\dfrac{-72}{8} = -9$    Check: $-9 \cdot 8 = -72$

**11.** $-100 \div (-50) = 2$    Check: $2(-50) = -100$

**13.** $-344 \div 8 = -43$    Check: $-43 \cdot 8 = -344$

**15.** $\dfrac{200}{-25} = -8$    Check: $-8(-25) = 200$

**17.** $\dfrac{-56}{0}$ is undefined.

**19.** $\dfrac{88}{-11} = -8$    Check: $-8(-11) = 88$

**21.** $-\dfrac{276}{12} = \dfrac{-276}{12} = -23$    Check: $-23 \cdot 12 = -276$

**23.** $\dfrac{0}{-2} = 0$    Check: $0 \cdot (-2) = 0$

**25.** $\dfrac{19}{-1} = -19$    Check: $-19(-1) = 19$

**27.** $-41 \div 1 = -41$    Check: $-41 \cdot 1 = -41$

**29.** $8 - 2 \cdot 3 - 9 = 8 - 6 - 9$    Multiplying
$$= 2 - 9 \qquad \text{Doing all additions and}$$
$$\qquad \text{subtractions in order}$$
$$= -7 \qquad \text{from left to right}$$

**31.** $8 - 2(3 - 9) = 8 - 2(-6)$    Subtracting inside
$$\qquad\qquad\qquad\qquad \text{parentheses}$$
$$= 8 + 12 \qquad \text{Multiplying}$$
$$= 20 \qquad \text{Adding}$$

**33.** $16 \cdot (-24) + 50 = -384 + 50$    Multiplying
$$= -334 \qquad \text{Adding}$$

**35.**
$$40 - 3^2 - 2^3$$
$$= 40 - 9 - 8 \quad \text{Evaluating the exponential expressions}$$
$$= 31 - 8 \quad \text{Doing all additions and subtractions}$$
$$= 23 \qquad \text{in order from left to right}$$

**37.**
$$4 \cdot (6 + 8)/(4 + 3)$$
$$= 4 \cdot 14/7 \quad \text{Adding inside parentheses}$$
$$= 56/7 \quad \text{Doing all multiplications and divisions}$$
$$= 8 \qquad \text{in order from left to right}$$

**39.** $4 \cdot 5 - 2 \cdot 6 + 4 = 20 - 12 + 4$    Multiplying
$$= 8 + 4$$
$$= 12$$

**41.**
$$\frac{9^2 - 1}{1 - 3^2}$$
$$= \frac{81 - 1}{1 - 9} \quad \text{Evaluating the exponential expressions}$$
$$= \frac{80}{-8} \quad \text{Subtracting in the numerator and in the denominator}$$
$$= -10$$

**43.** $8(-7) + 6(-5) = -56 - 30$    Multiplying
$$= -86$$

**45.** $20 \div 5(-3) + 3 = 4(-3) + 3$    Dividing
$$= -12 + 3 \quad \text{Multiplying}$$
$$= -9 \qquad \text{Adding}$$

**47.**
$$8 \div 2 \cdot 0 \div 6$$
$$= 4 \cdot 0 \div 6 \quad \text{Doing all multiplications}$$
$$= 0 \div 6 \qquad \text{and divisions in order}$$
$$= 0 \qquad \text{from left to right}$$

**49.**
$$4 \cdot 5^2 \div 10$$
$$= 4 \cdot 25 \div 10 \quad \text{Evaluating the exponential expression}$$
$$= 100 \div 10 \quad \text{Multiplying}$$
$$= 10 \qquad \text{Dividing}$$

**51.**
$$(3 - 8)^2 \div (-1)$$
$$= (-5)^2 \div (-1) \quad \text{Subtracting inside parentheses}$$
$$= 25 \div (-1) \quad \text{Evaluating the exponential expression}$$
$$= -25$$

**53.** $12 - 20^3 = 12 - 8000$
$$= -7988$$

**55.** $2 \times 10^3 - 5000 = 2 \times 1000 - 5000$
$$= 2000 - 5000$$
$$= -3000$$

**57.** $6[9 - (3 - 4)] = 6[9 - (-1)]$    Subtracting inside the
$$\qquad\qquad\qquad\qquad\qquad \text{innermost parentheses}$$
$$= 6[9 + 1]$$
$$= 6[10]$$
$$= 60$$

**59.** $-1000 \div (-100) \div 10 = 10 \div 10$    Doing the divisions in order
$$= 1 \qquad \text{from left to right}$$

**61.** $8 - |7 - 9| \cdot 3 = 8 - |-2| \cdot 3$
$= 8 - 2 \cdot 3$
$= 8 - 6$
$= 2$

**63.** $9 - |7 - 3^2| = 9 - |7 - 9|$
$= 9 - |-2|$
$= 9 - 2$
$= 7$

**65.**
$$\frac{(-5)^3 + 17}{10(2 - 6) - 2(5 + 2)}$$

$= \dfrac{-125 + 17}{10(2 - 6) - 2(5 + 2)}$    Evaluating the exponential expression

$= \dfrac{-125 + 17}{10(-4) - 2 \cdot 7}$    Doing the calculations within parentheses

$= \dfrac{-125 + 17}{-40 - 14}$    Multiplying

$= \dfrac{-108}{-54}$    Adding and subtracting

$= 2$

**67.**
$$\frac{2 \cdot 4^3 - 4 \cdot 32}{19^3 - 17^4}$$

$= \dfrac{2 \cdot 64 - 4 \cdot 32}{6859 - 83,521}$    Evaluating the exponential expressions

$= \dfrac{128 - 128}{6859 - 83,521}$    Multiplying

$= \dfrac{0}{-76,662}$    Subtracting

$= 0$    Dividing

**69.** *Familiarize.* We first make a drawing.

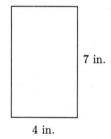

7 in.

4 in.

Let $A$ = the area.

*Translate.* Using the formula for area, we have $A = l \cdot w = 7 \cdot 4$.

*Solve.* We carry out the multiplication.

$A = 7 \cdot 4 = 28$

*Check.* We repeat the calculation.

*State.* The area of the ad was 28 in$^2$.

**71.** *Familiarize.* We let $g$ = the number of gallons needed to travel 350 miles. Think of a rectangular array consisting of 350 miles with 25 miles in each row. The number $g$ is the number of rows.

*Translate.*

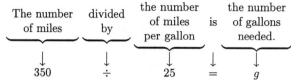

| The number of miles | divided by | the number of miles per gallon | is | the number of gallons needed. |
|---|---|---|---|---|
| ↓ | ↓ | ↓ | ↓ | ↓ |
| 350 | ÷ | 25 | = | $g$ |

*Solve.* We carry out the division.

```
      1 4
2 5 ⟌ 3 5 0
      2 5 0
      ─────
      1 0 0
      1 0 0
      ─────
          0
```

*Check.* We multiply the number of gallons by the number of miles per gallon:

$14 \cdot 25 = 350$

We get the number of miles to be traveled, so the answer checks.

*State.* It will take 14 gallons of gasoline to travel 350 miles.

**73.** *Familiarize.* We let $c$ = the number of calories in a 1-oz serving. Think of a rectangular array consisting of 1050 calories arranged in 7 rows. The number $c$ is the number of calories in each row.

*Translate.*

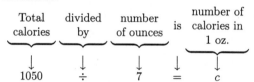

| Total calories | divided by | number of ounces | is | number of calories in 1 oz. |
|---|---|---|---|---|
| ↓ | ↓ | ↓ | ↓ | ↓ |
| 1050 | ÷ | 7 | = | $c$ |

*Solve.* We carry out the division.

```
      1 5 0
7 ⟌ 1 0 5 0
    7 0 0
    ─────
    3 5 0
    3 5 0
    ─────
        0
        0
      ───
        0
```

*Check.* We multiply the number of calories in 1 oz by 7:

$150 \cdot 7 = 1050$

The result checks

*State.* There are 150 calories in a 1-oz serving.

**75.**

**77.** ◈

**79.**
$$\frac{(7 - 8)^{37}}{7^2 - 8^2} \cdot (98 - 7^2 \cdot 2)$$

$= \dfrac{(-1)^{37}}{49 - 64} \cdot (98 - 49 \cdot 2)$

$= \dfrac{-1}{-15} \cdot (98 - 98)$

$= \dfrac{-1}{-15} \cdot 0$

$= 0$

**81.** $\dfrac{195 + (-15)^3}{195 - 7 \cdot 5^2} = \dfrac{195 - 3375}{195 - 7 \cdot 25}$

$= \dfrac{195 - 3375}{195 - 175}$

$= \dfrac{-3180}{20}$

$= -159$

**83.** $-n$ is negative and $-m$ is positive, so $\dfrac{-n}{-m}$ is the quotient of a negative and a positive number and, thus, is negative.

**85.** $n$ and $-m$ are both positive, so $\dfrac{n}{-m}$ is the quotient of two positive numbers and, thus, is positive. Then $-\left(\dfrac{n}{-m}\right)$ is the opposite of a positive number and, thus, is negative.

## Exercise Set 2.6

**1.** $7t = 7 \cdot 2 = 14\cent$

**3.** $\dfrac{x}{y} = \dfrac{9}{-3} = -3$

**5.** $\dfrac{3p}{q} = \dfrac{3 \cdot 2}{6} = \dfrac{6}{6} = 1$

**7.** $\dfrac{x+y}{5} = \dfrac{-10+20}{5} = \dfrac{10}{5} = 2$

**9.** $3 + 5 \cdot x = 3 + 5 \cdot 2 = 3 + 10 = 13$

**11.** $2l + 2w = 2 \cdot 3 + 2 \cdot 4 = 6 + 8 = 14$ ft

**13.** $2(l + w) = 2(3 + 4) = 2 \cdot 7 = 14$ ft

**15.** $7a - 7b = 7 \cdot 5 - 7 \cdot 2 = 35 - 14 = 21$

**17.** $7(a - b) = 7(5 - 2) = 7 \cdot 3 = 21$

**19.** $16t^2 = 16 \cdot 5^2 = 16 \cdot 25 = 400$ ft

**21.** $9m - m^2 = 9(-4) - (-4)^2$
$= 9(-4) - 16$
$= -36 - 16$
$= -52$

**23.** $a + (b - a)^2 = 6 + (4 - 6)^2$
$= 6 + (-2)^2$
$= 6 + 4$
$= 10$

**25.** $a + b - a^2 = 6 + 4 - 6^2$
$= 6 + 4 - 36$
$= 10 - 36$
$= -26$

**27.** $\dfrac{n^2 - n}{2} = \dfrac{9^2 - 9}{2}$

$= \dfrac{81 - 9}{2}$

$= \dfrac{72}{2}$

$= 36$

**29.** $-\dfrac{a}{b}, \dfrac{-a}{b},$ and $\dfrac{a}{-b}$ all represent the same number. Thus we can also write $-\dfrac{3}{a}$ as $\dfrac{-3}{a}$ and $\dfrac{3}{-a}$.

**31.** $-\dfrac{a}{b}, \dfrac{-a}{b},$ and $\dfrac{a}{-b}$ all represent the same number. Thus we can also write $\dfrac{-n}{b}$ as $-\dfrac{n}{b}$ and $\dfrac{n}{-b}$.

**33.** $-\dfrac{a}{b}, \dfrac{-a}{b},$ and $\dfrac{a}{-b}$ all represent the same number. Thus we can also write $\dfrac{9}{-p}$ as $-\dfrac{9}{p}$ and $\dfrac{-9}{p}$.

**35.** $-\dfrac{a}{b}, \dfrac{-a}{b},$ and $\dfrac{a}{-b}$ all represent the same number. Thus we can also write $\dfrac{-14}{w}$ as $-\dfrac{14}{w}$ and $\dfrac{14}{-w}$.

**37.** $\dfrac{-a}{b} = \dfrac{-35}{7} = -5;$

$\dfrac{a}{-b} = \dfrac{35}{-7} = -5;$

$-\dfrac{a}{b} = -\dfrac{35}{7} = -5$

**39.** $\dfrac{-a}{b} = \dfrac{-81}{3} = -27;$

$\dfrac{a}{-b} = \dfrac{81}{-3} = -27;$

$-\dfrac{a}{b} = -\dfrac{81}{3} = -27$

**41.** $(-3x)^2 = (-3 \cdot 7)^2 = (-21)^2 = 441;$
$-3x^2 = -3(7)^2 = -3 \cdot 49 = -147$

**43.** $5x^2 = 5(2)^2 = 5 \cdot 4 = 20;$
$5x^2 = 5(-2)^2 = 5 \cdot 4 = 20$

**45.** $x^3 = 6^3 = 6 \cdot 6 \cdot 6 = 216;$
$x^3 = (-6)^3 = (-6) \cdot (-6) \cdot (-6) = -216$

**47.** $x^6 = 1^6 = 1 \cdot 1 \cdot 1 \cdot 1 \cdot 1 \cdot 1 = 1;$
$x^6 = (-1)^6 = (-1) \cdot (-1) \cdot (-1) \cdot (-1) \cdot (-1) \cdot (-1) = 1$

**49.** $a^7 = 2^7 = 2 \cdot 2 \cdot 2 \cdot 2 \cdot 2 \cdot 2 \cdot 2 = 128;$
$a^7 = (-2)^7 = (-2)(-2)(-2)(-2)(-2)(-2)(-2) = -128$

**51.** $-m^2 + m = -(-4)^2 + (-4)$
$= -16 - 4$
$= -20$

**53.** $a - 3a^3 = -5 - 3(-5)^3$
$= -5 - 3(-125)$
$= -5 + 375$
$= 370$

**55.** $x^2 + 5x \div 2 = (-6)^2 + 5(-6) \div 2$
$= 36 + 5(-6) \div 2$
$= 36 - 30 \div 2$
$= 36 - 15$
$= 21$

**57.** $m^3 - m^2 = 5^3 - 5^2$
$= 125 - 25$
$= 100$

**59.** Twenty-three million, forty-three thousand, nine hundred twenty-one

**61.**
$$\begin{array}{r} 5\ 2\ 8\ 3 \\ -\ 2\ 4\ 7\ 5 \\ \hline \end{array} \qquad \begin{array}{r} 5\ 2\ 8\ 0 \\ -\ 2\ 4\ 8\ 0 \\ \hline 2\ 8\ 0\ 0 \end{array}$$

**63.** *Familiarize.* Since we are combining snowfall amounts, addition can be used. We let $s = $ the total snowfall.

*Translate.* We translate to an equation.
$$9 + 8 = s$$

*Solve.* We carry out the addition.
$$9 + 8 = 17$$

Thus, $17 = s$, or $s = 17$.

*Check.* We repeat the calculation.

*State.* It snowed 17 in. altogether.

**65.**

**67.**

**69.** Use a calculator.
$$x^2 - 23y + y^3 = 18^2 - 23(-21) + (-21)^3 = -8454$$

**71.** $x^{1492} - x^{1493} = (-1)^{1492} - (-1)^{1493}$
$$= 1 - (-1)$$
$$= 1 + 1$$
$$= 2$$

**73.** $5a^{3a-4} = 5 \cdot 2^{3 \cdot 2 - 4}$
$$= 5 \cdot 2^{6-4}$$
$$= 5 \cdot 2^2$$
$$= 5 \cdot 4$$
$$= 20$$

**75.** False; $2^3 = 8$; but $-2^3 = -8$.

**77.** True

## Exercise Set 2.7

**1.** $5(a + b) = 5 \cdot a + 5 \cdot b = 5a + 5b$

**3.** $4(x + 1) = 4 \cdot x + 4 \cdot 1 = 4x + 4$

**5.** $2(b + 5) = 2 \cdot b + 2 \cdot 5 = 2b + 10$

**7.** $7(1 - t) = 7 \cdot 1 - 7 \cdot t = 7 - 7t$

**9.** $6(5x + 2) = 6 \cdot 5x + 6 \cdot 2 = 30x + 12$

**11.** $8(x + 7 + 6y) = 8 \cdot x + 8 \cdot 7 + 8 \cdot 6y = 8x + 56 + 48y$

**13.** $-7(y - 2) = -7 \cdot y - (-7) \cdot 2 = -7y - (-14) = -7y + 14$

**15.** $-9(-5x - 6y + 8) = -9(-5x) - (-9)6y + (-9)8 =$
$45x - (-54y) + (-72) = 45x + 54y - 72$

**17.** $-4(x - 3y - 2z) = -4 \cdot x - (-4)3y - (-4)2z =$
$-4x - (-12y) - (-8z) = -4x + 12y + 8z$

**19.** $8(a - 3b + c) = 8 \cdot a - 8 \cdot 3b + 8 \cdot c =$
$8a - 24b + 8c$

**21.** $4(x - 3y - 7z) = 4 \cdot x - 4 \cdot 3y - 4 \cdot 7z =$
$4x - 12y - 28z$

**23.** $5(4a - 5b + c - 2d) = 5 \cdot 4a - 5 \cdot 5b + 5 \cdot c - 5 \cdot 2d =$
$20a - 25b + 5c - 10d$

**25.** $7a + 12a = (7 + 12)a = 19a$

**27.** $10a - a = 10a - 1 \cdot a = (10 - 1)a = 9a$

**29.** $2x + 6z + 9x = 2x + 9x + 6z$
$= (2 + 9)x + 6z$
$= 11x + 6z$

**31.** $27a + 70 - 40a - 8 = 27a - 40a + 70 - 8$
$= (27 - 40)a + 70 - 8$
$= -13a + 62$

**33.** $23 + 5t + 7y - t - y - 27$
$= 23 - 27 + 5t - 1 \cdot t + 7y - 1 \cdot y$
$= (23 - 27) + (5 - 1)t + (7 - 1)y$
$= -4 + 4t + 6y$

**35.** $5x - 12x = (5 - 12)x = -7x$

**37.** $y - 17y = (1 - 17)y = -16y$

**39.** $-8 + 11a - 5b + 6a - 7b + 7$
$= 11a + 6a - 5b - 7b - 8 + 7$
$= (11 + 6)a + (-5 - 7)b + (-8 + 7)$
$= 17a - 12b - 1$

**41.** $8x + 3y - 2x = 8x - 2x + 3y$
$= (8 - 2)x + 3y$
$= 6x + 3y$

**43.** $11x + 2y - 4x - y = 11x - 4x + 2y - y$
$= (11 - 4)x + (2 - 1)y$
$= 7x + y$

**45.** $a + 3b + 5a - 2 + b$
$= a + 5a + 3b + b - 2$
$= (1 + 5)a + (3 + 1)b - 2$
$= 6a + 4b - 2$

**47.** $6x^3 + 2x - 5x^3 + 7x = 6x^3 - 5x^3 + 2x + 7x$
$= (6 - 5)x^3 + (2 + 7)x$
$= x^3 + 9x$

**49.** $3a^2 + 7a^3 - a^2 + 5 + a^3$
$= 7a^3 + a^3 + 3a^2 - a^2 + 5$
$= (7 + 1)a^3 + (3 - 1)a^2 + 5$
$= 8a^3 + 2a^2 + 5$

**51.** $9xy + 4y^2 - 2xy + 2y^2 - 1$
$= 9xy - 2xy + 4y^2 + 2y^2 - 1$
$= (9 - 2)xy + (4 + 2)y^2 - 1$
$= 7xy + 6y^2 - 1$

**53.** $8a^2b - 3ab^2 - 4a^2b + 2ab$
$= 8a^2b - 4a^2b - 3ab^2 + 2ab$
$= (8 - 4)a^2b - 3ab^2 + 2ab$
$= 4a^2b - 3ab^2 + 2ab$

**55.**  $3x^4 - 2y^4 + 8x^4y^4 - 7x^4 + 8y^4$
$= 3x^4 - 7x^4 - 2y^4 + 8y^4 + 8x^4y^4$
$= (3 - 7)x^4 + (-2 + 8)y^4 + 8x^4y^4$
$= -4x^4 + 6y^4 + 8x^4y^4$

**57.**  Perimeter $= 4$ mm $+ 6$ mm $+ 7$ mm
$= (4 + 6 + 7)$ mm
$= 17$ mm

**59.**  Perimeter $= 4$ m $+ 4$ m $+ 4$ m $+ 5$ m $+ 1$ m
$= (4 + 4 + 4 + 5 + 1)$ m
$= 18$ m

**61.**  $P = 4s$        Perimeter of a square
$P = 4 \cdot 5$ in.
$P = 20$ in.

**63.**  The minimum length and width are 100 yards and 50 yards, respectively.
$P = 2l + 2w = 2 \cdot 100$ yd $+ 2 \cdot 50$ yd
$= 200$ yd $+ 100$ yd $= 300$ yd

**65.**  The penalty area is a rectangle with length 44 yd and width 18 yd.
$P = 2l + 2w = 2 \cdot 44$ yd $+ 2 \cdot 18$ yd
$= 88$ yd $+ 36$ yd $= 124$ yd

**67.**  $P = 2(l + w) = 2(14$ ft $+ 12$ ft$)$
$= 2 \cdot 26$ ft $= 52$ ft

**69.**  $P = 4s$
$= 4 \cdot 14$ in. $= 56$ in.

**71.**  $P = 4s$
$= 4 \cdot 75$ cm $= 300$ cm

**73.**  $P = 2(l + w) = 2(20$ ft $+ 12$ ft$)$
$= 2 \cdot 32$ ft $= 64$ ft

**75.**  *Familiarize.* Let $s =$ the number of servings of Grand Union Corn Flakes in one box. Visualize a rectangular array consisting of 510 grams with 30 grams in each row. Then $s$ is the number of rows.

*Translate.*

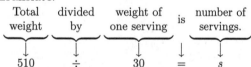

| Total weight | divided by | weight of one serving | is | number of servings. |
|---|---|---|---|---|
| ↓ | ↓ | ↓ | ↓ | ↓ |
| 510 | ÷ | 30 | = | $s$ |

*Solve.* We carry out the division.

$$
\begin{array}{r}
17 \\
30\overline{)510} \\
\underline{300} \\
210 \\
\underline{210} \\
0
\end{array}
$$

We have $17 = s$, or $s = 17$.

*Check.* We multiply the number of servings by the weight of a serving.

$17 \cdot 30 = 510$

We get 510 oz, the total weight of the corn flakes, so the answer checks.

*State.* There are 17 servings in a box of Grand Union Corn Flakes.

**77.**
$$
\begin{array}{r}
\overset{1}{5}\overset{5}{2}9 \\
\times \quad 6 \\
\hline
3174
\end{array}
$$

**79.**
$$
\begin{array}{r}
709 \\
5\overline{)3549} \\
\underline{3500} \\
49 \\
\underline{45} \\
4
\end{array}
$$

The answer is 709 R 4.

**81.**  ◈

**83.**  ◈

**85.**  $59 \boxed{\times} 17 \boxed{\vert} 59 \boxed{\times} 8 = 1475$

**87.**  $5(x + 3) + 2(x - 7) = 5x + 15 + 2x - 14 = 7x + 1$

**89.**  $2(3 - 4a) + 5(a - 7) = 6 - 8a + 5a - 35 = -3a - 29$

**91.**  $-5(2 + 3x + 4y) + 7(2x - y) =$
$-10 - 15x - 20y + 14x - 7y = -10 - x - 27y$

**93.**  *Familiarize.* First we will find the perimeter of each door and each window. Then we will find the total perimeter of all the doors and windows. Next we will determine how many sealant cartridges are needed and, finally, we will find the cost of the sealant.

*Translate.*

The perimeter of each door is given by
$$P = 2(l + w) = 2(3 \text{ ft} + 7 \text{ ft}).$$

The perimeter of each window is given by
$$P = 2(l + w) = 2(3 \text{ ft} + 4 \text{ ft}).$$

*Solve.* First we calculate the perimeters.

For each door: $P = 2(3 \text{ ft} + 7 \text{ ft}) = 2 \cdot 10$ ft $= 20$ ft

For each window: $P = 2(3 \text{ ft} + 4 \text{ ft}) = 2 \cdot 7$ ft $= 14$ ft

We multiply to find the perimeter of 3 doors and of 13 windows.

Doors: $3 \cdot 20$ ft $= 60$ ft

Windows: $13 \cdot 14$ ft $= 182$ ft

We add to find the total of the perimeters:
$$60 \text{ ft} + 182 \text{ ft} = 242 \text{ ft}$$

Next we divide to determine how many sealant cartridges are needed:

$$
\begin{array}{r}
4 \\
56\overline{)242} \\
\underline{224} \\
18
\end{array}
$$

The answer is 4 R 18. Since 18 ft will be left unsealed after 4 cartridges are used, Andrea should buy 5 sealant cartridges.

Finally, we multiply to find the cost of 5 sealant cartridges.
$$5 \cdot \$5.95 = \$29.75$$

*Check.* We repeat the calculations. The result checks.

*State.* It will cost Andrea $29.75 to seal the windows and doors.

## Exercise Set 2.8

**1.**
$$x - 7 = 5$$
$$x - 7 + 7 = 5 + 7 \quad \text{Adding 7 on both sides}$$
$$x + 0 = 12$$
$$x = 12$$
Check: $\dfrac{x - 7 = 5}{12 - 7 \; ? \; 5}$
$$\quad\quad 5 \; | \quad \text{TRUE}$$
The solution is 12.

**3.**
$$x - 6 = -9$$
$$x - 6 + 6 = -9 + 6 \quad \text{Adding 6 on both sides}$$
$$x + 0 = -3$$
$$x = -3$$
Check: $\dfrac{x - 6 = -9}{-3 - 6 \; ? \; -9}$
$$\quad\quad -9 \; | \quad \text{TRUE}$$
The solution is $-3$.

**5.**
$$t + 5 = 13$$
$$t + 5 - 5 = 13 - 5 \quad \text{Subtracting 5 on both sides}$$
$$t + 0 = 8$$
$$t = 8$$
The solution is 8.

**7.**
$$x + 9 = -3$$
$$x + 9 - 9 = -3 - 9 \quad \text{Subtracting 9 on both sides}$$
$$x + 0 = -12$$
$$x = -12$$
The solution is $-12$.

**9.**
$$17 = n - 6$$
$$17 + 6 = n - 6 + 6 \quad \text{Adding 6 on both sides}$$
$$23 = n + 0$$
$$23 = n$$
The solution is 23.

**11.**
$$-9 = x + 3$$
$$-9 - 3 = x + 3 - 3 \quad \text{Subtracting 3 on both sides}$$
$$-12 = x + 0$$
$$-12 = x$$
The solution is $-12$.

**13.**
$$-9 + t = 8$$
$$9 - 9 + t = 9 + 8 \quad \text{Adding 9 on both sides}$$
$$0 + t = 17$$
$$t = 17$$
The solution is 17.

**15.**
$$3 = 17 + x$$
$$3 - 17 = 17 - 17 + x \quad \text{Subtracting 17 on both sides}$$
$$-14 = 0 + x$$
$$-14 = x$$
The solution is $-14$.

**17.**
$$3x = 24$$
$$\frac{3x}{3} = \frac{24}{3} \quad \text{Dividing by 3 on both sides}$$
$$x = 8$$
Check: $\dfrac{3x = 24}{3 \cdot 8 \; ? \; 24}$
$$\quad\quad 24 \; | \quad \text{TRUE}$$
The solution is 8.

**19.**
$$-8t = 32$$
$$\frac{-8t}{-8} = \frac{32}{-8} \quad \text{Dividing by } -8 \text{ on both sides}$$
$$t = -4$$
Check: $\dfrac{-8t = 32}{-8(-4) \; ? \; 32}$
$$\quad\quad 32 \; | \quad \text{TRUE}$$
The solution is $-4$.

**21.**
$$-5n = -65$$
$$\frac{-5n}{-5} = \frac{-65}{-5} \quad \text{Dividing by } -5 \text{ on both sides}$$
$$n = 13$$
The solution is 13.

**23.**
$$64 = 8x$$
$$\frac{64}{8} = \frac{8x}{8} \quad \text{Dividing by 8 on both sides}$$
$$8 = x$$
The solution is 8.

**25.**
$$81 = -3t$$
$$\frac{81}{-3} = \frac{-3t}{-3} \quad \text{Dividing by } -3 \text{ on both sides}$$
$$-27 = t$$
The solution is $-27$.

**27.**
$$-x = 83$$
$$-1 \cdot x = 83$$
$$\frac{-1 \cdot x}{-1} = \frac{83}{-1} \quad \text{Dividing by } -1 \text{ on both sides}$$
$$x = -83$$
The solution is $-83$.

**29.**
$$n(-6) = -42$$
$$\frac{n(-6)}{-6} = \frac{-42}{-6} \quad \text{Dividing by } -6 \text{ on both sides}$$
$$n = 7$$
The solution is 7.

**31.**
$$-x = -475$$
$$-1 \cdot x = -475$$
$$\frac{-1 \cdot x}{-1} = \frac{-475}{-1} \quad \text{Dividing by } -1 \text{ on both sides}$$
$$x = 475$$

The solution is 475.

**33.** $t - 7 = -2$

To undo the addition of $-7$, or the subtraction of 7, we subtract $-7$, or simply add 7, on both sides.
$$t - 7 = -2$$
$$t - 7 + 7 = -2 + 7$$
$$t + 0 = 5$$
$$t = 5$$

The solution is 5.

**35.** $6x = -90$

To undo multiplication by 6, we divide by 6 on both sides.
$$6x = -90$$
$$\frac{6x}{6} = \frac{-90}{6}$$
$$x = -15$$

The solution is $-15$.

**37.** $8 + x = 43$

To undo the addition of 8, we subtract 8 on both sides.
$$8 + x = 43$$
$$-8 + 8 + x = -8 + 43$$
$$0 + x = 35$$
$$x = 35$$

The solution is 35.

**39.** $18 = x - 27$

To undo the addition of $-27$, or the subtraction of 27, we subtract $-27$, or simply add 27, on both sides.
$$18 = x - 27$$
$$18 + 27 = x - 27 + 27$$
$$45 = x + 0$$
$$45 = x$$

The solution is 45.

**41.** $35 = -5t$

To undo multiplication by $-5$, we divide by $-5$ on both sides.
$$35 = -5t$$
$$\frac{35}{-5} = \frac{-5t}{-5}$$
$$-7 = t$$

The solution is $-7$.

**43.** $19x = -171$

To undo multiplication by 19, we divide by 19 on both sides.
$$19x = -171$$
$$\frac{19x}{19} = \frac{-171}{19}$$
$$x = -9$$

The solution is $-9$.

**45.** $19 + x = -171$

To undo the addition of 19, we subtract 19 on both sides.
$$19 + x = -171$$
$$-19 + 19 + x = -19 - 171$$
$$0 + x = -190$$
$$x = -190$$

The solution is $-190$.

**47.** $-38 = t + 43$

To undo the addition of 43, we subtract 43 on both sides.
$$-38 = t + 43$$
$$-38 - 43 = t + 43 - 43$$
$$-81 = t + 0$$
$$-81 = t$$

The solution is $-81$.

**49.**
$$5 + 3 \cdot 2^3$$
$$= 5 + 3 \cdot 8 \quad \text{Evaluating the exponential expression}$$
$$= 5 + 24 \quad \text{Multiplying}$$
$$= 29 \quad \text{Adding}$$

**51.**
$$12 \div 3 \cdot 2$$
$$= 4 \cdot 2 \quad \text{Dividing and multiplying in order}$$
$$= 8 \quad \text{from left to right}$$

**53.**
$$15 - 3 \cdot 2 + 7$$
$$= 15 - 6 + 7 \quad \text{Multiplying}$$
$$= 9 + 7 \quad \text{Subtracting and adding in order}$$
$$= 16 \quad \text{from left to right}$$

**55.** ◈

**57.** ◈

**59.**
$$9 + x - 5 = 23$$
$$x + 4 = 23 \quad \text{Combining like terms on the left}$$
$$x + 4 - 4 = 23 - 4 \quad \text{Subtracting 4 on both sides}$$
$$x + 0 = 19$$
$$x = 19$$

The solution is 19.

**61.** $(-9)^2 = 2^3 t + (3 \cdot 6 + 1)t$

$(-9)^2 = 2^3 t + (18 + 1)t$    Doing the calculations

$(-9)^2 = 2^3 t + 19t$        inside the parentheses

$81 = 8t + 19t$   Evaluating the exponential
          expressions

$81 = 27t$      Combining like terms

$\dfrac{81}{27} = \dfrac{27t}{27}$    Dividing by 27 on both sides

$3 = t$

The solution is 3.

**63.** $(-17)^3 = 15^3 x$

$-4913 = 3375x$   Evaluating the exponential
           expressions

$\dfrac{-4913}{3375} = \dfrac{3375x}{3375}$   Dividing by 3375 on both sides

$-\dfrac{4913}{3375} = x$

The solution is $-\dfrac{4913}{3375}$.

**65.** $23^2 = x + 22^2$

$529 = x + 484$      Evaluating the exponential
                expression

$529 - 484 = x + 484 - 484$   Subtracting 484 on
                            both sides

$45 = x + 0$

$45 = x$

The solution is 45.

**67.** $4x + 3 = 31$

$4x + 3 - 3 = 31 - 3$   Subtracting 3 on both sides

$4x + 0 = 28$

$4x = 28$

$\dfrac{4x}{4} = \dfrac{28}{4}$    Dividing by 4 on both sides

$x = 7$

The solution is 7.

**69.** $8 + 3x = -22$

$-8 + 8 + 3x = -8 - 22$   Subtracting 8 on both sides

$0 + 3x = -30$

$3x = -30$

$\dfrac{3x}{3} = \dfrac{-30}{3}$   Dividing by 3 on both sides

$x = -10$

The solution is $-10$.

# Chapter 3

# Fractional Notation: Multiplication and Division

## Exercise Set 3.1

**1.** $1 \cdot 6 = 6$     $6 \cdot 6 = 36$
$2 \cdot 6 = 12$     $7 \cdot 6 = 42$
$3 \cdot 6 = 18$     $8 \cdot 6 = 48$
$4 \cdot 6 = 24$     $9 \cdot 6 = 54$
$5 \cdot 6 = 30$     $10 \cdot 6 = 60$

**3.** $1 \cdot 20 = 20$     $6 \cdot 20 = 120$
$2 \cdot 20 = 40$     $7 \cdot 20 = 140$
$3 \cdot 20 = 60$     $8 \cdot 20 = 160$
$4 \cdot 20 = 80$     $9 \cdot 20 = 180$
$5 \cdot 20 = 100$     $10 \cdot 20 = 200$

**5.** $1 \cdot 3 = 3$     $6 \cdot 3 = 18$
$2 \cdot 3 = 6$     $7 \cdot 3 = 21$
$3 \cdot 3 = 9$     $8 \cdot 3 = 24$
$4 \cdot 3 = 12$     $9 \cdot 3 = 27$
$5 \cdot 3 = 15$     $10 \cdot 3 = 30$

**7.** $1 \cdot 13 = 13$     $6 \cdot 13 = 78$
$2 \cdot 13 = 26$     $7 \cdot 13 = 91$
$3 \cdot 13 = 39$     $8 \cdot 13 = 104$
$4 \cdot 13 = 52$     $9 \cdot 13 = 117$
$5 \cdot 13 = 65$     $10 \cdot 13 = 130$

**9.** $1 \cdot 10 = 10$     $6 \cdot 10 = 60$
$2 \cdot 10 = 20$     $7 \cdot 10 = 70$
$3 \cdot 10 = 30$     $8 \cdot 10 = 80$
$4 \cdot 10 = 40$     $9 \cdot 10 = 90$
$5 \cdot 10 = 50$     $10 \cdot 10 = 100$

**11.** $1 \cdot 9 = 9$     $6 \cdot 9 = 54$
$2 \cdot 9 = 18$     $7 \cdot 9 = 63$
$3 \cdot 9 = 27$     $8 \cdot 9 = 72$
$4 \cdot 9 = 36$     $9 \cdot 9 = 81$
$5 \cdot 9 = 45$     $10 \cdot 9 = 90$

**13.** We divide 26 by 7.

$$
\begin{array}{r}
3 \\
7 \overline{)26} \\
21 \\
\hline
5
\end{array}
$$

Since the remainder is not 0, 26 is not divisible by 7.

**15.** We divide 1880 by 8.

$$
\begin{array}{r}
235 \\
8 \overline{)1880} \\
1600 \\
\hline
280 \\
240 \\
\hline
40 \\
40 \\
\hline
0
\end{array}
$$

The remainder of 0 indicates that 1880 is divisible by 8.

**17.** We divide 106 by 4.

$$
\begin{array}{r}
26 \\
4 \overline{)106} \\
80 \\
\hline
26 \\
24 \\
\hline
2
\end{array}
$$

Since the remainder is not 0, 106 is not divisible by 4.

**19.** We divide 4227 by 9.

$$
\begin{array}{r}
469 \\
9 \overline{)4227} \\
3600 \\
\hline
627 \\
540 \\
\hline
87 \\
81 \\
\hline
6
\end{array}
$$

Since the remainder is not 0, 4227 is not divisible by 9.

**21.** We divide 8650 by 16.

$$
\begin{array}{r}
540 \\
16 \overline{)8650} \\
8000 \\
\hline
650 \\
640 \\
\hline
10
\end{array}
$$

Since the remainder is not 0, 8650 is not divisible by 16.

**23.** A number is divisible by 2 if its <u>ones digit</u> is even.

4<u>6</u> is divisible by 2 because <u>6</u> is even.
22<u>4</u> is divisible by 2 because <u>4</u> is even.
1<u>9</u> is not divisible by 2 because <u>9</u> is not even.
55<u>5</u> is not divisible by 2 because <u>5</u> is not even.
30<u>0</u> is divisible by 2 because <u>0</u> is even.
3<u>6</u> is divisible by 2 because <u>6</u> is even.
45,27<u>0</u> is divisible by 2 because <u>0</u> is even.
444<u>4</u> is divisible by 2 because <u>4</u> is even.
8<u>5</u> is not divisible by 2 because <u>5</u> is not even.
71<u>1</u> is not divisible by 2 because <u>1</u> is not even.
13,25<u>1</u> is not divisible by 2 because <u>1</u> is not even.
254,76<u>5</u> is not divisible by 2 because <u>5</u> is not even.
25<u>6</u> is divisible by 2 because <u>6</u> is even.
806<u>4</u> is divisible by 2 because <u>4</u> is even.
186<u>7</u> is not divisible by 2 because <u>7</u> is not even.
21,56<u>8</u> is divisible by 2 because <u>8</u> is even.

**25.** A number is divisible by 10 if its ones digit is 0.

Of the numbers under consideration, only 300 and 45,270 have one digits of 0. Therefore, only 300 and 45,270 are divisible by 10.

**27.** For a number to be divisible by 6, the sum of the digits must be divisible by 3 and the ones digit must be 0, 2, 4, 6 or 8 (even). It is most efficient to determine if the ones digit is even first and then, if so, to determine if the sum of the digits is divisible by 3.

46 is not divisible by 6 because 46 is not divisible by 3.

$$4 + 6 = 10$$
$$\uparrow$$
Not divisible by 3

224 is not divisible by 6 because 224 is not divisible by 3.

$$2 + 2 + 4 = 8$$
$$\uparrow$$
Not divisible by 3

19 is not divisible by 6 because 19 is not even.

19
$\uparrow$
Not even

555 is not divisible by 6 because 555 is not even.

555
$\uparrow$
Not even

300 is divisible by 6.

300    $3 + 0 + 0 = 3$
$\uparrow$    $\uparrow$
Even    Divisible by 3

36 is divisible by 6.

36    $3 + 6 = 9$
$\uparrow$    $\uparrow$
Even    Divisible by 3

45,270 is divisible by 6.

45,270    $4 + 5 + 2 + 7 + 0 = 18$
$\uparrow$    $\uparrow$
Even    Divisible by 3

4444 is not divisible by 6 because 4444 is not divisible by 3.

$$4 + 4 + 4 + 4 = 16$$
$$\uparrow$$
Not divisible by 3

85 is not divisible by 6 because 85 is not even.

85
$\uparrow$
Not even

711 is not divisible by 6 because 711 is not even.

711
$\uparrow$
Not even

13,251 is not divisible by 6 because 13,251 is not even.

13,251
$\uparrow$
Not even

254,765 is not divisible by 6 because 254,765 is not even.

254,765
$\uparrow$
Not even

256 is not divisible by 6 because 256 is not divisible by 3.

$$2 + 5 + 6 = 13$$
$$\uparrow$$
Not divisible by 3

8064 is divisible by 6.

8064    $8 + 0 + 6 + 4 = 18$
$\uparrow$    $\uparrow$
Even    Divisible by 3

1867 is not divisible by 6 because 1867 is not even.

1867
$\uparrow$
Not even

21,568 is not divisible by 6 because 21,568 is not divisible by 3.

$$2 + 1 + 5 + 6 + 8 = 22$$
$$\uparrow$$
Not divisible by 3

**29.** A number is divisible by 3 if the sum of the digits is divisible by 3.

56 is not divisible by 3 because $5 + 6 = 11$ and 11 is not divisible by 3.

324 is divisible by 3 because $3 + 2 + 4 = 9$ and 9 is divisible by 3.

784 is not divisible by 3 because $7 + 8 + 4 = 19$ and 19 is not divisible by 3.

55,555 is not divisible by 3 because $5 + 5 + 5 + 5 + 5 = 25$ and 25 is not divisible by 3.

200 is not divisible by 3 because $2 + 0 + 0 = 2$ and 2 is not divisible by 3.

42 is divisible by 3 because $4 + 2 = 6$ and 6 is divisible by 3.

501 is divisible by 3 because $5 + 0 + 1 = 6$ and 6 is divisible by 3.

3009 is divisible by 3 because $3 + 0 + 0 + 9 = 12$ and 12 is divisible by 3.

75 is divisible by 3 because $7 + 5 = 12$ and 12 is divisible by 3.

812 is not divisible by 3 because $8 + 1 + 2 = 11$ and 11 is not divisible by 3.

2345 is not divisible by 3 because $2 + 3 + 4 + 5 = 14$ and 14 is not divisible by 3.

2001 is divisible by 3 because $2 + 0 + 0 + 1 = 3$ and 3 is divisible by 3.

35 is not divisible by 3 because $3 + 5 = 8$ and 8 is not divisible by 3.

402 is divisible by 3 because $4 + 0 + 2 = 6$ and 6 is divisible by 3.

111,111 is divisible by 3 because $1 + 1 + 1 + 1 + 1 + 1 = 6$ and 6 is divisible by 3.

1005 is divisible by 3 because $1 + 0 + 0 + 5 = 6$ and 6 is divisible by 3.

**31.** A number is divisible by 5 if the ones digit is 0 or 5.

5<u>6</u> is not divisible by 5 because the ones digit (6) is not 0 or 5.

32<u>4</u> is not divisible by 5 because the ones digit (4) is not 0 or 5.

78<u>4</u> is not divisible by 5 because the ones digit (4) is not 0 or 5.

55,55<u>5</u> is divisible by 5 because the ones digit is 5.

20<u>0</u> is divisible by 5 because the ones digit is 0.

4<u>2</u> is not divisible by 5 because the ones digit (2) is not 0 or 5.

50<u>1</u> is not divisible by 5 because the ones digit (1) is not 0 or 5.

300<u>9</u> is not divisible by 5 because the ones digit (9) is not 0 or 5.

7<u>5</u> is divisible by 5 because the ones digit is 5.

81<u>2</u> is not divisible by 5 because the ones digit (2) is not 0 or 5.

234<u>5</u> is divisible by 5 because the ones digit is 5.

200<u>1</u> is not divisible by 5 because the ones digit (1) is not 0 or 5.

3<u>5</u> is divisible by 5 because the ones digit is 5.

40<u>2</u> is not divisible by 5 because the ones digit (2) is not 0 or 5.

111,11<u>1</u> is not divisible by 5 because the ones digit (1) is not 0 or 5.

100<u>5</u> is divisible by 5 because the ones digit is 5.

**33.** A number is divisible by 9 if the sum of the digits is divisible by 9.

56 is not divisible by 9 because $5 + 6 = 11$ and 11 is not divisible by 9.

324 is divisible by 9 because $3 + 2 + 4 = 9$ and 9 is divisible by 9.

784 is not divisible by 9 because $7 + 8 + 4 = 19$ and 19 is not divisible by 9.

55,555 is not divisible by 9 because $5 + 5 + 5 + 5 + 5 = 25$ and 25 is not divisible by 9.

200 is not divisible by 9 because $2 + 0 + 0 = 2$ and 2 is not divisible by 9.

42 is not divisible by 9 because $4 + 2 = 6$ and 6 is not divisible by 9.

501 is not divisible by 9 because $5 + 0 + 1 = 6$ and 6 is not divisible by 9.

3009 is not divisible by 9 because $3 + 0 + 0 + 9 = 12$ and 12 is not divisible by 9.

75 is not divisible by 9 because $7 + 5 = 12$ and 12 is not divisible by 9.

812 is not divisible by 9 because $8 + 1 + 2 = 11$ and 11 is not divisible by 9.

2345 is not divisible by 9 because $2 + 3 + 4 + 5 = 14$ and 14 is not divisible by 9.

2001 is not divisible by 9 because $2 + 0 + 0 + 1 = 3$ and 3 is not divisible by 9.

35 is not divisible by 9 because $3 + 5 = 8$ and 8 is not divisible by 9.

402 is not divisible by 9 because $4 + 0 + 2 = 6$ and is not divisible by 9.

111,111 is not divisible by 9 because $1+1+1+1+1+1 = 6$ and 6 is not divisible by 9.

1005 is not divisible by 9 because $1 + 0 + 0 + 5 = 6$ and 6 is not divisible by 9.

**35.** $16 \cdot t = 848$

$$\frac{16 \cdot t}{16} = \frac{848}{16} \quad \text{Dividing by 16 on both sides}$$

$$t = 53$$

The solution is 53.

**37.**
$$23 + x = 15$$
$$23 + x - 23 = 15 - 23 \quad \text{Subtracting 23 on both sides}$$
$$x = -8$$

The solution is $-8$.

**39.** *Familiarize*. This is a multistep problem. Find the total cost of the shirts and the total cost of the pants and then find the sum of the two.

We let $s =$ the total cost of the shirts and $t =$ the total cost of the pants.

*Translate*. We write two equations.

| Number of shirts | times | Cost of one shirt | is | Total cost of shirts |
|---|---|---|---|---|
| ↓ | ↓ | ↓ | ↓ | ↓ |
| 12 | · | 37 | = | s |

| Number of pairs of pants | times | Cost of one pair | is | Total cost of pants |
|---|---|---|---|---|
| ↓ | ↓ | ↓ | ↓ | ↓ |
| 4 | · | 59 | = | t |

*Solve*. We carry out the multiplication.

$$12 \cdot 37 = s$$
$$444 = s \quad \text{Doing the multiplication}$$

The total cost of the 12 shirts is $444.

$$4 \cdot 59 = t$$
$$236 = t \quad \text{Doing the multiplication}$$

The total cost of the 4 pairs of pants is $236.

Now we find the total amount spent. We let $a =$ this amount.

| Total cost of shirts | plus | Total cost of pants | is | Total amount spent |
|---|---|---|---|---|
| ↓ | ↓ | ↓ | ↓ | ↓ |
| 444 | + | 236 | = | a |

To solve the equation, carry out the addition.

$$\begin{array}{r} 4\,4\,4 \\ +\,2\,3\,6 \\ \hline 6\,8\,0 \end{array}$$

*Check*.  We can repeat the calculations.  The answer checks.

*State*.  The total cost is $680.

**41.**

**43.**

**45.** When we use a calculator to divide the largest five-digit number, 99,999, by 47 we get 2127.638298. This tells us that 99,999 is not divisible by 47 but that $2127 \times 47$, or 99,969, is divisible by 47 and that it is the largest such five-digit number.

**47.** We list multiples of 2, 3, and 5 and find the smallest number that is on all 3 lists.

Multiples of 2:  2, 4, 6, 8, 10, 12, 14, 16, 18, 20, 22, 24, 26, 28, <u>30</u>, 32, · · ·

Multiples of 3:  3, 6, 9, 12, 15, 18, 21, 24, 27, <u>30</u>, 33, · · ·

Multiples of 5:  5, 10, 15, 20, 25, <u>30</u>, 35, · · ·

The smallest number that is simultaneously a multiple of 2, 3, and 5 is 30.

**49.** We list multiples of 4, 6, and 10 and find the smallest number that is on all 3 lists.

Multiples of 4:  4, 8, 12, 16, 20, 24, 28, 32, 36, 40, 44, 48, 52, 56, <u>60</u>, 64, · · ·

Multiples of 6:  6, 12, 18, 24, 30, 36, 42, 48, 54, <u>60</u>, 66, · · ·

Multiples of 10:  10, 20, 30, 40, 50, <u>60</u>, 70, · · ·

The smallest number that is simultaneously a multiple of 4, 6, and 10 is 60.

**51.** The number must be a multiple of 11, so we try numbers of the form $11 \cdot n$ where $n$ is greater than 6. The smallest multiple that meets the criteria is $11 \cdot 11$, or 121. This is the driver's number.

## Exercise Set 3.2

**1.** Since 18 is even, we know that 2 is a factor.  Since the sum of the digits is 9 and 9 is divisible by both 3 and 9, we know that 3 and 9 are both factors.  Also, since 2 and 3 are factors, 6 is a factor as well.  We write a list of factorizations.

$18 = 1 \cdot 18$ $\qquad 18 = 3 \cdot 6$
$18 = 2 \cdot 9$

Factors: 1, 2, 3, 6, 9, 18

**3.** Since 54 is even, we know that 2 is a factor.  Since the sum of the digits is 9 and 9 is divisible by both 3 and 9, we know that 3 and 9 are both factors.  Also, since 2 and 3 are factors, 6 is a factor as well.  We write a list of factorizations.

$54 = 1 \cdot 54$ $\qquad 54 = 3 \cdot 18$
$54 = 2 \cdot 27$ $\qquad 54 = 6 \cdot 9$

Factors: 1, 2, 3, 6, 9, 18, 27, 54

**5.** Since 4 is even, we know that 2 is a factor.  We write a list of factorizations:

$4 = 1 \cdot 4$ $\qquad\qquad 4 = 2 \cdot 2$

Factors: 1, 2, 4

**7.** The only factorization is $7 = 1 \cdot 7$.

Factors: 1, 7

**9.** The only factorization is $1 = 1 \cdot 1$.

Factor: 1

**11.** Since 98 is even, we now that 2 is a factor.  Using other tests for divisibility, we determine that 3, 5, 6, 9, and 10 are not factors.  We write a list of factorizations.

$98 = 1 \cdot 98$ $\qquad 98 = 7 \cdot 14$
$98 = 2 \cdot 49$

Factors: 1, 2, 7, 14, 49, 98

**13.** Using tests for divisibility we determine that 2, 3, and 6 are factors.  We write a list of factorizations.

$42 = 1 \cdot 42$ $\qquad 42 = 3 \cdot 14$
$42 = 2 \cdot 21$ $\qquad 42 = 6 \cdot 7$

Factors: 1, 2, 3, 6, 7, 14, 21, 42

**15.** Using tests for divisibility we determine that 5 is a factor.  We write a list of factorizations.

$385 = 1 \cdot 385$ $\qquad 385 = 7 \cdot 55$
$385 = 5 \cdot 77$ $\qquad 385 = 11 \cdot 35$

Factors: 1, 5, 7, 11, 35, 55, 77, 385

**17.** Using tests for divisibility we determine that 2, 3, 6, and 9 are factors.  We write a list of factorizations.

$36 = 1 \cdot 36$ $\qquad 36 = 4 \cdot 9$
$36 = 2 \cdot 18$ $\qquad 36 = 6 \cdot 6$
$36 = 3 \cdot 12$

Factors: 1, 2, 3, 4, 6, 9, 12, 18, 36

**19.** Using tests for divisibility we determine that 3, 5, and 9 are factors.  We write a list of factorizations.

$225 = 1 \cdot 225$ $\qquad 225 = 9 \cdot 25$
$225 = 3 \cdot 75$ $\qquad 225 = 15 \cdot 15$
$225 = 5 \cdot 45$

Factors: 1, 3, 5, 9, 15, 25, 45, 75, 225

**21.** The number 17 is prime. It has only the factors 1 and 17.

**23.** The number 22 has factors 1, 2, 11, and 22. Since it has at least one factor other than itself and 1, it is composite.

**25.** The number 48 has factors 1, 2, 3, 4, 6, 8, 12, 16, 24, and 48. Since it has at least one factor other than itself and 1, it is composite.

**27.** The number 31 is prime. It has only the factors 1 and 31.

**29.** 1 is neither prime nor composite.

**31.** The number 9 has factors 1, 3, and 9.

Since it has at least one factor other than itself and 1, it is composite.

**33.** The number 47 is prime. It has only the factors 1 and 47.

**35.** The number 29 is prime. It has only the factors 1 and 29.

**37.**
$$\begin{array}{r} 2 \quad \leftarrow \quad 2 \text{ is prime.} \\ 2\overline{\smash{\big)}\,4} \\ 2\overline{\smash{\big)}\,8} \\ 2\overline{\smash{\big)}\,16} \end{array}$$

$16 = 2 \cdot 2 \cdot 2 \cdot 2$

**39.**
$$\begin{array}{r} 7 \quad \leftarrow \quad 7 \text{ is prime.} \\ 2\overline{\smash{\big)}\,14} \end{array}$$

$14 = 2 \cdot 7$

**41.**
$$\begin{array}{r} 11 \quad \leftarrow \quad 11 \text{ is prime.} \\ 2\overline{\smash{\big)}\,22} \end{array}$$

$22 = 2 \cdot 11$

**43.**
$$\begin{array}{r} 5 \quad \leftarrow \quad 5 \text{ is prime.} \\ 5\overline{\smash{\big)}\,25} \end{array}$$
(25 is not divisible by 2 or 3. We move to 5.)

$25 = 5 \cdot 5$

**45.**
$$\begin{array}{r} 31 \quad \leftarrow \quad 31 \text{ is prime.} \\ 2\overline{\smash{\big)}\,62} \end{array}$$

$62 = 2 \cdot 31$

**47.**
$$\begin{array}{r} 7 \quad \leftarrow \quad 7 \text{ is prime.} \\ 5\overline{\smash{\big)}\,35} \\ 2\overline{\smash{\big)}\,70} \\ 2\overline{\smash{\big)}\,140} \end{array}$$

$140 = 2 \cdot 2 \cdot 5 \cdot 7$

**49.**
$$\begin{array}{r} 5 \quad \leftarrow \quad 5 \text{ is prime.} \\ 5\overline{\smash{\big)}\,25} \\ 2\overline{\smash{\big)}\,50} \\ 2\overline{\smash{\big)}\,100} \end{array}$$
(25 is not divisible by 2 or 3. We move to 5.)

$100 = 2 \cdot 2 \cdot 5 \cdot 5$

We can also use a factor tree.

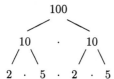

$$\begin{array}{ccccc} & & 100 & & \\ & 10 & \cdot & 10 & \\ 2 & \cdot & 5 & \cdot \quad 2 & \cdot \quad 5 \end{array}$$

**51.**
$$\begin{array}{r} 7 \quad \leftarrow \quad 7 \text{ is prime.} \\ 5\overline{\smash{\big)}\,35} \end{array}$$
(35 is not divisible by 2 or 3. We move to 5.)

$35 = 5 \cdot 7$

**53.**
$$\begin{array}{r} 13 \quad \leftarrow \quad 13 \text{ is prime.} \\ 3\overline{\smash{\big)}\,39} \\ 2\overline{\smash{\big)}\,78} \end{array}$$
(39 is not divisible by 2. We move to 3.)

$78 = 2 \cdot 3 \cdot 13$

**55.**
$$\begin{array}{r} 11 \quad \leftarrow \quad 11 \text{ is prime.} \\ 7\overline{\smash{\big)}\,77} \end{array}$$
(77 is not divisible by 2, 3, or 5. We move to 7.)

$77 = 7 \cdot 11$

**57.**
$$\begin{array}{r} 7 \quad \leftarrow \quad 7 \text{ is prime.} \\ 2\overline{\smash{\big)}\,14} \\ 2\overline{\smash{\big)}\,28} \\ 2\overline{\smash{\big)}\,56} \\ 2\overline{\smash{\big)}\,112} \end{array}$$

$112 = 2 \cdot 2 \cdot 2 \cdot 2 \cdot 7$

We can also use a factor tree.

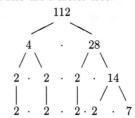

$$\begin{array}{ccccccc} & & & 112 & & & \\ & & 4 & \cdot & 28 & & \\ & 2 \cdot 2 & & & 2 & \cdot & 14 \\ 2 & \cdot \quad 2 & \cdot & 2 & \cdot \quad 2 & \cdot & 7 \end{array}$$

**59.**
$$\begin{array}{r} 5 \quad \leftarrow \quad 5 \text{ is prime.} \\ 5\overline{\smash{\big)}\,25} \\ 3\overline{\smash{\big)}\,75} \\ 2\overline{\smash{\big)}\,150} \\ 2\overline{\smash{\big)}\,300} \end{array}$$

$300 = 2 \cdot 2 \cdot 3 \cdot 5 \cdot 5$

We can also use a factor tree.

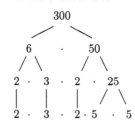

$$\begin{array}{ccccccc} & & & 300 & & & \\ & & 6 & \cdot & 50 & & \\ & 2 \cdot 3 & & & 2 & \cdot & 25 \\ 2 & \cdot \quad 3 & \cdot & 2 & \cdot \quad 5 & \cdot & 5 \end{array}$$

**61.** $-2 \cdot 13 = -26$ (The signs are different, so the answer is negative.)

**63.** $-17 + 25$ The absolute values are 17 and 25. The difference is 8. The positive number has the larger absolute value, so the answer is positive. $-17 + 25 = 8$

**65.** $0 \div 22 = 0$ (0 divided by a nonzero number is 0.)

**67.** ◈

**69.** ◈

**71.** Use a calculator to perform successive divisions by prime numbers.

$28,502,923 = 53 \cdot 53 \cdot 73 \cdot 139$

**73.**
$$\begin{array}{r} 7 \quad \leftarrow \quad 7 \text{ is prime.} \\ 5\overline{\smash{\big)}\,35} \\ 3\overline{\smash{\big)}\,105} \\ 3\overline{\smash{\big)}\,315} \\ 2\overline{\smash{\big)}\,630} \\ 2\overline{\smash{\big)}\,1260} \\ 2\overline{\smash{\big)}\,2520} \end{array}$$

$2520 = 2 \cdot 2 \cdot 2 \cdot 3 \cdot 3 \cdot 5 \cdot 7$

**75.**
$$3\overline{\smash{\big)}\,111} \leftarrow 37 \text{ is prime.}$$
$$3\overline{\smash{\big)}\,333}$$
$$3\overline{\smash{\big)}\,999}$$
$$2\overline{\smash{\big)}\,1998}$$

$1998 = 2 \cdot 3 \cdot 3 \cdot 3 \cdot 37$

**77.** Answers may vary. One arrangement is a 3-dimensional rectangular array consisting of 2 tiers of 12 objects each where each tier consists of a rectangular array of 4 rows with 3 objects each.

## Exercise Set 3.3

**1.** The top number is the numerator, and the bottom number is the denominator.
$$\frac{3}{4} \quad \begin{array}{l} \leftarrow \text{Numerator} \\ \leftarrow \text{Denominator} \end{array}$$

**3.** $\dfrac{7}{-9} \quad \begin{array}{l} \leftarrow \text{Numerator} \\ \leftarrow \text{Denominator} \end{array}$

**5.** $\dfrac{2x}{3z} \quad \begin{array}{l} \leftarrow \text{Numerator} \\ \leftarrow \text{Denominator} \end{array}$

**7.** The dollar is divided into 4 parts of the same size, and 3 of them are shaded. This is $3 \cdot \dfrac{1}{4}$ or $\dfrac{3}{4}$. Thus, $\$\dfrac{3}{4}$ (three-fourths of a dollar) is shaded.

**9.** We have 2 miles, each of which is divided into 8 parts of the same size. Two of those parts are shaded. This is $2 \cdot \dfrac{1}{8}$, or $\dfrac{2}{8}$ mi.

**11.** We have 2 liters, each divided into three parts of the same size. Four of those parts are shaded. This is $4 \cdot \dfrac{1}{3}$, or $\dfrac{4}{3}$ L.

**13.** The acre is divided into 4 parts of the same size, and 3 of them are shaded. This is $3 \cdot \dfrac{1}{4}$ or $\dfrac{3}{4}$ acre.

**15.** The pound is divided into 16 parts of the same size, and 8 of them are shaded. This is $8 \cdot \dfrac{1}{16}$, or $\dfrac{8}{16}$ lb.

**17.** There are 12 marbles and 5 are shaded. Thus, $\dfrac{5}{12}$ of the marbles are shaded.

**19.** There are 4 sailboats and 1 of them is in the shade. Thus, $\dfrac{1}{4}$ of the sailboats are shaded.

**21.** Remember: $\dfrac{0}{n} = 0$, for any integer $n$ that is not 0.
$$\frac{0}{17} = 0$$
Think of dividing an object into 17 parts and taking none of them. We get 0.

**23.** Remember: $\dfrac{n}{1} = n$.
$$\frac{15}{1} = 15$$
Think of taking 15 objects and dividing them into 1 part. (We do not divide them.) We have 15 objects.

**25.** Remember: $\dfrac{n}{n} = 1$, for any integer $n$ that is not 0.
$$\frac{20}{20} = 1$$
If we divide an object into 20 parts and take 20 of them, we get all of the object (1 whole object).

**27.** Remember: $\dfrac{n}{n} = 1$, for any integer $n$ that is not 0.
$$\frac{-14}{-14} = 1$$

**29.** Remember: $\dfrac{0}{n} = 0$, for any integer $n$ that is not 0.
$$\frac{0}{-234} = 0$$

**31.** Remember: $\dfrac{n}{n} = 1$, for any integer $n$ that is not 0.
$$\frac{3n}{3n} = 1$$

**33.** Remember: $\dfrac{n}{n} = 1$, for any integer $n$ that is not 0.
$$\frac{9x}{9x} = 1$$

**35.** Remember: $\dfrac{n}{1} = n$
$$\frac{-63}{1} = -63$$

**37.** Remember: $\dfrac{0}{n} = 0$, for any integer $n$ that is not 0.
$$\frac{0}{2a} = 0$$

**39.** Remember: $\dfrac{n}{0}$ is not defined.
$$\frac{52}{0} \text{ is not defined.}$$

**41.** Remember: $\dfrac{n}{1} = n$
$$\frac{7n}{1} = 7n$$

**43.** $\dfrac{6}{7-7} = \dfrac{6}{0}$

Remember: $\dfrac{n}{0}$ is not defined. Thus, $\dfrac{6}{7-7}$ is not defined.

**45.** $-7(30) = -210$
(The signs are different, so the answer is negative.)

**47.** $(-71)(-12)0 = -71 \cdot 0 = 0$
(We might have observed at the outset that the answer is 0 since one of the factors is 0.)

**49.** *Familiarize.* We let $a$ = the amount by which the average income in Connecticut exceeds the average income in Mississippi.

*Translate.* This is "how-much-more" situation.

| Average income in Mississippi | plus | Excess income in Connecticut | is | Average income in Connecticut |
|:---:|:---:|:---:|:---:|:---:|
| ↓ | ↓ | ↓ | ↓ | ↓ |
| 16,531 | + | $a$ | = | 30,303 |

*Solve.* We solve the equation.

$$16,531 + a = 30,303$$
$$16,531 + a - 16,531 = 30,303 - 16,531$$
$$a = 13,772$$

*Check.* We add the answer to the average income in Mississippi: $\$13,772 + \$16,531 = \$30,303$. This is the average income in Connecticut, so our answer checks.

*State.* On average, people living in Connecticut make $13,772 more than those living in Mississippi.

**51.** ◈

**53.** ◈

**55.** $365 = 52 \cdot 7 + 1$, so in one year there are 52 full weeks plus one additional day. Since 1999 started on a Friday, the additional day was not a Monday. (It is a Friday.) Thus, of the 365 days in 1999, 52 were Mondays, so $\dfrac{52}{365}$ were Mondays.

**57.** The couple's 3 sons had a total of $3 \cdot 3$, or 9 daughters. The 9 daughters had a total of $9 \cdot 3$, or 27 sons. Altogether the couple had $3 + 9 + 27$, or 39 descendants. Nine of them were female, so $\dfrac{9}{39}$ are female.

---

## Exercise Set 3.4

**1.** $3 \cdot \dfrac{1}{7} = \dfrac{3 \cdot 1}{7} = \dfrac{3}{7}$

**3.** $(-5) \times \dfrac{1}{6} = \dfrac{-5 \times 1}{6} = \dfrac{-5}{6}$, or $-\dfrac{5}{6}$

**5.** $\dfrac{2}{3} \cdot 7 = \dfrac{2 \cdot 7}{3} = \dfrac{14}{3}$

**7.** $(-1)\dfrac{7}{9} = \dfrac{(-1)7}{9} = \dfrac{-7}{9}$, or $-\dfrac{7}{9}$

**9.** $\dfrac{2}{5} \cdot x = \dfrac{2 \cdot x}{5} = \dfrac{2x}{5}$

**11.** $\dfrac{2}{5}(-3) = \dfrac{2(-3)}{5} = \dfrac{-6}{5}$, or $-\dfrac{6}{5}$

**13.** $a \cdot \dfrac{3}{4} = \dfrac{a \cdot 3}{4} = \dfrac{3a}{4}$

**15.** $17 \times \dfrac{m}{6} = \dfrac{17 \times m}{6} = \dfrac{17m}{6}$

**17.** $-3 \cdot \dfrac{-2}{5} = \dfrac{-3}{1} \cdot \dfrac{-2}{5} = \dfrac{-3(-2)}{1 \cdot 5} = \dfrac{6}{5}$

**19.** $-\dfrac{2}{7}(-x) = \dfrac{-2}{7} \cdot \dfrac{-x}{1} = \dfrac{-2(-x)}{7 \cdot 1} = \dfrac{2x}{7}$

**21.** $\dfrac{1}{2} \cdot \dfrac{1}{5} = \dfrac{1 \cdot 1}{2 \cdot 5} = \dfrac{1}{10}$

**23.** $\left(-\dfrac{1}{4}\right) \times \dfrac{1}{10} = -\dfrac{1 \times 1}{4 \times 10} = -\dfrac{1}{40}$, or $\dfrac{-1}{40}$

**25.** $\dfrac{2}{3} \times \dfrac{1}{5} = \dfrac{2 \times 1}{3 \times 5} = \dfrac{2}{15}$

**27.** $\dfrac{2}{y} \cdot \dfrac{x}{5} = \dfrac{2 \cdot x}{y \cdot 5} = \dfrac{2x}{5y}$

**29.** $\left(-\dfrac{3}{4}\right)\left(-\dfrac{3}{4}\right) = \dfrac{(-3)(-3)}{4 \cdot 4} = \dfrac{9}{16}$

**31.** $\dfrac{2}{3} \cdot \dfrac{7}{13} = \dfrac{2 \cdot 7}{3 \cdot 13} = \dfrac{14}{39}$

**33.** $\dfrac{1}{10}\left(\dfrac{-3}{5}\right) = \dfrac{1(-3)}{10 \cdot 5} = \dfrac{-3}{50}$, or $-\dfrac{3}{50}$

**35.** $\dfrac{7}{8} \cdot \dfrac{a}{8} = \dfrac{7 \cdot a}{8 \cdot 8} = \dfrac{7a}{64}$

**37.** $\dfrac{1}{y} \cdot \dfrac{1}{100} = \dfrac{1 \cdot 1}{y \cdot 100} = \dfrac{1}{100y}$

**39.** $\dfrac{-14}{15} \cdot \dfrac{13}{19} = \dfrac{-14 \cdot 13}{15 \cdot 19} = \dfrac{-182}{285}$, or $-\dfrac{182}{285}$

**41.** *Familiarize.* Recall that area is length times width. We draw a picture. We will let A = the area of the table top.

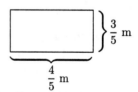

*Translate.* Then we translate.

Area is length times width
↓ ↓ ↓ ↓ ↓
$$A = \dfrac{4}{5} \times \dfrac{3}{5}$$

*Solve.* The sentence tells us what to do. We multiply.

$$\dfrac{4}{5} \times \dfrac{3}{5} = \dfrac{4 \times 3}{5 \times 5} = \dfrac{12}{25}$$

*Check.* We repeat the calculation. The answer checks.

*State.* The area is $\dfrac{12}{25}$ m².

**43.** *Familiarize.* Let $t$ = the number of gallons of two-cycle oil in a freshly filled chainsaw.

*Translate.* The multiplication sentence $\dfrac{1}{16} \cdot \dfrac{1}{5} = t$ corresponds to this situation.

*Solve*. We carry out the multiplication.

$$\frac{1}{16} \cdot \frac{1}{5} = \frac{1 \cdot 1}{16 \cdot 5} = \frac{1}{80}$$

*Check*. We repeat the calculation. The answer checks.

*State*. There is $\frac{1}{80}$ gal of two-cycle oil in a freshly filled chainsaw.

**45.** *Familiarize*. We draw a picture. We let $m =$ the amount of molasses in $\frac{3}{4}$ of a batch.

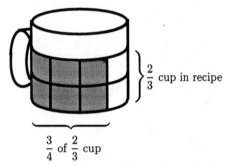

$$\frac{3}{4} \text{ of } \frac{2}{3} \text{ cup}$$

*Translate*. The multiplication sentence $\frac{3}{4} \cdot \frac{2}{3} = m$ corresponds to the situation.

*Solve*. We carry out the multiplication:

$$\frac{3}{4} \cdot \frac{2}{3} = \frac{3 \cdot 2}{4 \cdot 3} = \frac{6}{12}$$

*Check*. We repeat the calculation or determine what fractional part of the drawing is shaded. The answer checks.

*State*. $\frac{6}{12}$ cup of molasses is needed to make $\frac{3}{4}$ of a batch.

**47.** *Familiarize*. Two of every 3 tons of municipal waste are dumped in landfills, so $\frac{2}{3}$ of all municipal waste is dumped in landfills. We also know that $\frac{1}{10}$ of the waste in landfills is yard trimmings. We let $y =$ the fractional part of municipal waste that is yard trimmings.

*Translate*. The multiplication sentence $\frac{1}{10} \cdot \frac{2}{3} = y$ corresponds to this situation.

*Solve*. We carry out the multiplication.

$$\frac{1}{10} \cdot \frac{2}{3} = \frac{1 \cdot 2}{10 \cdot 3} = \frac{2}{30}$$

*Check*. We repeat the calculation. The answer checks.

*State*. $\frac{2}{30}$ of municipal waste is yard trimmings.

**49.**  $5 - 3^2 = 5 - 9$   Evaluating the exponential
                          expression
      $\quad\quad\; = -4$   Subtracting

**51.**  $\quad 8 \cdot 12 - (7 + 13)$
      $= 8 \cdot 12 - 20$   Adding inside parentheses
      $= 96 - 20$   Multiplying
      $= 76$   Subtracting

**53.** 4, $\boxed{6}$ 7 8, 9 5 2

The digit 6 means 6 hundred thousands.

**55.** ◈

**57.** ◈

**59.** Use a calculator.

$$\left(-\frac{57}{61}\right)^3 = -\frac{185,193}{226,981}$$

**61.**  $\left(-\frac{1}{2}\right)^5 \left(\frac{3}{5}\right) = -\frac{1}{32}\left(\frac{3}{5}\right) = -\frac{3}{160}$

**63.**  $-\frac{2}{3}xy = -\frac{2}{3} \cdot \frac{2}{5}\left(-\frac{1}{7}\right)$

$$= -\frac{4}{15}\left(-\frac{1}{7}\right)$$

$$= \frac{4}{105}$$

## Exercise Set 3.5

**1.** Since $10 \div 2 = 5$, we multiply by $\frac{5}{5}$.

$$\frac{1}{2} = \frac{1}{2} \cdot \frac{5}{5} = \frac{1 \cdot 5}{2 \cdot 5} = \frac{5}{10}$$

**3.** Since $-48 \div 4 = -12$, we multiply by $\frac{-12}{-12}$.

$$\frac{3}{4} = \frac{3}{4}\left(\frac{-12}{-12}\right) = \frac{3(-12)}{4(-12)} == \frac{-36}{-48}$$

**5.** Since $30 \div 10 = 3$, we multiply by $\frac{3}{3}$.

$$\frac{9}{10} = \frac{9}{10} \cdot \frac{3}{3} = \frac{9 \cdot 3}{10 \cdot 3} = \frac{27}{30}$$

**7.** Since $5t \div 5 = t$, we multiply by $\frac{t}{t}$.

$$\frac{11}{5} = \frac{11}{5} \cdot \frac{t}{t} = \frac{11 \cdot t}{5 \cdot t} = \frac{11t}{5t}$$

**9.** Since $48 \div 12 = 4$, we multiply by $\frac{4}{4}$.

$$\frac{5}{12} = \frac{5}{12} \cdot \frac{4}{4} = \frac{5 \cdot 4}{12 \cdot 4} = \frac{20}{48}$$

**11.** Since $54 \div 18 = 3$, we multiply by $\frac{3}{3}$.

$$-\frac{17}{18} = -\frac{17}{18} \cdot \frac{3}{3} = -\frac{17 \cdot 3}{18 \cdot 3} = -\frac{51}{54}$$

**13.** Since $-25 \div -5 = 5$, we multiply by $\frac{5}{5}$.

$$\frac{2}{-5} = \frac{2}{-5} \cdot \frac{5}{5} = \frac{2 \cdot 5}{-5 \cdot 5} = \frac{10}{-25}$$

**15.** Since $132 \div 22 = 6$, we multiply by $\frac{6}{6}$.

$$\frac{-7}{22} = \frac{-7}{22} \cdot \frac{6}{6} = \frac{-7 \cdot 6}{22 \cdot 6} = \frac{-42}{132}$$

**17.** Since $8x \div 8 = x$, we multiply by $\dfrac{x}{x}$.

$$\frac{5}{8} = \frac{5}{8} \cdot \frac{x}{x} = \frac{5x}{8x}$$

**19.** Since $11m \div 11 = m$, we multiply by $\dfrac{m}{m}$.

$$\frac{7}{11} \cdot \frac{m}{m} = \frac{7m}{11m}$$

**21.** Since $9ab \div 9 = ab$, we multiply by $\dfrac{ab}{ab}$.

$$\frac{4}{9} \cdot \frac{ab}{ab} = \frac{4ab}{9ab}$$

**23.** Since $27b \div 9 = 3b$, we multiply by $\dfrac{3b}{3b}$.

$$\frac{4}{9} = \frac{4}{9} \cdot \frac{3b}{3b} = \frac{12b}{27b}$$

**25.**
$$\begin{aligned}
\frac{2}{4} &= \frac{1 \cdot 2}{2 \cdot 2} && \longleftarrow \text{Factor the numerator} \\
&&& \longleftarrow \text{Factor the denominator} \\
&= \frac{1}{2} \cdot \frac{2}{2} && \longleftarrow \text{Factor the fraction} \\
&= \frac{1}{2} \cdot 1 && \longleftarrow \tfrac{2}{2} = 1 \\
&= \frac{1}{2} && \longleftarrow \text{Removing a factor of 1}
\end{aligned}$$

**27.**
$$\begin{aligned}
-\frac{6}{9} &= -\frac{2 \cdot 3}{3 \cdot 3} && \longleftarrow \text{Factor the numerator} \\
&&& \longleftarrow \text{Factor the denominator} \\
&= -\frac{2}{3} \cdot \frac{3}{3} && \longleftarrow \text{Factor the fraction} \\
&= -\frac{2}{3} \cdot 1 && \longleftarrow \tfrac{3}{3} = 1 \\
&= -\frac{2}{3} && \longleftarrow \text{Removing a factor of 1}
\end{aligned}$$

**29.**
$$\begin{aligned}
\frac{10}{25} &= \frac{2 \cdot 5}{5 \cdot 5} && \longleftarrow \text{Factor the numerator} \\
&&& \longleftarrow \text{Factor the denominator} \\
&= \frac{2}{5} \cdot \frac{5}{5} && \longleftarrow \text{Factor the fraction} \\
&= \frac{2}{5} \cdot 1 && \longleftarrow \tfrac{5}{5} = 1 \\
&= \frac{2}{5} && \longleftarrow \text{Removing a factor of 1}
\end{aligned}$$

**31.** $\dfrac{24}{-8} = \dfrac{3 \cdot 8}{-1 \cdot 8} = \dfrac{3}{-1} \cdot \dfrac{8}{8} = \dfrac{3}{-1} \cdot 1 = \dfrac{3}{-1} = -3$

**33.** $\dfrac{27}{36} = \dfrac{9 \cdot 3}{9 \cdot 4} = \dfrac{9}{9} \cdot \dfrac{3}{4} = 1 \cdot \dfrac{3}{4} = \dfrac{3}{4}$

**35.** $-\dfrac{24}{14} = -\dfrac{12 \cdot 2}{7 \cdot 2} = -\dfrac{12}{7} \cdot \dfrac{2}{2} = -\dfrac{12}{7}$

**37.** $\dfrac{16n}{48n} = \dfrac{1 \cdot 16n}{3 \cdot 16n} = \dfrac{1}{3} \cdot \dfrac{16n}{16n} = \dfrac{1}{3}$

**39.** $\dfrac{-17}{51} = \dfrac{-1 \cdot 17}{3 \cdot 17} = \dfrac{-1}{3} \cdot \dfrac{17}{17} = \dfrac{-1}{3}$

**41.**
$$\begin{aligned}
\frac{420}{480} &= \frac{2 \cdot 2 \cdot 3 \cdot 5 \cdot 7}{2 \cdot 2 \cdot 2 \cdot 2 \cdot 3 \cdot 5} \\
&= \frac{2}{2} \cdot \frac{2}{2} \cdot \frac{3}{3} \cdot \frac{5}{5} \cdot \frac{7}{2 \cdot 2 \cdot 2} \\
&= \frac{7}{2 \cdot 2 \cdot 2} \\
&= \frac{7}{8}
\end{aligned}$$

**43.**
$$\begin{aligned}
\frac{136}{153} &= \frac{2 \cdot 2 \cdot 2 \cdot 17}{3 \cdot 3 \cdot 17} \\
&= \frac{2 \cdot 2 \cdot 2}{3 \cdot 3} \cdot \frac{17}{17} \\
&= \frac{2 \cdot 2 \cdot 2}{3 \cdot 3} \\
&= \frac{8}{9}
\end{aligned}$$

**45.** $\dfrac{3ab}{8ab} = \dfrac{3 \cdot a \cdot b}{8 \cdot a \cdot b} = \dfrac{3}{8} \cdot \dfrac{a}{a} \cdot \dfrac{b}{b} = \dfrac{3}{8}$

**47.** $\dfrac{9xy}{6x} = \dfrac{3 \cdot 3 \cdot x \cdot y}{2 \cdot 3 \cdot x} = \dfrac{3 \cdot y}{2} \cdot \dfrac{3}{3} \cdot \dfrac{x}{x} = \dfrac{3y}{2}$

**49. *Familiarize*.** We make a drawing. We let $A$ = the area.

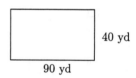

*Translate*. Using the formula for area, we have
$$A = l \cdot w = 90 \cdot 40.$$

*Solve*. We carry out the multiplication.
$$\begin{array}{r} 4\,0 \\ \times\ 9\,0 \\ \hline 3\,6\,0\,0 \end{array}$$

Thus, $A = 3600$.

*Check*. We repeat the calculation. The answer checks.

*State*. The area is $3600$ yd$^2$.

**51.** $34 - 39 = 34 + (-39) = -5$

**53.**
$$\begin{array}{r} {\scriptstyle 7\ 9\ 13} \\ \cancel{8\,0\,3} \\ -\ 6\,1\,7 \\ \hline 1\,8\,6 \end{array}$$

**55.**
$$\begin{aligned}
30x &= -150 \\
\frac{30x}{30} &= \frac{-150}{30} \\
x &= -5
\end{aligned}$$

The solution is $-5$.

**57.** ◈

**59.** ◈

**61.** $\dfrac{209ab}{247ac} = \dfrac{11 \cdot 19 \cdot a \cdot b}{13 \cdot 19 \cdot a \cdot c}$

$\qquad = \dfrac{11 \cdot b}{13 \cdot c} \cdot \dfrac{19 \cdot a}{19 \cdot a}$

$\qquad = \dfrac{11b}{13c}$

**63.** $-\dfrac{187a}{287b} = -\dfrac{11 \cdot 17 \cdot a}{17 \cdot 17 \cdot b}$

$\qquad = -\dfrac{11 \cdot a}{17 \cdot b} \cdot \dfrac{17}{17}$

$\qquad = -\dfrac{11a}{17b}$

**65.** $\dfrac{3473}{3197} = \dfrac{23 \cdot 151}{23 \cdot 139} = \dfrac{23}{23} \cdot \dfrac{151}{139} = \dfrac{151}{139}$

**67.** $\dfrac{3}{20} = \dfrac{?}{460}$

$460 \div 20 = 23$, so we have $\dfrac{3}{20} \cdot \dfrac{23}{23} = \dfrac{69}{460}$. You would

expect 69 people in a crowd of 460 to be left-handed.

**69.** (a) Find the amount Pam will pay:

$\qquad \dfrac{1}{2} \cdot 14{,}400 = \dfrac{14{,}400}{2} = \dfrac{2 \cdot 7200}{2 \cdot 1}$

$\qquad\qquad = \dfrac{2}{2} \cdot \dfrac{7200}{1} = 7200$

Find the amount Sam will pay:

$\qquad \dfrac{1}{4} \cdot 14{,}400 = \dfrac{14{,}400}{4} = \dfrac{4 \cdot 3600}{4 \cdot 1}$

$\qquad\qquad = \dfrac{4}{4} \cdot \dfrac{3600}{1} = 3600$

Find the amount Jan will pay:

$\qquad \dfrac{1}{6} \cdot 14{,}400 = \dfrac{14{,}400}{6} = \dfrac{6 \cdot 2400}{6 \cdot 1}$

$\qquad\qquad = \dfrac{6}{6} \cdot \dfrac{2400}{1} = 2400$

Find the amount Nan will pay:

$\qquad 14{,}400 - (7200 + 3600 + 2400)$

$\qquad = 14{,}400 - (13{,}200)$

$\qquad = 1200$

Nan will pay \$1200.

(b) $\dfrac{1200}{14{,}400} = \dfrac{1 \cdot 12 \cdot 100}{12 \cdot 12 \cdot 100} = \dfrac{1}{12} \cdot \dfrac{12 \cdot 100}{12 \cdot 100} = \dfrac{1}{12}$

Nan will pay $\dfrac{1}{12}$ of the cost.

## Exercise Set 3.6

**1.** $\dfrac{3}{8} \cdot \dfrac{5}{3} = \dfrac{3 \cdot 5}{8 \cdot 3} = \dfrac{3}{3} \cdot \dfrac{5}{8} = 1 \cdot \dfrac{5}{8} = \dfrac{5}{8}$

**3.** $\dfrac{7}{8} \cdot \dfrac{-1}{7} = \dfrac{7(-1)}{8 \cdot 7} = \dfrac{7}{7} \cdot \dfrac{-1}{8} = \dfrac{-1}{8}$, or $-\dfrac{1}{8}$

**5.** $\dfrac{1}{8} \cdot \dfrac{6}{5} = \dfrac{1 \cdot 6}{8 \cdot 5} = \dfrac{1 \cdot 2 \cdot 3}{2 \cdot 4 \cdot 5} = \dfrac{2}{2} \cdot \dfrac{1 \cdot 3}{4 \cdot 5} = \dfrac{3}{20}$

**7.** $\dfrac{1}{6} \cdot \dfrac{4}{3} = \dfrac{1 \cdot 4}{6 \cdot 3} = \dfrac{1 \cdot 2 \cdot 2}{2 \cdot 3 \cdot 3} = \dfrac{2}{2} \cdot \dfrac{1 \cdot 2}{3 \cdot 3} = \dfrac{2}{9}$

**9.** $\dfrac{12}{-5} \cdot \dfrac{9}{8} = \dfrac{12 \cdot 9}{-5 \cdot 8} = \dfrac{4 \cdot 3 \cdot 9}{-5 \cdot 2 \cdot 4} = \dfrac{4}{4} \cdot \dfrac{3 \cdot 9}{-5 \cdot 2} = \dfrac{3 \cdot 9}{-5 \cdot 2} =$

$\dfrac{27}{-10}$, or $-\dfrac{27}{10}$

**11.** $\dfrac{5x}{9} \cdot \dfrac{7}{5} = \dfrac{5x \cdot 7}{9 \cdot 5} = \dfrac{5 \cdot x \cdot 7}{9 \cdot 5} = \dfrac{5}{5} \cdot \dfrac{x \cdot 7}{9} = \dfrac{7x}{9}$

**13.** $\dfrac{1}{4} \cdot 8 = \dfrac{1 \cdot 8}{4} = \dfrac{8}{4} = \dfrac{4 \cdot 2}{4 \cdot 1} = \dfrac{4}{4} \cdot \dfrac{2}{1} = \dfrac{2}{1} = 2$

**15.** $15 \cdot \dfrac{1}{3} = \dfrac{15 \cdot 1}{3} = \dfrac{15}{3} = \dfrac{3 \cdot 5}{3 \cdot 1} = \dfrac{3}{3} \cdot \dfrac{5}{1} = \dfrac{5}{1} = 5$

**17.** $-12 \cdot \dfrac{3}{4} = \dfrac{-12 \cdot 3}{4} = \dfrac{4(-3) \cdot 3}{4 \cdot 1} = \dfrac{4}{4} \cdot \dfrac{-3 \cdot 3}{1} =$

$\dfrac{-3 \cdot 3}{1} = \dfrac{-9}{1} = -9$

**19.** $\dfrac{3}{8} \cdot 8a = \dfrac{3 \cdot 8a}{8} = \dfrac{3 \cdot 8 \cdot a}{8 \cdot 1} = \dfrac{8}{8} \cdot \dfrac{3 \cdot a}{1} = \dfrac{3a}{1} = 3a$

**21.** $\left(-\dfrac{3}{7}\right)\left(-\dfrac{7}{3}\right) = \dfrac{3 \cdot 7}{7 \cdot 3} = \dfrac{3 \cdot 7}{3 \cdot 7} = 1$

**23.** $\dfrac{a}{b} \cdot \dfrac{b}{a} = \dfrac{a \cdot b}{b \cdot a} = \dfrac{a \cdot b}{a \cdot b} = 1$

**25.** $\dfrac{1}{27} \cdot 360a = \dfrac{1 \cdot 360a}{27} = \dfrac{1 \cdot 9 \cdot 40 \cdot a}{9 \cdot 3} =$

$\dfrac{9}{9} \cdot \dfrac{1 \cdot 40 \cdot a}{3} = \dfrac{40a}{3}$

**27.** $176\left(\dfrac{1}{-6}\right) = \dfrac{176 \cdot 1}{-6} = \dfrac{2 \cdot 88 \cdot 1}{-3 \cdot 2} = \dfrac{2}{2} \cdot \dfrac{88 \cdot 1}{-3} =$

$\dfrac{88}{-3}$, or $-\dfrac{88}{3}$

**29.** $7x \cdot \dfrac{1}{7x} = \dfrac{7x \cdot 1}{7x \cdot 1} = 1$

**31.** $\dfrac{2x}{9} \cdot \dfrac{27}{2x} = \dfrac{2x \cdot 27}{9 \cdot 2x} = \dfrac{2x \cdot 9 \cdot 3}{9 \cdot 2x \cdot 1} =$

$\dfrac{2x \cdot 9}{2x \cdot 9} \cdot \dfrac{3}{1} = 3$

**33.** $\dfrac{7}{10} \cdot \dfrac{34}{150} = \dfrac{7 \cdot 34}{10 \cdot 150} = \dfrac{7 \cdot 2 \cdot 17}{2 \cdot 5 \cdot 150} = \dfrac{2}{2} \cdot \dfrac{7 \cdot 17}{5 \cdot 150} = \dfrac{119}{750}$

**35.** $\dfrac{36}{85} \cdot \dfrac{25}{-99} = \dfrac{9 \cdot 4 \cdot 5 \cdot 5}{5 \cdot 17 \cdot 9(-11)} = \dfrac{9 \cdot 5}{9 \cdot 5} \cdot \dfrac{4 \cdot 5}{17(-11)} =$

$\dfrac{20}{-187}$, or $-\dfrac{20}{187}$

**37.** $\dfrac{-98}{99} \cdot \dfrac{27a}{175a} = \dfrac{7(-14) \cdot 9 \cdot 3 \cdot a}{9 \cdot 11 \cdot 7 \cdot 25 \cdot a} =$

$\dfrac{7 \cdot 9 \cdot a}{7 \cdot 9 \cdot a} \cdot \dfrac{-14 \cdot 3}{11 \cdot 25} = \dfrac{-42}{275}$, or $-\dfrac{42}{275}$

**39.** $\dfrac{110}{33} \cdot \dfrac{-24}{25} = \dfrac{2 \cdot 5 \cdot 11 \cdot 3(-8)}{3 \cdot 11 \cdot 5 \cdot 5} = \dfrac{3 \cdot 5 \cdot 11}{3 \cdot 5 \cdot 11} \cdot \dfrac{2(-8)}{5} =$

$\dfrac{-16}{5}$, or $-\dfrac{16}{5}$

**41.** $\left(-\dfrac{11}{24}\right)\dfrac{3}{5} = -\dfrac{11 \cdot 3}{24 \cdot 5} = -\dfrac{11 \cdot 3}{3 \cdot 8 \cdot 5} = \dfrac{3}{3}\left(-\dfrac{11}{8 \cdot 5}\right) =$

$-\dfrac{11}{40}$

**43.** $\dfrac{10a}{21} \cdot \dfrac{3}{4a} = \dfrac{10a \cdot 3}{21 \cdot 4a} = \dfrac{2 \cdot 5 \cdot a \cdot 3}{3 \cdot 7 \cdot 2 \cdot 2 \cdot a}$

$= \dfrac{2 \cdot 3 \cdot a}{2 \cdot 3 \cdot a} \cdot \dfrac{5}{2 \cdot 7} = \dfrac{5}{14}$

**45.** *Familiarize.* We visualize the situation. We let $a$ = the amount received for working $\dfrac{3}{4}$ of a day.

| 1 day $56 | |
|---|---|
| 3/4 day $a$ | |

*Translate.* We write an equation.

Pay for 3/4 of a day is $\dfrac{3}{4}$ of $56.

$$a = \dfrac{3}{4} \cdot 56$$

*Solve.* We carry out the multiplication.

$a = \dfrac{3}{4} \cdot 56 = \dfrac{3 \cdot 56}{4}$

$= \dfrac{3 \cdot 14 \cdot 4}{1 \cdot 4} = \dfrac{3 \cdot 14}{1} \cdot \dfrac{4}{4}$

$= 42$

*Check.* We can repeat the calculation. We can also determine that the answer seems reasonable since we multiplied 56 by a number less than 1 and the result is less than 56. The answer checks.

*State.* $42 is received for working $\dfrac{3}{4}$ of a day.

**47.** *Familiarize.* Visualize 30 addends of $\dfrac{2}{5}$. Let $s$ = the number of pounds of salmon needed.

*Translate.*

| Amount per person | times | number of people | is | amount needed. |
|---|---|---|---|---|
| $\dfrac{2}{5}$ | $\cdot$ | 30 | $=$ | $s$ |

*Solve.* We carry out the multiplication.

$\dfrac{2}{5} \cdot 30 = \dfrac{2 \cdot 30}{5} = \dfrac{2 \cdot 5 \cdot 6}{5 \cdot 1}$

$= \dfrac{5}{5} \cdot \dfrac{2 \cdot 6}{1}$

$= 12$

We have $12 = s$, or $s = 12$.

*Check.* We repeat the calculation. The answer checks.

*State.* 12 lb of salmon will be needed.

**49.** *Familiarize.* Let $s$ = the number of people who might be shy.

*Translate.*

$\dfrac{2}{5}$ of 650 is number of shy people.

$$\dfrac{2}{5} \cdot 650 = s$$

*Solve.* We carry out the multiplication.

$\dfrac{2}{5} \cdot 650 = \dfrac{2 \cdot 650}{5} = \dfrac{2 \cdot 5 \cdot 130}{5 \cdot 1}$

$= \dfrac{5}{5} \cdot \dfrac{2 \cdot 130}{1}$

$= 260$

We have $260 = s$, or $s = 260$.

*Check.* We repeat the calculation. The answer checks.

*State.* 260 people might be shy.

**51.** *Familiarize.* We draw a picture.

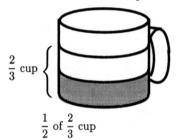

$\dfrac{2}{3}$ cup

$\dfrac{1}{2}$ of $\dfrac{2}{3}$ cup

We let $n$ = the amount of flour the chef should use.

*Translate.* The multiplication sentence

$$\dfrac{1}{2} \cdot \dfrac{2}{3} = n$$

corresponds to the situation.

*Solve.* We multiply and simplify:

$n = \dfrac{1}{2} \cdot \dfrac{2}{3} = \dfrac{1 \cdot 2}{2 \cdot 3} = \dfrac{2}{2} \cdot \dfrac{1}{3} = \dfrac{1}{3}$

*Check.* We can repeat the calculation. We can also determine that the answer seems reasonable since we multiplied $\dfrac{2}{3}$ by a number less than 1 and the result is less than $\dfrac{2}{3}$. The answer checks.

*State.* The chef should use $\dfrac{1}{3}$ cup of flour.

**53.** *Familiarize.* We visualize the situation. Let $a$ = the assessed value of the house.

| Value of house $124,000 | |
|---|---|
| 3/4 of the value $a | |

*Translate.* We write an equation.

Assessed value is $\dfrac{3}{4}$ of the value of the house

$$a = \dfrac{3}{4} \cdot 124,000$$

*Solve*. We carry out the multiplication.

$$a = \frac{3}{4} \cdot 124{,}000 = \frac{3 \cdot 124{,}000}{4}$$

$$= \frac{3 \cdot 4 \cdot 31{,}000}{4 \cdot 1} = \frac{4}{4} \cdot \frac{3 \cdot 31{,}000}{1}$$

$$= 93{,}000$$

*Check*. We can repeat the calculation. We can also determine that the answer seems reasonable since we multiplied 124,000 by a number less than 1 and the result is less than 124,000. The answer checks.

*State*. The assessed value of the house is $93,000.

**55. Familiarize**. We draw a picture.

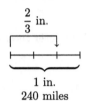

$\frac{2}{3}$ in.

1 in.
240 miles

We let $n =$ the number of miles represented by $\frac{2}{3}$ in.

*Translate*. The multiplication sentence

$$n = \frac{2}{3} \cdot 240$$

corresponds to the situation.

*Solve*. We multiply and simplify:

$$n = \frac{2}{3} \cdot 240 = \frac{2 \cdot 240}{3} = \frac{2 \cdot 3 \cdot 80}{1 \cdot 3}$$

$$= \frac{3}{3} \cdot \frac{2 \cdot 80}{1} = \frac{2 \cdot 80}{1}$$

$$= 160$$

*Check*. We can repeat the calculation. We can also determine that the answer seems reasonable since we multiplied 240 by a number less than 1 and the result is less than 240.

*State*. $\frac{2}{3}$ in. on the map represents 160 miles.

**57. Familiarize**. This is a multistep problem. First we find the amount of each of the given expenses. Then we find the total of these expenses and take it away from the annual income to find how much is spent for other expenses.

| $27,000 | $27,000 | $27,000 |
|---|---|---|
| $\frac{1}{4}$ for food | $\frac{1}{5}$ for housing | $\frac{1}{10}$ for clothing |

| $27,000 | $27,000 |
|---|---|
| $\frac{1}{9}$ for savings | $\frac{1}{4}$ for taxes |

We let $f$, $h$, $c$, $s$, and $t$ represent the amounts spent on food, housing, clothing, savings, and taxes, respectively.

*Translate*. The following multiplication sentences correspond to the situation.

$$\frac{1}{4} \cdot 27{,}000 = f \qquad \frac{1}{9} \cdot 27{,}000 = s$$

$$\frac{1}{5} \cdot 27{,}000 = h \qquad \frac{1}{4} \cdot 27{,}000 = t$$

$$\frac{1}{10} \cdot 27{,}000 = c$$

*Solve*. We multiply and simplify.

$$f = \frac{1}{4} \cdot 27{,}000 = \frac{27{,}000}{4} = \frac{4 \cdot 6750}{4 \cdot 1} = \frac{4}{4} \cdot \frac{6750}{1} =$$
6750

$$h = \frac{1}{5} \cdot 27{,}000 = \frac{27{,}000}{5} = \frac{5 \cdot 5400}{5 \cdot 1} = \frac{5}{5} \cdot \frac{5400}{1} =$$
5400

$$c = \frac{1}{10} \cdot 27{,}000 = \frac{27{,}000}{10} = \frac{10 \cdot 2700}{10 \cdot 1} = \frac{10}{10} \cdot \frac{2700}{1} =$$
2700

$$s = \frac{1}{9} \cdot 27{,}000 = \frac{27{,}000}{9} = \frac{9 \cdot 3000}{9 \cdot 1} = \frac{9}{9} \cdot \frac{3000}{1} =$$
3000

$$t = \frac{1}{4} \cdot 27{,}000 = \frac{27{,}000}{4} = \frac{4 \cdot 6750}{4 \cdot 1} = \frac{4}{4} \cdot \frac{6750}{1} =$$
6750

We add to find the total of these expenses.

```
       2 1
   $ 6 7 5 0
     5 4 0 0
     2 7 0 0
     3 0 0 0
     6 7 5 0
   ─────────
   $ 2 4, 6 0 0
```

We let $m =$ the amount spent on other expenses and subtract to find this amount.

| Annual income | minus | Total of itemized expenses | is | Total spent on other expenses |
|---|---|---|---|---|
| ↓ | ↓ | ↓ | ↓ | ↓ |
| $27,000 | − | $24,600 | = | $m$ |
| | | $2400 | = | $m$     Subtracting |

*Check*. We repeat the calculations. The results check.

*State*. $6750 is spent for food, $5400 for housing, $2700 for clothing, $3000 for savings, $6750 for taxes, and $2400 for other expenses.

**59.** $A = \frac{1}{2} \cdot b \cdot h$ 	Area of a triangle

$A = \frac{1}{2} \cdot 15$ in. $\cdot 8$ in. 	Substituting 15 in. for $b$ and 8 in. for $h$

$A = \frac{15 \cdot 8}{2}$ in$^2$

$A = 60$ in$^2$

**61.**  $A = \frac{1}{2} \cdot b \cdot h$          Area of a triangle

$A = \frac{1}{2} \cdot 5 \text{ mm} \cdot \frac{7}{2} \text{ mm}$  Substituting 5 mm for $b$

and $\frac{7}{2}$ mm for $h$

$A = \frac{5 \cdot 7}{2 \cdot 2} \text{ mm}^2$

$A = \frac{35}{4} \text{ mm}^2$

**63.**  $A = \frac{1}{2} \cdot b \cdot h$          Area of a triangle

$A = \frac{1}{2} \cdot \frac{9}{2} \text{ m} \cdot \frac{7}{2} \text{ m}$  Substituting $\frac{9}{2}$ m for $b$ and

$\frac{7}{2}$ m for $h$

$A = \frac{9 \cdot 7}{2 \cdot 2 \cdot 2} \text{ m}^2$

$A = \frac{63}{8} \text{ m}^2$

**65.** *Familiarize.* We look for figures whose areas we can calculate using area formulas we already know.

*Translate.* The figure consists of a rectangle with a length of 10 mi and a width of 8 mi and of a triangle with a base of $13 - 10$, or 3 mi, and a height of 8 mi. We use the formula $A = l \cdot w$ for the area of a rectangle and the formula $A = \frac{1}{2} \cdot b \cdot h$ for the area of a triangle and add the two areas.

*Solve.* For the rectangle: $A = l \cdot w = 10 \text{ mi} \cdot 8 \text{ mi} = 80 \text{ mi}^2$

For the triangle: $A = \frac{1}{2} \cdot b \cdot h = \frac{1}{2} \cdot 3 \text{ mi} \cdot 8 \text{ mi} = 12 \text{ mi}^2$

Then we add: $80 \text{ mi}^2 + 12 \text{ mi}^2 = 92 \text{ mi}^2$

*Check.* We repeat the calculations.

*State.* The area of the figure is $92 \text{ mi}^2$.

**67.** *Familiarize.* We look for figures whose areas we can calculate using area formulas we already know.

*Translate.* Each of the 2 ends of the building consists of a rectangle with a length of 50 ft and a width of 25 ft and of a triangle with a base of 50 ft and a height of 11 ft. Each of the two sides is a rectangle with a length of 75 ft and a width of 25 ft. We use the formula $A = l \cdot w$ for the area of a rectangle twice and the formula $A = \frac{1}{2} \cdot b \cdot h$ for the area of a triangle, multiply each area by 2, and then add the two results.

*Solve.*

For each end:

Rectangle: $A = l \cdot w = 50 \text{ ft} \cdot 25 \text{ ft} = 1250 \text{ ft}^2$

Triangle: $A = \frac{1}{2} \cdot b \cdot h = \frac{1}{2} \cdot 50 \text{ ft} \cdot 11 \text{ ft} = 275 \text{ ft}^2$

For each side:

$A = l \cdot w = 75 \text{ ft} \cdot 25 \text{ ft} = 1875 \text{ ft}^2$

We multiply each area by 2:

$2 \cdot 1250 \text{ ft}^2 = 2500 \text{ ft}^2$

$2 \cdot 275 \text{ ft}^2 = 550 \text{ ft}^2$

$2 \cdot 1875 \text{ ft}^2 = 3750 \text{ ft}^2$

Now we add:

$2500 \text{ ft}^2 + 550 \text{ ft}^2 + 3750 \text{ ft}^2 = 6800 \text{ ft}^2$

*Check.* We repeat the calculations.

*State.* The total area of the sides and ends of the building is $6800 \text{ ft}^2$.

**69.**  $48 \cdot t = 1680$

$\frac{48 \cdot t}{48} = \frac{1680}{48}$

$t = 35$

The solution is 35.

**71.**  $747 = x + 270$

$747 - 270 = x + 270 - 270$

$477 = x$

The solution is 477.

**73.** We add absolute values and make the result negative.

$(-39) + (-72) = -111$

**75.** ◈

**77.** ◈

**79.**  $\frac{5767}{3763} \cdot \frac{159}{395} = \frac{5767 \cdot 159}{3763 \cdot 395} = \frac{73 \cdot 79 \cdot 3 \cdot 53}{53 \cdot 71 \cdot 5 \cdot 79} =$

$\frac{79 \cdot 53}{79 \cdot 53} \cdot \frac{73 \cdot 3}{71 \cdot 5} = \frac{219}{355}$

**81.** *Familiarize.* We look for figures whose areas we can calculate using area formulas we already know.

*Translate.* Each of the 8 triangular portions of the steeple has a base of 4 ft and a height of 15 ft. We use the formula $A = \frac{1}{2} \cdot b \cdot h$ for the area of a triangle and multiply the area by 8. Each of the 8 rectangular portions has a length of 6 ft and a width of 4 ft. We use the formula $A = l \cdot w$ for the area of a rectangle and multiply the area by 8. Then we add the areas of the triangles and the rectangles.

*Solve.*

For each triangular portion:

$A = \frac{1}{2} \cdot b \cdot h = \frac{1}{2} \cdot 4 \text{ ft} \cdot 15 \text{ ft} = 30 \text{ ft}^2$

We multiply this area by 8.

$8 \cdot 30 \text{ ft}^2 = 240 \text{ ft}^2$

For each rectangular portion:

$A = l \cdot w = 6 \text{ ft} \cdot 4 \text{ ft} = 24 \text{ ft}^2$

We multiply this area by 8:

$8 \cdot 24 \text{ ft}^2 = 192 \text{ ft}^2$

Finally we add the areas:

$240 \text{ ft}^2 + 192 \text{ ft}^2 = 432 \text{ ft}^2$

*Check.* We repeat the calculations.

*State.* The total area of the steeple is $432 \text{ ft}^2$.

**83. *Familiarize*.** We are told that $\frac{2}{3}$ of $\frac{7}{8}$ of the students are high school graduates who are older than 20, and $\frac{1}{7}$ of this fraction are left-handed. Thus, we want to find $\frac{1}{7}$ of $\frac{2}{3}$ of $\frac{7}{8}$. We let $f$ represent this fraction.

***Translate*.** The multiplication sentence
$$f = \frac{1}{7} \cdot \frac{2}{3} \cdot \frac{7}{8}$$
corresponds to this situation.

***Solve*.** We multiply and simplify.
$$f = \frac{1}{7} \cdot \frac{2}{3} \cdot \frac{7}{8} = \frac{1 \cdot 2}{7 \cdot 3} \cdot \frac{7}{8} = \frac{1 \cdot 2 \cdot 7}{7 \cdot 3 \cdot 8} = \frac{1 \cdot 2 \cdot 7}{7 \cdot 3 \cdot 2 \cdot 4} =$$
$$\frac{2 \cdot 7}{2 \cdot 7} \cdot \frac{1}{3 \cdot 4} = \frac{1}{3 \cdot 4} = \frac{1}{12}$$

***Check*.** We repeat the calculation. The result checks.

***State*.** $\frac{1}{12}$ of the students are left-handed high school graduates over the age of 20.

**85. *Familiarize*.** If we divide the group of entering students into 8 equal parts and take 7 of them, we have the fractional part of the students that completed high school. Then the 1 part remaining, or $\frac{1}{8}$ of the students, did not graduate from high school. Similarly, if we divide the group of entering students into 3 equal parts and take 2 of them, we have the fractional part of the students that is older than 20. Then the 1 part remaining, or $\frac{1}{3}$ of the students, are 20 years old or younger. From Exercise 79 we know that $\frac{1}{7}$ of the students are left-handed. Thus, we want to find $\frac{1}{7}$ of $\frac{1}{3}$ of $\frac{1}{8}$. We let $f$ = this fraction.

***Translate*.** The multiplication sentence
$$f = \frac{1}{7} \cdot \frac{1}{3} \cdot \frac{1}{8}$$
corresponds to this situation.

***Solve*.** We multiply.
$$f = \frac{1}{7} \cdot \frac{1}{3} \cdot \frac{1}{8} = \frac{1 \cdot 1 \cdot 1}{7 \cdot 3 \cdot 8} = \frac{1}{168}$$

***Check*.** We repeat the calculation. The result checks.

***State*.** $\frac{1}{168}$ of the students did not graduate from high school, are 20 years old or younger, and are left-handed.

## Exercise Set 3.7

**1.** $\frac{7}{3}$   Interchange the numerator and denominator.

The reciprocal of $\frac{7}{3}$ is $\frac{3}{7}$.   $\left( \frac{7}{3} \cdot \frac{3}{7} = \frac{21}{21} = 1 \right)$

**3.** Think of 4 as $\frac{4}{1}$.

$\frac{4}{1}$   Interchange the numerator and denominator.

The reciprocal of 4 is $\frac{1}{4}$.   $\left( \frac{4}{1} \cdot \frac{1}{4} = \frac{4}{4} = 1 \right)$

**5.** $\frac{1}{6}$   Interchange the numerator and denominator.

The reciprocal of $\frac{1}{6}$ is 6.   $\left( \frac{6}{1} = 6; \frac{1}{6} \cdot \frac{6}{1} = \frac{6}{6} = 1 \right)$

**7.** $-\frac{10}{3}$   Interchange the numerator and denominator.

The reciprocal of $-\frac{10}{3}$ is $-\frac{3}{10}$.   $\left( -\frac{10}{3}\left(-\frac{3}{10}\right) = \frac{30}{30} = 1 \right)$

**9.** $\frac{2}{21}$   Interchange the numerator and denominator.

The reciprocal of $\frac{2}{21}$ is $\frac{21}{2}$.   $\left( \frac{2}{21} \cdot \frac{21}{2} = \frac{42}{42} = 1 \right)$

**11.** $\frac{-3n}{m}$   Interchange the numerator and denominator.

The reciprocal of $\frac{-3n}{m}$ is $\frac{m}{-3n}$.
$\left( \frac{-3n}{m} \cdot \frac{m}{-3n} = \frac{-3mn}{-3mn} = 1 \right)$

**13.** $\frac{7}{-15}$   Interchange the numerator and denominator.

The reciprocal of $\frac{7}{-15}$ is $\frac{-15}{7}$.   $\left( \frac{7}{-15}\left(\frac{-15}{7}\right) = \frac{-105}{-105} = 1 \right)$

**15.** Think of $7m$ as $\frac{7m}{1}$.

$\frac{7m}{1}$   Interchange the numerator and denominator.

The reciprocal of $7m$ is $\frac{1}{7m}$.   $\left( \frac{7m}{1} \cdot \frac{1}{7m} = \frac{7m}{7m} = 1 \right)$

**17.** $\frac{3}{5} \div \frac{3}{4} = \frac{3}{5} \cdot \frac{4}{3}$   Multiplying by the reciprocal of the divisor

$= \frac{3 \cdot 4}{5 \cdot 3}$   Multiplying numerators and denominators

$= \frac{3}{3} \cdot \frac{4}{5} = \frac{4}{5}$   Removing a factor equal to 1

**19.** $\frac{7}{6} \div \frac{5}{-3} = \frac{7}{6} \cdot \frac{-3}{5}$   Multiplying by the reciprocal of the divisor

$= \frac{7(-1)(3)}{2 \cdot 3 \cdot 5}$   Factoring

$= \frac{3}{3} \cdot \frac{7(-1)}{2 \cdot 5}$

$= \frac{-7}{10}$, or $-\frac{7}{10}$

**21.** $\frac{4}{3} \div \frac{1}{3} = \frac{4}{3} \cdot 3 = \frac{4 \cdot 3}{3} = \frac{3}{3} \cdot 4 = 4$

**23.** $\left(-\dfrac{1}{3}\right) \div \dfrac{1}{6} = -\dfrac{1}{3} \cdot 6 = -\dfrac{1 \cdot 2 \cdot 3}{1 \cdot 3} =$

$-\dfrac{1 \cdot 3}{1 \cdot 3} \cdot 2 = -2$

**25.** $\dfrac{3}{8} \div 24 = \dfrac{3}{8} \cdot \dfrac{1}{24} = \dfrac{3 \cdot 1}{8 \cdot 3 \cdot 8} = \dfrac{3}{3} \cdot \dfrac{1}{8 \cdot 8} = \dfrac{1}{64}$

**27.** $\dfrac{12}{7} \div (4x) = \dfrac{12}{7} \cdot \dfrac{1}{4x} = \dfrac{4 \cdot 3 \cdot 1}{7 \cdot 4 \cdot x} = \dfrac{4}{4} \cdot \dfrac{3 \cdot 1}{7 \cdot x} = \dfrac{3}{7x}$

**29.** $(-12) \div \dfrac{3}{2} = -12 \cdot \dfrac{2}{3} = -\dfrac{3 \cdot 4 \cdot 2}{3 \cdot 1}$

$= -\dfrac{3}{3} \cdot \dfrac{4 \cdot 2}{1} = -\dfrac{8}{1} = -8$

**31.** $28 \div \dfrac{4}{5a} = 28 \cdot \dfrac{5a}{4} = \dfrac{28 \cdot 5a}{4} = \dfrac{4 \cdot 7 \cdot 5 \cdot a}{4 \cdot 1} = \dfrac{4}{4} \cdot \dfrac{7 \cdot 5 \cdot a}{1}$

$= 35a$

**33.** $\left(-\dfrac{5}{8}\right) \div \left(-\dfrac{5}{8}\right) = -\dfrac{5}{8}\left(-\dfrac{8}{5}\right) = \dfrac{5 \cdot 8}{8 \cdot 5} = \dfrac{5 \cdot 8}{5 \cdot 8} = 1$

**35.** $\dfrac{-8}{15} \div \dfrac{4}{5} = \dfrac{-8}{15} \cdot \dfrac{5}{4} = \dfrac{-8 \cdot 5}{15 \cdot 4} = \dfrac{-2 \cdot 4 \cdot 5}{3 \cdot 5 \cdot 4} =$

$\dfrac{4 \cdot 5}{4 \cdot 5} \cdot \dfrac{-2}{3} = \dfrac{-2}{3},$ or $-\dfrac{2}{3}$

**37.** $\dfrac{77}{64} \div \dfrac{49}{18} = \dfrac{77}{64} \cdot \dfrac{18}{49} = \dfrac{7 \cdot 11 \cdot 2 \cdot 9}{2 \cdot 32 \cdot 7 \cdot 7} =$

$\dfrac{2 \cdot 7}{2 \cdot 7} \cdot \dfrac{11 \cdot 9}{32 \cdot 7} = \dfrac{99}{224}$

**39.** $120a \div \dfrac{45}{14} = 120a \cdot \dfrac{14}{45} = \dfrac{8 \cdot 15 \cdot a \cdot 14}{3 \cdot 15} =$

$\dfrac{15}{15} \cdot \dfrac{8 \cdot a \cdot 14}{3} = \dfrac{112a}{3}$

**41.** **Familiarize.** We draw a picture. Let $t$ = the number of times Benny will be able to brush his teeth.

$$t \text{ brushings}$$

**Translate.** The multiplication that corresponds to the situation is

$$\dfrac{5}{4} \cdot t = 110.$$

**Solve.** We solve the equation by dividing on both sides by $\dfrac{5}{4}$ and carrying out the division:

$$t = 110 \div \dfrac{5}{4} = 110 \cdot \dfrac{4}{5} = \dfrac{5 \cdot 22 \cdot 4}{5 \cdot 1} = \dfrac{5}{5} \cdot \dfrac{22 \cdot 4}{1} = 88$$

**Check.** We repeat the calculation. The answer checks.

**State.** Benny can brush his teeth 88 times with a 110-g tube of toothpaste.

**43.** **Familiarize.** We make a drawing.

} How long for $\dfrac{3}{4}$ mi?

$\dfrac{1}{12}$ mi each day

We let $d$ = the number of days it will take to repave a $\dfrac{3}{4}$-mi stretch of road.

**Translate.** The problem translates to the following equation:

$$d = \dfrac{3}{4} \div \dfrac{1}{12}$$

**Solve.** We carry out the division.

$$d = \dfrac{3}{4} \div \dfrac{1}{12}$$

$$= \dfrac{3}{4} \cdot \dfrac{12}{1}$$

$$= \dfrac{3 \cdot 3 \cdot 4}{4 \cdot 1} = \dfrac{4}{4} \cdot \dfrac{3 \cdot 3}{1}$$

$$= 9$$

**Check.** If the crew repaves $\dfrac{1}{12}$ mi of road each day for 9 days, a total of

$$\dfrac{1}{12} \cdot 9 = \dfrac{1 \cdot 9}{12} = \dfrac{1 \cdot 3 \cdot 3}{3 \cdot 4} = \dfrac{1 \cdot 3}{4},$$

or $\dfrac{3}{4}$ mi of road will be repaved. Our answer checks.

**State.** It will take 9 days to repave a $\dfrac{3}{4}$-mi stretch of road.

**45.** **Familiarize.** We make a drawing. Let $p$ = the number of packages that can be made from 15 lb of cheese.

$$p \text{ packages}$$

**Translate.** The problem translates to the following equation:

$$p = 15 \div \dfrac{3}{4}$$

**Solve.** We carry out the division.

$$p = 15 \div \dfrac{3}{4}$$

$$= 15 \cdot \dfrac{4}{3}$$

$$= \dfrac{3 \cdot 5 \cdot 4}{3 \cdot 1} = \dfrac{3}{3} \cdot \dfrac{5 \cdot 4}{1}$$

$$= 20$$

**Check.** If 20 packages, each containing $\dfrac{3}{4}$ lb of cheese, are made, a total of

$$20 \cdot \dfrac{3}{4} = \dfrac{5 \cdot 4 \cdot 3}{4 \cdot 1} = \dfrac{4}{4} \cdot \dfrac{5 \cdot 3}{1} = 15,$$

or 15 lb of cheese is used. The answer checks.

**State.** 20 packages can be made.

**47. Familiarize.** We make a drawing.

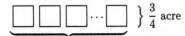

$$\left.\begin{array}{}\end{array}\right\} \frac{3}{4} \text{ acre}$$

How much for each
of 16 plots?

We let $p$ = the size of each plot, in acres.

**Translate.** The problem translates to the following equation:

$$p = \frac{3}{4} \div 16$$

**Solve.** We carry out the division.

$$p = \frac{3}{4} \div 16$$
$$= \frac{3}{4} \cdot \frac{1}{16} = \frac{3}{64}$$

**Check.** If each of 16 plots is $\frac{3}{64}$ acre, then the total acreage is

$$16 \cdot \frac{3}{64} = \frac{16 \cdot 3}{16 \cdot 4} = \frac{16}{16} \cdot \frac{3}{4} = \frac{3}{4} \text{ acre}$$

The answer checks.

**State.** Each plot will be $\frac{3}{64}$ acre.

**49. Familiarize.** We draw a picture.

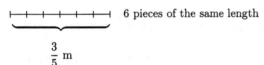

6 pieces of the same length

$$\frac{3}{5} \text{ m}$$

We let $n$ = the length of each piece.

**Translate.** The problem translates to the following equation:

$$n = \frac{3}{5} \div 6$$

**Solve.** We carry out the division.

$$n = \frac{3}{5} \div 6$$
$$= \frac{3}{5} \cdot \frac{1}{6}$$
$$= \frac{3 \cdot 1}{5 \cdot 2 \cdot 3} = \frac{3}{3} \cdot \frac{1}{5 \cdot 2}$$
$$= \frac{1}{10}$$

**Check.** If each of 6 pieces of wire is $\frac{1}{10}$ m long, there is a total of

$$6 \cdot \frac{1}{10} = \frac{6 \cdot 1}{10} = \frac{2 \cdot 3 \cdot 1}{2 \cdot 5} = \frac{2}{2} \cdot \frac{3 \cdot 1}{5},$$

or $\frac{3}{5}$ m of wire. Our answer checks.

**State.** Each piece is $\frac{1}{10}$ m.

**51. Familiarize.** We draw a picture.

$$\left.\begin{array}{}\end{array}\right\} \text{ 24 yd makes how many pairs?}$$

$\frac{3}{4}$ yd per pair

We let $s$ = the number of pairs of basketball shorts that can be made.

**Translate.** The problem translates to the following equation:

$$s = 24 \div \frac{3}{4}$$

**Solve.** We carry out the division.

$$s = 24 \div \frac{3}{4}$$
$$= 24 \cdot \frac{4}{3}$$
$$= \frac{3 \cdot 8 \cdot 4}{1 \cdot 3} = \frac{3}{3} \cdot \frac{8 \cdot 4}{1}$$
$$= 32$$

**Check.** If each of 32 pairs of shorts requires $\frac{3}{4}$ yd of nylon, a total of

$$32 \cdot \frac{3}{4} = \frac{32 \cdot 3}{4} = \frac{4 \cdot 8 \cdot 3}{4} = 8 \cdot 3,$$

or 24 yd of nylon is needed. Our answer checks.

**State.** 32 pairs of basketball shorts can be made from 24 yd of nylon.

**53. Familiarize.** Let $s$ = the number of stitches required for a 12-in. row.

**Translate.** The problem translates to the following equation:

$$s = 12 \div \frac{3}{8}$$

**Solve.** We carry out the division.

$$s = 12 \div \frac{3}{8}$$
$$= 12 \cdot \frac{8}{3}$$
$$= \frac{3 \cdot 4 \cdot 8}{3 \cdot 1} = \frac{3}{3} \cdot \frac{4 \cdot 8}{1}$$
$$= 32$$

**Check.** If 32 $\frac{3}{8}$-in. stitches are made, the length of the row is

$$32 \cdot \frac{3}{8} = \frac{4 \cdot 8 \cdot 3}{8 \cdot 1} = \frac{8}{8} \cdot \frac{4 \cdot 3}{1} = 12 \text{ in.}$$

The answer checks.

**State.** Brianna will need 32 stitches for a 12-in. row.

**55.** $(-17)(-30) = 510$ (The signs are the same, so the answer is positive.)

**57.** $x^3 = 3^3 = 3 \cdot 3 \cdot 3 = 27;$
$x^3 = (-3)^3 = (-3)(-3)(-3) = -27$

**59.** $3x^2 = 3 \cdot 7^2 = 3 \cdot 49 = 147;$
$3x^2 = 3(-7)^2 = 3 \cdot 49 = 147$

**61.** ◈

**63.** ◈

**65.** $\dfrac{\left(-\dfrac{3}{7}\right)^2 \div \dfrac{12}{5}}{\left(\dfrac{-2}{9}\right)\left(\dfrac{9}{2}\right)} = \dfrac{\dfrac{9}{49} \div \dfrac{12}{5}}{\left(\dfrac{-2}{9}\right)\left(\dfrac{9}{2}\right)}$

$= \dfrac{\dfrac{9}{49} \cdot \dfrac{5}{12}}{\dfrac{-2 \cdot 9}{9 \cdot 2}}$

$= \dfrac{\dfrac{3 \cdot 3 \cdot 5}{7 \cdot 7 \cdot 3 \cdot 4}}{\dfrac{-1 \cdot 2 \cdot 9}{9 \cdot 2}}$

$= \dfrac{\dfrac{3}{3} \cdot \dfrac{3 \cdot 5}{7 \cdot 7 \cdot 4}}{\dfrac{2 \cdot 9}{2 \cdot 9} \cdot (-1)}$

$= \dfrac{\dfrac{3 \cdot 5}{7 \cdot 7 \cdot 4}}{-1}$

$= -\dfrac{15}{196}$

**67.** $\left(\dfrac{10}{9}\right)^2 \div \dfrac{35}{27} \cdot \dfrac{49}{44} = \dfrac{100}{81} \div \dfrac{35}{27} \cdot \dfrac{49}{44}$

$= \dfrac{100}{81} \cdot \dfrac{27}{35} \cdot \dfrac{49}{44}$

$= \dfrac{4 \cdot 5 \cdot 5 \cdot 27 \cdot 7 \cdot 7}{3 \cdot 27 \cdot 5 \cdot 7 \cdot 4 \cdot 11}$

$= \dfrac{4 \cdot 5 \cdot 7 \cdot 27}{4 \cdot 5 \cdot 7 \cdot 27} \cdot \dfrac{5 \cdot 7}{3 \cdot 11}$

$= \dfrac{35}{33}$

**69.** $\dfrac{8633}{7387} \div \dfrac{485}{581} = \dfrac{8633}{7387} \cdot \dfrac{581}{485} = \dfrac{89 \cdot 97 \cdot 7 \cdot 83}{83 \cdot 89 \cdot 5 \cdot 97} =$

$\dfrac{89 \cdot 97 \cdot 83}{89 \cdot 97 \cdot 83} \cdot \dfrac{7}{5} = \dfrac{7}{5}$

---

## Exercise Set 3.8

**1.** $\dfrac{8}{5}x = 16$

$\dfrac{5}{8} \cdot \dfrac{8}{5}x = \dfrac{5}{8} \cdot 16$    The reciprocal of $\dfrac{8}{5}$ is $\dfrac{5}{8}$.

$1x = \dfrac{5 \cdot 2 \cdot 8}{8}$

$x = 10$    Removing the factor $\dfrac{8}{8}$

Check:    $\dfrac{8}{5}x = 16$

$$\dfrac{8}{5} \cdot 10 \ ? \ 16$$

$$\dfrac{8 \cdot 2 \cdot 5}{5 \cdot 1}$$

$$16 \ \big| \ 16 \quad \text{TRUE}$$

The solution is 10.

**3.** $\dfrac{7}{3}a = 21$

$\dfrac{3}{7} \cdot \dfrac{7}{3}a = \dfrac{3}{7} \cdot 21$    The reciprocal of $\dfrac{7}{3}$ is $\dfrac{3}{7}$.

$1a = \dfrac{3 \cdot 3 \cdot 7}{7}$

$a = 9$    Removing the factor $\dfrac{7}{7}$

Check:    $\dfrac{7}{3}a = 21$

$$\dfrac{7}{3} \cdot 9 \ ? \ 21$$

$$\dfrac{7 \cdot 3 \cdot 3}{3 \cdot 1}$$

$$21 \ \big| \ 21 \quad \text{TRUE}$$

The solution is 9.

**5.** $\dfrac{3}{7}x = -18$

$\dfrac{7}{3} \cdot \dfrac{3}{7}x = \dfrac{7}{3}(-18)$    The reciprocal of $\dfrac{3}{7}$ is $\dfrac{7}{3}$.

$1x = -\dfrac{7 \cdot 3 \cdot 6}{3}$

$x = -42$    Removing the factor $\dfrac{3}{3}$

Check:    $\dfrac{3}{7}x = -18$

$$\dfrac{3}{7}(-42) \ ? \ -18$$

$$-\dfrac{3 \cdot 6 \cdot 7}{7 \cdot 1}$$

$$-18 \ \big| \ -18 \quad \text{TRUE}$$

The solution is $-42$.

**7.** $6a = \dfrac{12}{17}$

$\dfrac{1}{6} \cdot 6a = \dfrac{1}{6} \cdot \dfrac{12}{17}$    The reciprocal of 6 is $\dfrac{1}{6}$.

$1a = \dfrac{2 \cdot 6}{6 \cdot 17}$

$a = \dfrac{2}{17}$    Removing the factor $\dfrac{6}{6}$

Check:    $6a = \dfrac{12}{17}$

$$6 \cdot \dfrac{2}{17} \ ? \ \dfrac{12}{17}$$

$$\dfrac{12}{17} \ \big| \ \dfrac{12}{17} \quad \text{TRUE}$$

The solution is $\dfrac{12}{17}$.

**9.** $\frac{3}{5}x = \frac{2}{7}$

$\frac{5}{3} \cdot \frac{3}{5}x = \frac{5}{3} \cdot \frac{2}{7}$

$x = \frac{5 \cdot 2}{3 \cdot 7}$

$x = \frac{10}{21}$

$\frac{10}{21}$ checks and is the solution.

**11.** $\frac{3}{2}t = -\frac{8}{7}$

$\frac{2}{3} \cdot \frac{3}{2}t = \frac{2}{3}\left(-\frac{8}{7}\right)$

$t = -\frac{16}{21}$

$-\frac{16}{21}$ checks and is the solution.

**13.** $\frac{4}{5} = -10a$

$-\frac{1}{10} \cdot \frac{4}{5} = -\frac{1}{10}(-10a)$

$-\frac{2 \cdot 2}{2 \cdot 5 \cdot 5} = a$

$-\frac{2}{25} = a$

$-\frac{2}{25}$ checks and is the solution.

**15.** $\frac{9}{5}x = \frac{3}{10}$

$\frac{5}{9} \cdot \frac{9}{5}x = \frac{5}{9} \cdot \frac{3}{10}$

$x = \frac{5 \cdot 3 \cdot 1}{3 \cdot 3 \cdot 2 \cdot 5}$

$x = \frac{1}{6}$

$\frac{1}{6}$ checks and is the solution.

**17.** $-\frac{3}{10}x = 8$

$-\frac{10}{3}\left(-\frac{3}{10}x\right) = -\frac{10}{3} \cdot 8$

$x = -\frac{10 \cdot 8}{3}$

$x = -\frac{80}{3}$

$-\frac{80}{3}$ checks and is the solution.

**19.** $a \cdot \frac{9}{7} = -\frac{3}{14}$

$a \cdot \frac{9}{7} \cdot \frac{7}{9} = -\frac{3}{14} \cdot \frac{7}{9}$

$a \cdot 1 = -\frac{3 \cdot 7 \cdot 1}{2 \cdot 7 \cdot 3 \cdot 3}$

$a = -\frac{1}{6}$

$-\frac{1}{6}$ checks and is the solution.

**21.** $-x = \frac{9}{13}$

$-1(-x) = -1 \cdot \frac{9}{13}$

$x = -\frac{9}{13}$

$-\frac{9}{13}$ checks and is the solution.

**23.** $-x = -\frac{27}{31}$

$-1(-x) = -1\left(-\frac{27}{31}\right)$

$x = \frac{27}{31}$

$\frac{27}{31}$ checks and is the solution.

**25.** $7t = 5$

$\frac{1}{7} \cdot 7t = \frac{1}{7} \cdot 5$

$t = \frac{5}{7}$

$\frac{5}{7}$ checks and is the solution.

**27.** $-24 = -10a$

$-\frac{1}{10}(-24) = -\frac{1}{10}(-10a)$

$\frac{2 \cdot 12}{2 \cdot 5} = a$

$\frac{12}{5} = a$

$\frac{12}{5}$ checks and is the solution.

**29.** $-\frac{15}{7} = \frac{3}{2}t$

$\frac{2}{3}\left(-\frac{15}{7}\right) = \frac{2}{3} \cdot \frac{3}{2}t$

$-\frac{2 \cdot 3 \cdot 5}{3 \cdot 7} = t$

$-\frac{10}{7} = t$

$-\frac{10}{7}$ checks and is the solution.

**31.**
$$x \cdot \frac{5}{16} = \frac{15}{14}$$
$$x \cdot \frac{5}{16} \cdot \frac{16}{5} = \frac{15}{14} \cdot \frac{16}{5}$$
$$x \cdot 1 = \frac{3 \cdot 5 \cdot 2 \cdot 8}{2 \cdot 7 \cdot 5}$$
$$x = \frac{24}{7}$$
$\frac{24}{7}$ checks and is the solution.

**33.**
$$-\frac{3}{20}x = -\frac{21}{10}$$
$$-\frac{20}{3}\left(-\frac{3}{20}x\right) = -\frac{20}{3}\left(-\frac{21}{10}\right)$$
$$x = \frac{2 \cdot 10 \cdot 3 \cdot 7}{3 \cdot 10}$$
$$x = 14$$
14 checks and is the solution.

**35.**
$$-\frac{25}{17} = -\frac{35}{34}a$$
$$-\frac{34}{35}\left(-\frac{25}{17}\right) = -\frac{34}{35}\left(-\frac{35}{34}a\right)$$
$$\frac{2 \cdot 17 \cdot 5 \cdot 5}{5 \cdot 7 \cdot 17} = a$$
$$\frac{10}{7} = a$$
$\frac{10}{7}$ checks and is the solution.

**37.**
$$36 \div (-3)^2 \times (7 - 2) = 36 \div (-3)^2 \times 5$$
$$= 36 \div 9 \times 5$$
$$= 4 \times 5$$
$$= 20$$

**39.** $13x + 4x = (13 + 4)x = 17x$

**41.**
$$2a + 3 + 5a = 2a + 5a + 3$$
$$= (2 + 5)a + 3$$
$$= 7a + 3$$

**43.** ◈

**45.** ◈

**47.**
$$\left(-\frac{4}{7}\right)^2 = \left(\frac{2^3 - 9}{3}\right)^3 x$$
$$\left(-\frac{4}{7}\right)^2 = \left(\frac{8 - 9}{3}\right)^3 x$$
$$\left(-\frac{4}{7}\right)^2 = \left(\frac{-1}{3}\right)^3 x$$
$$\frac{16}{49} = -\frac{1}{27}x$$
$$-27 \cdot \frac{16}{49} = -27\left(-\frac{1}{27}x\right)$$
$$-\frac{432}{49} = x$$
$-\frac{432}{49}$ checks and is the solution.

**49.** *Familiarize.* This is a multistep problem. First we find the length of the total trip. Then we find how many kilometers were left to drive. We draw a picture. We let $n =$ the length of the total trip.

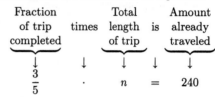

$\frac{3}{5}$ of the trip

240 km

$n$ km

*Translate.* We translate to an equation.

| Fraction of trip completed | times | Total length of trip | is | Amount already traveled |
|---|---|---|---|---|
| ↓ | ↓ | ↓ | ↓ | ↓ |
| $\frac{3}{5}$ | $\cdot$ | $n$ | $=$ | 240 |

*Solve.* We solve the equation.
$$\frac{3}{5} \cdot n = 240$$
$$\frac{5}{3} \cdot \frac{3}{5}n = \frac{5}{3} \cdot 240$$
$$n = \frac{5 \cdot 3 \cdot 80}{3}$$
$$n = 400$$

The total trip was 400 km.

Now we find how many kilometers were left to travel. Let $t =$ this number.

| Length of total trip | minus | Distance traveled | is | Distance left to travel |
|---|---|---|---|---|
| ↓ | ↓ | ↓ | ↓ | ↓ |
| 400 | $-$ | 240 | $=$ | $t$ |

We carry out the subtraction:
$$400 - 240 = t$$
$$160 = t$$

*Check.* We repeat the calculations. The results check.

*State.* The total trip was 400 km. There were 160 km left to travel.

**51.** *Familiarize.* We draw a picture. Let $r =$ the length of the race.

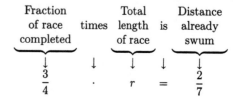

$\frac{3}{4}$ of the race

$\frac{2}{7}$ mi

$r$ mi

*Translate.* We translate to an equation.

| Fraction of race completed | times | Total length of race | is | Distance already swum |
|---|---|---|---|---|
| ↓ | ↓ | ↓ | ↓ | ↓ |
| $\frac{3}{4}$ | $\cdot$ | $r$ | $=$ | $\frac{2}{7}$ |

**Solve**. We solve the equation.

$$\frac{3}{4} \cdot r = \frac{2}{7}$$

$$\frac{4}{3} \cdot \frac{3}{4}r = \frac{4}{3} \cdot \frac{2}{7}$$

$$r = \frac{8}{21}$$

**Check**. Since $\frac{3}{4} \cdot \frac{8}{21} = \frac{3 \cdot 2 \cdot 4}{4 \cdot 3 \cdot 7} = \frac{2}{7}$ mi, the answer checks.

**State**. The race was $\frac{8}{21}$ mi long.

# Chapter 4

# Fractional Notation: Addition and Subtraction

## Exercise Set 4.1

In this section we will find the LCM using the multiples method in Exercises 1 - 19 and the factorization method in Exercises 21 - 39.

**1.** 1. 4 is the larger number and is a multiple of 2, so it is the LCM.

The LCM = 4.

**3.** 1. 25 is the larger number, but it is not a multiple of 10.

2. Check multiples of 25:

$2 \cdot 25 = 50$    A multiple of 10

The LCM = 50.

**5.** 1. 40 is the larger number and is a multiple of 20, so it is the LCM.

The LCM = 40.

**7.** 1. 27 is the larger number, but it is not a multiple of 18.

2. Check multiples of 27:

$2 \cdot 27 = 54$    A multiple of 18

The LCM = 54.

**9.** 1. 50 is the larger number, but it is not a multiple of 30.

2. Check multiples of 50:

$2 \cdot 50 = 100$    Not a multiple of 30
$3 \cdot 50 = 150$    A multiple of 30

The LCM = 150.

**11.** 1. 40 is the larger number, but it is not a multiple of 30.

2. Check multiples of 40:

$2 \cdot 40 = 80$    Not a multiple of 30
$3 \cdot 40 = 120$    A multiple of 30

The LCM = 120.

**13.** 1. 24 is the larger number, but it is not a multiple of 18.

2. Check multiples of 24:

$2 \cdot 24 = 48$    Not a multiple of 18
$3 \cdot 24 = 72$    A multiple of 18

The LCM = 72.

**15.** 1. 70 is the larger number, but it is not a multiple of 60.

2. Check multiples of 70:

$2 \cdot 70 = 140$    Not a multiple of 60
$3 \cdot 70 = 210$    Not a multiple of 60
$4 \cdot 70 = 280$    Not a multiple of 60
$5 \cdot 70 = 350$    Not a multiple of 60
$6 \cdot 70 = 420$    A multiple of 60

The LCM = 420.

**17.** 1. 36 is the larger number, but it is not a multiple of 16.

2. Check multiples of 36:

$2 \cdot 36 = 72$    Not a multiple of 16
$3 \cdot 36 = 108$    Not a multiple of 16
$4 \cdot 36 = 144$    A multiple of 16

The LCM = 144.

**19.** 1. 36 is the larger number, but it is not a multiple of 32.

2. Check multiples of 36:

$2 \cdot 36 = 72$    Not a multiple of 32
$3 \cdot 36 = 108$    Not a multiple of 32
$4 \cdot 36 = 144$    Not a multiple of 32
$5 \cdot 36 = 180$    Not a multiple of 32
$6 \cdot 36 = 216$    Not a multiple of 32
$7 \cdot 36 = 252$    Not a multiple of 32
$8 \cdot 36 = 288$    A multiple of 32

The LCM = 288.

**21.** 1. Write the prime factorization of each number. Because 2, 3, and 5 are all prime we write $2 = 2$, $3 = 3$, and $5 = 5$.

2. a) None of the factorizations contains the other two.

b) We begin with 2. Since 3 contains a factor of 3, we multiply by 3:

$2 \cdot 3$

Next we multiply $2 \cdot 3$ by 5, the factor of 5 that is missing:

$2 \cdot 3 \cdot 5$

The LCM is $2 \cdot 3 \cdot 5$, or 30.

3. To check, note that 2, 3, and 5 appear in the LCM the greatest number of times that each appears as a factor of 2, 3, or 5. The LCM is $2 \cdot 3 \cdot 5$, or 30.

**23.** 1. Write the prime factorization of each number. Because 3, 5, and 7 are all prime we write $3 = 3$, $5 = 5$, and $7 = 7$.

2. a) None of the factorizations contains the other two.

b) We begin with 3. Since 5 contains a factor of 5, we multiply by 5:

$3 \cdot 5$

Next we multiply $3 \cdot 5$ by 7, the factor of 7 that is missing:

$3 \cdot 5 \cdot 7$

The LCM is $3 \cdot 5 \cdot 7$, or 105.

3. To check, note that $3, 5$, and $7$ appear in the LCM the greatest number of times that each appears as a factor of $3, 5$, or $7$. The LCM is $3 \cdot 5 \cdot 7$, or $105$.

**25.** 1. Write the prime factorization of each number.

$$24 = 2 \cdot 2 \cdot 2 \cdot 3$$
$$36 = 2 \cdot 2 \cdot 3 \cdot 3$$
$$12 = 2 \cdot 2 \cdot 3$$

2. a) None of the factorizations contains the other two.

   b) We begin with the factorization of $24$, $2 \cdot 2 \cdot 2 \cdot 3$. Since $36$ contains a second factor of $3$, we multiply by another factor of $3$:

   $$2 \cdot 2 \cdot 2 \cdot 3 \cdot 3$$

   Next we look for factors of $12$ that are still missing. There are none. The LCM is

   $2 \cdot 2 \cdot 2 \cdot 3 \cdot 3$, or $72$.

3. To check, note that $2$ and $3$ appear in the LCM the greatest number of times that each appears as a factor of $24, 36$, or $12$. The LCM is $2 \cdot 2 \cdot 2 \cdot 3 \cdot 3$, or $72$.

**27.** 1. Write the prime factorization of each number.

$$5 = 5$$
$$12 = 2 \cdot 2 \cdot 3$$
$$15 = 3 \cdot 5$$

2. a) None of the factorizations contains the other two.

   b) We begin with the factorization of $12$, $2 \cdot 2 \cdot 3$. Since $5$ contains a factor of $5$, we multiply by $5$:

   $$2 \cdot 2 \cdot 3 \cdot 5$$

   Next we look for factors of $15$ that are still missing. There are none. The LCM is $2 \cdot 2 \cdot 3 \cdot 5$, or $60$.

3. The result checks.

**29.** 1. Write the prime factorization of each number.

$$9 = 3 \cdot 3$$
$$12 = 2 \cdot 2 \cdot 3$$
$$6 = 2 \cdot 3$$

2. a) None of the factorizations contains the other two.

   b) We begin with the factorization of $12$, $2 \cdot 2 \cdot 3$. Since $9$ contains a second factor of $3$, we multiply by another factor of $3$:

   $$2 \cdot 2 \cdot 3 \cdot 3$$

   Next we look for factors of $6$ that are still missing. There are none. The LCM is $2 \cdot 2 \cdot 3 \cdot 3$, or $36$.

3. The result checks.

**31.** 1. Write the prime factorization of each number.

$$180 = 2 \cdot 2 \cdot 3 \cdot 3 \cdot 5$$
$$100 = 2 \cdot 2 \cdot 5 \cdot 5$$
$$450 = 2 \cdot 3 \cdot 3 \cdot 5 \cdot 5$$

2. a) None of the factorizations contains the other two.

   b) We begin with the factorization of $450$, $2 \cdot 3 \cdot 3 \cdot 5 \cdot 5$. Since $180$ contains another factor of $2$, we multiply by $2$:

   $$2 \cdot 3 \cdot 3 \cdot 5 \cdot 5 \cdot 2$$

   Next we look for factors of $100$ that are still missing. There are none. The LCM is $2 \cdot 3 \cdot 3 \cdot 5 \cdot 5 \cdot 2$, or $900$.

3. The result checks.

**33.** 1. Write the prime factorization of each number.

$$75 = 3 \cdot 5 \cdot 5$$
$$100 = 2 \cdot 2 \cdot 5 \cdot 5$$

2. a) Neither factorization contains the other.

   b) We begin with the factorization of $100$, $2 \cdot 2 \cdot 5 \cdot 5$. Since $75$ contains a factor of $3$, we multiply by $3$:

   $$2 \cdot 2 \cdot 5 \cdot 5 \cdot 3$$

   The LCM is $2 \cdot 2 \cdot 5 \cdot 5 \cdot 3$, or $300$.

3. The result checks.

**35.** 1. We have the following factorizations:

$$ab = a \cdot b$$
$$bc = b \cdot c$$

2. a) Neither factorization contains the other.

   b) Consider the factorization of $ab$, $a \cdot b$. Since $bc$ contains a factor of $c$, we multiply by $c$.

   $$a \cdot b \cdot c$$

   The LCM is $a \cdot b \cdot c$, or $abc$.

3. The result checks.

**37.** 1. We have the following factorizations:

$$3x = 3 \cdot x$$
$$9x^2 = 3 \cdot 3 \cdot x \cdot x$$

2. a) One factorization, $3 \cdot 3 \cdot x \cdot x$, contains the other. Thus the LCM is $3 \cdot 3 \cdot x \cdot x$, or $9x^2$.

**39.** 1. We have the following factorizations:

$$4x^3 = 2 \cdot 2 \cdot x \cdot x \cdot x$$
$$x^2 y = x \cdot x \cdot y$$

2. a) Neither factorization contains the other.

   b) Consider the factorization of $4x^3$, $2 \cdot 2 \cdot x \cdot x \cdot x$. Since $x^2 y$ contains a factor of $y$, we multiply by $y$.

   $$2 \cdot 2 \cdot x \cdot x \cdot x \cdot y$$

   The LCM is $2 \cdot 2 \cdot x \cdot x \cdot x \cdot y$, or $4x^3 y$.

3. The result checks.

**41.** We find the LCM of the number of years it takes Jupiter and Saturn to make a complete revolution around the sun.

Jupiter: $12 = 2 \cdot 2 \cdot 3$

Saturn: $30 = 2 \cdot 3 \cdot 5$

The LCM $= 2 \cdot 2 \cdot 3 \cdot 5$, or 60. Thus, Jupiter and Saturn will appear in the same direction in the night sky once every 60 years.

**43.** $-38 + 52$

The absolute values are 38 and 52. The difference is 14. The positive number has the larger absolute value, so the answer is positive.

$$-38 + 52 = 14$$

**45.**
$$\begin{array}{r} \overset{\scriptstyle 1}{\phantom{0}} \\ \overset{\scriptstyle 1\ 1}{\phantom{0}} \\ 3\ 4\ 5 \\ \times\quad 2\ 3 \\ \hline 1\ 0\ 3\ 5 \\ 6\ 9\ 0\ 0 \\ \hline 7\ 9\ 3\ 5 \end{array}$$

**47.** $\dfrac{4}{5} \div \left( -\dfrac{7}{10} \right) = \dfrac{4}{5} \cdot \left( -\dfrac{10}{7} \right) = -\dfrac{4 \cdot 10}{5 \cdot 7} = -\dfrac{4 \cdot 2 \cdot 5}{5 \cdot 7} =$

$-\dfrac{4 \cdot 2}{7} \cdot \dfrac{5}{5} = -\dfrac{8}{7}$

**49.** ◈

**51.** ◈

**53.** a) 7800 is the larger number, but it is not a multiple of 2700.

b) Check multiples, using a calculator:

| | |
|---|---|
| $2 \cdot 7800 = 15,600$ | Not a multiple of 2700 |
| $3 \cdot 7800 = 23,400$ | Not a multiple of 2700 |
| $4 \cdot 7800 = 31,200$ | Not a multiple of 2700 |
| $5 \cdot 7800 = 39,000$ | Not a multiple of 2700 |
| $6 \cdot 7800 = 46,800$ | Not a multiple of 2700 |
| $7 \cdot 7800 = 54,600$ | Not a multiple of 2700 |
| $8 \cdot 7800 = 62,400$ | Not a multiple of 2700 |
| $9 \cdot 7800 = 70,200$ | A multiple of 2700 |

c) The LCM is 70,200.

**55.** 1. From Example 6 we know that the LCM of 18 and 21 is $2 \cdot 3 \cdot 3 \cdot 7$. From Example 7 we know that the LCM of 24 and 36 is $2 \cdot 2 \cdot 2 \cdot 3 \cdot 3$. Also $63 = 3 \cdot 3 \cdot 7$, $56 = 2 \cdot 2 \cdot 2 \cdot 7$, and $20 = 2 \cdot 2 \cdot 5$.

2. a) None of the factorizations contains all of the others.

b) Begin with the LCM of 18 and 21. We multiply by two factors of 2, the prime factors of the LCM of 24 and 36 that are missing. We have $2 \cdot 3 \cdot 3 \cdot 7 \cdot 2 \cdot 2$. There are no prime factors of either 63 or 56 that are missing in this factorization. We multiply by 5, the prime factor of 20 that is missing. The LCM is $2 \cdot 3 \cdot 3 \cdot 7 \cdot 2 \cdot 2 \cdot 5$, or 2520.

3. The result checks.

**57.** a) This is not the LCM, because 8 is not a factor of $2 \cdot 2 \cdot 3 \cdot 3$.

b) This is not the LCM, because 8 is not a factor of $2 \cdot 2 \cdot 3$.

c) This is not the LCM, because neither 8 nor 12 is a factor of $2 \cdot 3 \cdot 3$.

d) This is the LCM, because both 8 and 12 are factors of $2 \cdot 2 \cdot 2 \cdot 3$ and this is the smallest such number.

**59.** Answers may vary. Since $54 = 2 \cdot 3 \cdot 3 \cdot 3$, we use appropriate combinations of these factors to find the desired numbers. One pair is 2 and $3 \cdot 3 \cdot 3$, or 2 and 27. Another is $2 \cdot 3$ and $3 \cdot 3 \cdot 3$, or 6 and 27. A third pair is $2 \cdot 3 \cdot 3$ and $3 \cdot 3 \cdot 3$, or 18 and 27.

---

## Exercise Set 4.2

**1.** $\dfrac{4}{9} + \dfrac{5}{9} = \dfrac{4+5}{9} = \dfrac{9}{9} = 1$

**3.** $\dfrac{1}{8} + \dfrac{5}{8} = \dfrac{1+5}{8} = \dfrac{6}{8} = \dfrac{3 \cdot 2}{4 \cdot 2} = \dfrac{3}{4} \cdot \dfrac{2}{2} = \dfrac{3}{4} \cdot 1 = \dfrac{3}{4}$

**5.** $\dfrac{7}{10} + \dfrac{3}{-10} = \dfrac{7}{10} + \dfrac{-3}{10} = \dfrac{7 + (-3)}{10} = \dfrac{4}{10} = \dfrac{2 \cdot 2}{2 \cdot 5} =$

$\dfrac{2}{2} \cdot \dfrac{2}{5} = 1 \cdot \dfrac{2}{5} = \dfrac{2}{5}$

**7.** $\dfrac{9}{a} + \dfrac{4}{a} = \dfrac{9+4}{a} = \dfrac{13}{a}$

**9.** $\dfrac{-5}{11} + \dfrac{3}{11} = \dfrac{-5+3}{11} = \dfrac{-2}{11}$, or $-\dfrac{2}{11}$

**11.** $\dfrac{2}{9}x + \dfrac{5}{9}x = \left( \dfrac{2}{9} + \dfrac{5}{9} \right)x = \dfrac{7}{9}x$

**13.** $\dfrac{5}{32} + \dfrac{3}{32}t + \dfrac{7}{32} + \dfrac{13}{32}t$

$= \dfrac{5}{32} + \dfrac{7}{32} + \dfrac{3}{32}t + \dfrac{13}{32}t$

$= \left( \dfrac{5}{32} + \dfrac{7}{32} \right) + \left( \dfrac{3}{32} + \dfrac{13}{32} \right)t$

$= \dfrac{12}{32} + \dfrac{16}{32}t$

$= \dfrac{4 \cdot 3}{4 \cdot 8} + \dfrac{16 \cdot 1}{16 \cdot 2}t$

$= \dfrac{4}{4} \cdot \dfrac{3}{8} + \dfrac{16}{16} \cdot \dfrac{1}{2}t$

$= \dfrac{3}{8} + \dfrac{1}{2}t$

**15.** $-\dfrac{3}{x} + \left( -\dfrac{7}{x} \right) = \dfrac{-3}{x} + \dfrac{-7}{x} = \dfrac{-3 + (-7)}{x} = \dfrac{-10}{x}$,

or $-\dfrac{10}{x}$

**17.** $\dfrac{1}{8} + \dfrac{1}{6}$ $\qquad$ $8 = 2 \cdot 2 \cdot 2$ and $6 = 2 \cdot 3$, so the LCD is $2 \cdot 2 \cdot 2 \cdot 3$, or 24

$= \dfrac{1}{8} \cdot \dfrac{3}{3} + \dfrac{1}{6} \cdot \dfrac{4}{4}$

$\qquad$ Think: $6 \times \square = 24$. The answer is 4, so we multiply by 1, using $\dfrac{4}{4}$.

$\qquad$ Think: $8 \times \square = 24$. The answer is 3, so we multiply by 1, using $\dfrac{3}{3}$.

$= \dfrac{3}{24} + \dfrac{4}{24}$

$= \dfrac{7}{24}$

**19.** $\dfrac{-4}{5} + \dfrac{7}{10}$ $\qquad$ 5 is a factor of 10, so the LCD is 10.

$= \dfrac{-4}{5} \cdot \dfrac{2}{2} + \dfrac{7}{10}$ ← This fraction already has the LCD as denominator.

$\qquad$ Think: $5 \times \square = 10$. The answer is 2, so we multiply by 1, using $\dfrac{2}{2}$.

$= \dfrac{-8}{10} + \dfrac{7}{10}$

$= \dfrac{-1}{10}$, or $-\dfrac{1}{10}$

**21.** $\dfrac{5}{12} + \dfrac{3}{8}$ $\qquad$ $12 = 2 \cdot 2 \cdot 3$ and $8 = 2 \cdot 2 \cdot 2$, so the LCD is $2 \cdot 2 \cdot 2 \cdot 3$, or 24.

$= \dfrac{5}{12} \cdot \dfrac{2}{2} + \dfrac{3}{8} \cdot \dfrac{3}{3}$

$\qquad$ Think: $8 \times \square = 24$. The answer is 3, so we multiply by 1, using $\dfrac{3}{3}$.

$\qquad$ Think: $12 \times \square = 24$. The answer is 2, so we multiply by 1, using $\dfrac{2}{2}$.

$= \dfrac{10}{24} + \dfrac{9}{24} = \dfrac{19}{24}$

**23.** $\dfrac{3}{20} + 4$

$= \dfrac{3}{20} + \dfrac{4}{1}$ $\qquad$ Rewriting 4 in fractional notation

$= \dfrac{3}{20} + \dfrac{4}{1} \cdot \dfrac{20}{20}$ $\qquad$ The LCD is 20.

$= \dfrac{3}{20} + \dfrac{80}{20}$

$= \dfrac{83}{20}$

**25.** $\dfrac{5}{-8} + \dfrac{5}{6}$

$= \dfrac{-5}{8} + \dfrac{5}{6}$ $\qquad$ Recall that $\dfrac{m}{-n} = \dfrac{-m}{n}$. The LCD is 24. (See Exercise 17.)

$= \dfrac{-5}{8} \cdot \dfrac{3}{3} + \dfrac{5}{6} \cdot \dfrac{4}{4}$

$= \dfrac{-15}{24} + \dfrac{20}{24}$

$= \dfrac{5}{24}$

**27.** $\dfrac{3}{10}x + \dfrac{7}{100}x$

$= \dfrac{3}{10} \cdot \dfrac{10}{10} \cdot x + \dfrac{7}{100}x$ $\qquad$ 10 is a factor of 100, so the LCD is 100.

$= \dfrac{30}{100}x + \dfrac{7}{100}x$

$= \dfrac{37}{100}x$

**29.** $\dfrac{5}{12} + \dfrac{4}{15}$ $\qquad$ $12 = 2 \cdot 2 \cdot 3$ and $15 = 3 \cdot 5$, so the LCM is $2 \cdot 2 \cdot 3 \cdot 5$, or 60.

$= \dfrac{5}{12} \cdot \dfrac{5}{5} + \dfrac{4}{15} \cdot \dfrac{4}{4}$

$= \dfrac{25}{60} + \dfrac{16}{60} = \dfrac{41}{60}$

**31.** $\dfrac{9}{10} + \dfrac{-99}{100}$ $\qquad$ 10 is a factor of 100, so the LCD is 100.

$= \dfrac{9}{10} \cdot \dfrac{10}{10} + \dfrac{-99}{100}$

$= \dfrac{90}{100} + \dfrac{-99}{100} = \dfrac{-9}{100}$, or $-\dfrac{9}{100}$

**33.** $5 + \dfrac{7}{12}$

$= \dfrac{5}{1} + \dfrac{7}{12}$ $\qquad$ The LCD is 12.

$= \dfrac{5}{1} \cdot \dfrac{12}{12} + \dfrac{7}{12}$

$= \dfrac{60}{12} + \dfrac{7}{12}$

$= \dfrac{67}{12}$

**35.** $-5t + \dfrac{2}{7}t$

$= \dfrac{-5}{1}t + \dfrac{2}{7}t$ $\qquad$ The LCD is 7.

$= \dfrac{-5}{1} \cdot \dfrac{7}{7} \cdot t + \dfrac{2}{7}t$

$= \dfrac{-35}{7}t + \dfrac{2}{7}t$

$= \dfrac{-33}{7}t$, or $-\dfrac{33}{7}t$

**37.** $-\dfrac{5}{12} + \dfrac{7}{-24}$

$\dfrac{-5}{12} + \dfrac{-7}{24}$     12 is a factor of 24, so the LCD is 24.

$= \dfrac{-5}{12} \cdot \dfrac{2}{2} + \dfrac{-7}{24}$

$= \dfrac{-10}{24} + \dfrac{-7}{24}$

$= \dfrac{-17}{24}$, or $-\dfrac{17}{24}$

**39.** $\dfrac{4}{10} + \dfrac{3}{100} + \dfrac{7}{1000}$     10 and 100 are factors of 1000, so the LCD is 1000.

$= \dfrac{4}{10} \cdot \dfrac{100}{100} + \dfrac{3}{100} \cdot \dfrac{10}{10} + \dfrac{7}{1000}$

$= \dfrac{400}{1000} + \dfrac{30}{1000} + \dfrac{7}{1000}$

$= \dfrac{437}{1000}$

**41.** $\dfrac{3}{10} + \dfrac{5}{12} + \dfrac{8}{15}$

$= \dfrac{3}{2 \cdot 5} + \dfrac{5}{2 \cdot 2 \cdot 3} + \dfrac{8}{3 \cdot 5}$     Factoring the denominators
The LCD is $2 \cdot 5 \cdot 2 \cdot 3$.

$= \dfrac{3}{2 \cdot 5} \cdot \dfrac{2 \cdot 3}{2 \cdot 3} + \dfrac{5}{2 \cdot 2 \cdot 3} \cdot \dfrac{5}{5} + \dfrac{8}{3 \cdot 5} \cdot \dfrac{2 \cdot 2}{2 \cdot 2}$

In each case we multiply by 1 to obtain the LCD.

$= \dfrac{3 \cdot 2 \cdot 3}{2 \cdot 5 \cdot 2 \cdot 3} + \dfrac{5 \cdot 5}{2 \cdot 2 \cdot 3 \cdot 5} + \dfrac{8 \cdot 2 \cdot 2}{3 \cdot 5 \cdot 2 \cdot 2}$

$= \dfrac{18}{2 \cdot 5 \cdot 2 \cdot 3} + \dfrac{25}{2 \cdot 5 \cdot 2 \cdot 3} + \dfrac{32}{2 \cdot 5 \cdot 2 \cdot 3}$

$= \dfrac{75}{2 \cdot 5 \cdot 2 \cdot 3}$

$= \dfrac{3 \cdot 5 \cdot 5}{2 \cdot 5 \cdot 2 \cdot 3} = \dfrac{3 \cdot 5}{3 \cdot 5} \cdot \dfrac{5}{2 \cdot 2}$

$= \dfrac{5}{4}$

**43.** $\dfrac{5}{6} + \dfrac{25}{52} + \dfrac{7}{4}$

$= \dfrac{5}{2 \cdot 3} + \dfrac{25}{2 \cdot 2 \cdot 13} + \dfrac{7}{2 \cdot 2}$     LCD is $2 \cdot 3 \cdot 2 \cdot 13$.

$= \dfrac{5}{2 \cdot 3} \cdot \dfrac{2 \cdot 13}{2 \cdot 13} + \dfrac{25}{2 \cdot 2 \cdot 13} \cdot \dfrac{3}{3} + \dfrac{7}{2 \cdot 2} \cdot \dfrac{3 \cdot 13}{3 \cdot 13}$

$= \dfrac{5 \cdot 2 \cdot 13}{2 \cdot 3 \cdot 2 \cdot 13} + \dfrac{25 \cdot 3}{2 \cdot 2 \cdot 13 \cdot 3} + \dfrac{7 \cdot 3 \cdot 13}{2 \cdot 2 \cdot 3 \cdot 13}$

$= \dfrac{130}{2 \cdot 3 \cdot 2 \cdot 13} + \dfrac{75}{2 \cdot 3 \cdot 2 \cdot 13} + \dfrac{273}{2 \cdot 3 \cdot 2 \cdot 13}$

$= \dfrac{478}{2 \cdot 3 \cdot 2 \cdot 13}$

$= \dfrac{2 \cdot 239}{2 \cdot 3 \cdot 2 \cdot 13} = \dfrac{2}{2} \cdot \dfrac{239}{3 \cdot 2 \cdot 13}$

$= \dfrac{239}{78}$

**45.** $\dfrac{2}{9} + \dfrac{7}{10} + \dfrac{-4}{15}$

$= \dfrac{2}{3 \cdot 3} + \dfrac{7}{2 \cdot 5} + \dfrac{-4}{3 \cdot 5}$     LCD is $3 \cdot 3 \cdot 2 \cdot 5$.

$= \dfrac{2}{3 \cdot 3} \cdot \dfrac{2 \cdot 5}{2 \cdot 5} + \dfrac{7}{2 \cdot 5} \cdot \dfrac{3 \cdot 3}{3 \cdot 3} + \dfrac{-4}{3 \cdot 5} \cdot \dfrac{3 \cdot 2}{3 \cdot 2}$

$= \dfrac{2 \cdot 2 \cdot 5}{3 \cdot 3 \cdot 2 \cdot 5} + \dfrac{7 \cdot 3 \cdot 3}{2 \cdot 5 \cdot 3 \cdot 3} + \dfrac{-4 \cdot 3 \cdot 2}{3 \cdot 5 \cdot 3 \cdot 2}$

$= \dfrac{20}{3 \cdot 3 \cdot 2 \cdot 5} + \dfrac{63}{3 \cdot 3 \cdot 2 \cdot 5} + \dfrac{-24}{3 \cdot 3 \cdot 2 \cdot 5}$

$= \dfrac{59}{3 \cdot 3 \cdot 2 \cdot 5}$

$= \dfrac{59}{90}$

**47.** Since there is a common denominator, compare the numerators.

$$5 < 6, \text{ so } \dfrac{5}{8} < \dfrac{6}{8}.$$

**49.** The LCD is 6. We multiply $\dfrac{2}{3}$ by 1 to make the denominators the same.

$$\dfrac{2}{3} \cdot \dfrac{2}{2} = \dfrac{4}{6}$$

The denominator of $\dfrac{5}{6}$ is the LCD.

Since $4 < 5$, it follows that $\dfrac{4}{6} < \dfrac{5}{6}$, so $\dfrac{2}{3} < \dfrac{5}{6}$.

**51.** The LCD is 21. We multiply by 1 to make the denominators the same.

$$\dfrac{-2}{3} \cdot \dfrac{7}{7} = \dfrac{-14}{21}$$

$$\dfrac{-5}{7} \cdot \dfrac{3}{3} = \dfrac{-15}{21}$$

Since $-14 > -15$, it follows that $\dfrac{-14}{21} > \dfrac{-15}{21}$, so

$\dfrac{-2}{3} > \dfrac{-5}{7}$.

**53.** The LCD is 30. We multiply by 1 to make the denominators the same.

$$\dfrac{11}{15} \cdot \dfrac{2}{2} = \dfrac{22}{30}$$

$$\dfrac{7}{10} \cdot \dfrac{3}{3} = \dfrac{21}{30}$$

Since $22 > 21$, it follows that $\dfrac{22}{30} > \dfrac{21}{30}$, so $\dfrac{11}{15} > \dfrac{7}{10}$.

**55.** Express $-\dfrac{1}{5}$ as $\dfrac{-1}{5}$. The LCD is 20. Multiply by 1 to make the denominators the same.

$$\dfrac{3}{4} \cdot \dfrac{5}{5} = \dfrac{15}{20}$$

$$\dfrac{-1}{5} \cdot \dfrac{4}{4} = \dfrac{-4}{20}$$

Since $15 > -4$, it follows that $\dfrac{15}{20} > \dfrac{-4}{20}$, so $\dfrac{3}{4} > -\dfrac{1}{5}$. We might have observed at the outset that one number is positive and the other is negative and it follows that the positive number is greater than the negative number.

**57.** The LCD is 60. We multiply by 1 to make the denominators the same.

$$\frac{-7}{20} \cdot \frac{3}{3} = \frac{-21}{60}$$

$$\frac{-6}{15} \cdot \frac{4}{4} = \frac{-24}{60}$$

Since $-21 > -24$, it follows that $\frac{-21}{60} > \frac{-24}{60}$, so $\frac{-7}{20} > \frac{-6}{15}$.

**59.** The LCD is 60. We multiply by 1 to make the denominators the same.

$$\frac{3}{10} \cdot \frac{6}{6} = \frac{18}{60}$$

$$\frac{5}{12} \cdot \frac{5}{5} = \frac{25}{60}$$

$$\frac{4}{15} \cdot \frac{4}{4} = \frac{16}{60}$$

Since $16 < 18$ and $18 < 25$, when we arrange $\frac{18}{60}, \frac{25}{60}$, and $\frac{16}{60}$ from smallest to largest we have $\frac{16}{60}, \frac{18}{60}, \frac{25}{60}$. Then it follows that when we arrange the original fractions from smallest to largest we have $\frac{4}{15}, \frac{3}{10}, \frac{5}{12}$.

**61.** *Familiarize*. We draw a picture. We let $p =$ the number of pounds of tea Rose bought.

| $\frac{1}{3}$ lb | $\frac{1}{2}$ lb |
|---|---|
| $p$ | |

*Translate*. The problem can be translated to an equation as follows:

| Pounds of orange pekoe | plus | Pounds of English cinnamon | is | Total pounds of tea |
|---|---|---|---|---|
| ↓ | ↓ | ↓ | ↓ | ↓ |
| $\frac{1}{3}$ | $+$ | $\frac{1}{2}$ | $=$ | $p$ |

*Solve*. We carry out the addition. The LCM of the denominators is $3 \cdot 2$, or 6.

$$\frac{1}{3} + \frac{1}{2} = p$$

$$\frac{1}{3} \cdot \frac{2}{2} + \frac{1}{2} \cdot \frac{3}{3} = p$$

$$\frac{2}{6} + \frac{3}{6} = p$$

$$\frac{5}{6} = p$$

*Check*. We check by repeating the calculation. We also note that the sum is larger than either of the individual weights, so the answer seems reasonable.

*State*. Rose bought $\frac{5}{6}$ lb of tea.

**63.** *Familiarize*. We draw a picture. We let $D =$ the total distance walked.

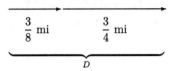

*Translate*. The problem can be translated to an equation as follows:

| Distance to friend's dorm | plus | Distance to class | is | Total distance |
|---|---|---|---|---|
| ↓ | ↓ | ↓ | ↓ | ↓ |
| $\frac{3}{8}$ | $+$ | $\frac{3}{4}$ | $=$ | $D$ |

*Solve*. To solve the equation, carry out the addition. Since 8 is a multiple of 4, the LCM of the denominators is 8.

$$\frac{3}{8} + \frac{3}{4} = D$$

$$\frac{3}{8} + \frac{3}{4} \cdot \frac{2}{2} = D$$

$$\frac{3}{8} + \frac{6}{8} = D$$

$$\frac{9}{8} = D$$

*Check*. We repeat the calculation. We also note that the sum is larger than either of the original distances, so the answer seems reasonable.

*State*. Ruwanda walked $\frac{9}{8}$ mi.

**65.** *Familiarize*. We draw a picture and let $c =$ the total number of quarts of liquid ingredients.

| $\frac{1}{2}$ qt | $\frac{1}{3}$ qt | $\frac{1}{16}$ qt |
|---|---|---|
| $c$ | | |

*Translate*. The problem can be translated to an equation as follows:

| Amount of buttermilk | plus | Amount of skim milk | plus | Amount of oil | is | Amount of liquid |
|---|---|---|---|---|---|---|
| ↓ | ↓ | ↓ | ↓ | ↓ | ↓ | ↓ |
| $\frac{1}{2}$ | $+$ | $\frac{1}{3}$ | $+$ | $\frac{1}{16}$ | $=$ | $c$ |

*Solve*. We carry out the addition. Since $2 = 2$, $3 = 3$, and $16 = 2 \cdot 2 \cdot 2 \cdot 2$, the LCM of the denominators is $2 \cdot 2 \cdot 2 \cdot 2 \cdot 3$, or 48.

$$\frac{1}{2} + \frac{1}{3} + \frac{1}{16} = c$$

$$\frac{1}{2} \cdot \frac{24}{24} + \frac{1}{3} \cdot \frac{16}{16} + \frac{1}{16} \cdot \frac{3}{3} = c$$

$$\frac{24}{48} + \frac{16}{48} + \frac{3}{48} = c$$

$$\frac{43}{48} = c$$

*Check*. We repeat the calculation. We also note that the sum is larger than any of the individual amounts, as expected.

*State*. The recipe calls for $\frac{43}{48}$ qt of liquid ingredients.

**67.** *Familiarize*. This is a multistep problem. First we find the total weight of the cubic meter of concrete mix. We visualize the situation, letting $w$ = the total weight.

| 420 kg | 150 kg | 120 kg |
|---|---|---|
| | $w$ | |

*Translate*. We translate to an equation.

| Weight of cement | plus | Weight of stone | plus | Weight of sand | is | Total weight |
|---|---|---|---|---|---|---|
| ↓ | ↓ | ↓ | ↓ | ↓ | ↓ | ↓ |
| 420 | + | 150 | + | 120 | = | $w$ |

*Solve*. We carry out the addition.

$$420 + 150 + 120 = w$$
$$690 = w$$

Since the mix contains 420 kg of cement, the part that is cement is $\frac{420}{690} = \frac{14 \cdot 30}{23 \cdot 30} = \frac{14}{23} \cdot \frac{30}{30} = \frac{14}{23}$.

Since the mix contains 150 kg of stone, the part that is stone is $\frac{150}{690} = \frac{5 \cdot 30}{23 \cdot 30} = \frac{5}{23} \cdot \frac{30}{30} = \frac{5}{23}$.

Since the mix contains 120 kg of sand, the part that is sand is $\frac{120}{690} = \frac{4 \cdot 30}{23 \cdot 30} = \frac{4}{23} \cdot \frac{30}{30} = \frac{4}{23}$.

We add these amounts: $\frac{14}{23} + \frac{5}{23} + \frac{4}{23} = \frac{14+5+4}{23} = \frac{23}{23} = 1$.

*Check*. We repeat the calculations. We also note that since the total of the fractional parts is 1, the answer is probably correct.

*State*. The total weight of the cubic meter of concrete mix is 690 kg. Of this, $\frac{14}{23}$ is cement, $\frac{5}{23}$ is stone, and $\frac{4}{23}$ is sand. The result when we add these amounts is 1.

**69.** *Familiarize*. We draw a picture and let $d$ = the number of miles the ranger hikes.

Lookout   Nest  Campsite

| $\frac{3}{5}$ mi | $\frac{3}{10}$ mi | $\frac{3}{4}$ mi |
|---|---|---|

$d$

*Translate*. We translate to an equation as follows:

| Miles to lookout | plus | Miles to nest | plus | Miles to campsite | is | Total miles |
|---|---|---|---|---|---|---|
| ↓ | ↓ | ↓ | ↓ | ↓ | ↓ | ↓ |
| $\frac{3}{5}$ | + | $\frac{3}{10}$ | + | $\frac{3}{4}$ | = | $d$ |

*Solve*. We carry out the addition. Since $5 = 5$, $10 = 2 \cdot 5$, and $4 = 2 \cdot 2$, the LCM of the denominators is $2 \cdot 2 \cdot 5$, or 20.

$$\frac{3}{5} + \frac{3}{10} + \frac{3}{4} = d$$
$$\frac{3}{5} \cdot \frac{4}{4} + \frac{3}{10} \cdot \frac{2}{2} + \frac{3}{4} \cdot \frac{5}{5} = d$$
$$\frac{12}{20} + \frac{6}{20} + \frac{15}{20} = d$$
$$\frac{33}{20} = d$$

*Check*. We repeat the calculation. We also note that the sum is larger than any of the individual distances, as expected.

*State*. The ranger hiked a total of $\frac{33}{20}$ mi.

**71.** *Familiarize*. We draw a picture. We let $t$ = the total thickness.

| $\frac{5}{8}$ in. |
|---|
| $\frac{7}{8}$ in. |

← $\frac{3}{32}$ in. } $t$

*Translate*. We translate to an equation.

| Thickness of one board | plus | Thickness of glue | plus |
|---|---|---|---|
| ↓ | ↓ | ↓ | ↓ |
| $\frac{5}{8}$ | + | $\frac{3}{32}$ | + |

| Thickness of second board | is | Total thickness |
|---|---|---|
| ↓ | ↓ | ↓ |
| $\frac{7}{8}$ | = | $t$ |

*Solve*. We carry out the addition. The LCD is 32 since 8 is a factor of 32.

$$\frac{5}{8} + \frac{3}{32} + \frac{7}{8} = t$$
$$\frac{5}{8} \cdot \frac{4}{4} + \frac{3}{32} + \frac{7}{8} \cdot \frac{4}{4} = t$$
$$\frac{20}{32} + \frac{3}{32} + \frac{28}{32} = t$$
$$\frac{51}{32} = t$$

*Check*. We repeat the calculation. We also note that the sum is larger than any of the individual thicknesses, as expected.

*State*. The result is $\frac{51}{32}$ in. thick.

**73.** $-7 - 6 = -7 + (-6) = -13$

**75.** $9 - 17 = 9 + (-17) = -8$

**77.** $\dfrac{x-y}{3} = \dfrac{7-(-3)}{3} = \dfrac{7+3}{3} = \dfrac{10}{3}$

**79.** ◈

**81.** ◈

**83.**
$$\frac{2}{9} + \frac{4}{21}x + \frac{4}{15} + \frac{3}{14}x$$
$$= \frac{2}{9} + \frac{4}{15} + \frac{4}{21}x + \frac{3}{14}x$$
$$= \frac{2}{9} \cdot \frac{5}{5} + \frac{4}{15} \cdot \frac{3}{3} + \frac{4}{21} \cdot \frac{2}{2} \cdot x + \frac{3}{14} \cdot \frac{3}{3} \cdot x$$
$$= \frac{10}{45} + \frac{12}{45} + \frac{8}{42}x + \frac{9}{42}x$$
$$= \frac{22}{45} + \frac{17}{42}x$$

**85.** Use a calculator to do this exercise. First, add on the left.
$$\frac{12}{97} + \frac{67}{137} = \frac{8143}{13,289}$$
Now compare $\dfrac{8143}{13,289}$ and $\dfrac{8144}{13,289}$.

The denominators are the same. Since $8143 < 8144$, it follows that $\dfrac{8143}{13,289} < \dfrac{8144}{13,289}$, so $\dfrac{12}{97} + \dfrac{67}{137} < \dfrac{8144}{13,289}$.

**87.** *Familiarize.* First we find the fractional part of the band's pay that the guitarist received. We let $f =$ this fraction.

*Translate.* We translate to an equation.

One-third of one-half plus one-fifth of one-half is fractional part
$$\frac{1}{3} \cdot \frac{1}{2} + \frac{1}{5} \cdot \frac{1}{2} = f$$

*Solve.* We carry out the calculation.
$$\frac{1}{3} \cdot \frac{1}{2} + \frac{1}{5} \cdot \frac{1}{2} = f$$
$$\frac{1}{6} + \frac{1}{10} = f \qquad \text{LCD is 30.}$$
$$\frac{1}{6} \cdot \frac{5}{5} + \frac{1}{10} \cdot \frac{3}{3} = f$$
$$\frac{5}{30} + \frac{3}{30} = f$$
$$\frac{8}{30} = f$$
$$\frac{4}{15} = f$$

Now we find how much of the $1200 received by the band was paid to the guitarist. We let $p =$ the amount.

Four-fifteenths of $1200 = guitarist's pay
$$\frac{4}{15} \cdot 1200 = p$$

We solve the equation.
$$\frac{4}{15} \cdot 1200 = p$$
$$\frac{4 \cdot 1200}{15} = p$$
$$\frac{4 \cdot 3 \cdot 5 \cdot 80}{3 \cdot 5} = p$$
$$320 = p$$

*Check.* We repeat the calculations.

*State.* The guitarist received $\dfrac{4}{15}$ of the band's pay. This was $320.

---

## Exercise Set 4.3

**1.** When denominators are the same, subtract the numerators and keep the denominator.
$$\frac{5}{6} - \frac{1}{6} = \frac{5-1}{6} = \frac{4}{6} = \frac{2 \cdot 2}{2 \cdot 3} = \frac{2}{2} \cdot \frac{2}{3} = \frac{2}{3}$$

**3.** When denominators are the same, subtract the numerators and keep the denominator.
$$\frac{11}{16} - \frac{15}{16} = \frac{11-15}{16} = \frac{-4}{16} = \frac{-1 \cdot 4}{4 \cdot 4} = \frac{-1}{4} \cdot \frac{4}{4} = \frac{-1}{4},$$
or $-\dfrac{1}{4}$

**5.** $\dfrac{7}{a} - \dfrac{3}{a} = \dfrac{7-3}{a} = \dfrac{4}{a}$

**7.** $\dfrac{10}{3t} - \dfrac{4}{3t} = \dfrac{10-4}{3t} = \dfrac{6}{3t} = \dfrac{3}{3} \cdot \dfrac{2}{t} = \dfrac{2}{t}$

**9.** $\dfrac{3}{5a} - \dfrac{7}{5a} = \dfrac{3-7}{5a} = \dfrac{-4}{5a}$, or $-\dfrac{4}{5a}$

**11.** The LCM of 8 and 16 is 16.
$$\frac{7}{8} - \frac{1}{16} = \frac{7}{8} \cdot \frac{2}{2} - \frac{1}{16} \quad \leftarrow \text{This fraction already has the LCM as the denominator.}$$

Think: $8 \times \Box = 16$. The answer is 2, so we multiply by 1, using $\dfrac{2}{2}$.

$$= \frac{14}{16} - \frac{1}{16} = \frac{13}{16}$$

**13.** The LCM of 15 and 5 is 15.

$$\frac{7}{15} - \frac{4}{5} = \frac{7}{15} - \frac{4}{5} \cdot \frac{3}{2}$$

└ Think: $5 \times \boxed{\phantom{x}} = 15$. The answer is 3, so we multiply by 1, using $\frac{3}{3}$.

This fraction already has the LCM as the denominator.

$$= \frac{7}{15} - \frac{12}{15}$$

$$= \frac{-5}{15} = \frac{5}{5} \cdot \frac{-1}{3}$$

$$= \frac{-1}{3}, \text{ or } -\frac{1}{3}$$

**15.** The LCM of 4 and 20 is 20.

$$\frac{3}{4} - \frac{1}{20} = \frac{3}{4} \cdot \frac{5}{5} - \frac{1}{20}$$

$$= \frac{15}{20} - \frac{1}{20} = \frac{14}{20}$$

$$= \frac{2 \cdot 7}{2 \cdot 10} = \frac{2}{2} \cdot \frac{7}{10}$$

$$= \frac{7}{10}$$

**17.** The LCM of 15 and 12 is 60.

$$\frac{2}{15} - \frac{5}{12} = \frac{2}{15} \cdot \frac{4}{4} - \frac{5}{12} \cdot \frac{5}{5}$$

$$= \frac{8}{60} - \frac{25}{60} = \frac{8 - 25}{60}$$

$$= \frac{-17}{60}, \text{ or } -\frac{17}{60}$$

**19.** The LCM of 10 and 100 is 100.

$$\frac{6}{10} - \frac{7}{100} = \frac{6}{10} \cdot \frac{10}{10} - \frac{7}{100}$$

$$= \frac{60}{100} - \frac{7}{100} = \frac{53}{100}$$

**21.** The LCM of 15 and 25 is 75.

$$\frac{7}{15} - \frac{3}{25} = \frac{7}{15} \cdot \frac{5}{5} - \frac{3}{25} \cdot \frac{3}{3}$$

$$= \frac{35}{75} - \frac{9}{75} = \frac{26}{75}$$

**23.** The LCM of 10 and 100 is 100.

$$\frac{69}{100} - \frac{9}{10} = \frac{69}{100} - \frac{9}{10} \cdot \frac{10}{10}$$

$$= \frac{69}{100} - \frac{90}{100} = \frac{69 - 90}{100}$$

$$= \frac{-21}{100}, \text{ or } -\frac{21}{100}$$

**25.** The LCM of 3 and 8 is 24.

$$\frac{2}{3} - \frac{1}{8} = \frac{2}{3} \cdot \frac{8}{8} - \frac{1}{8} \cdot \frac{3}{3}$$

$$= \frac{16}{24} - \frac{3}{24}$$

$$= \frac{13}{24}$$

**27.** The LCM of 5 and 2 is 10.

$$\frac{3}{5} - \frac{1}{2} = \frac{3}{5} \cdot \frac{2}{2} - \frac{1}{2} \cdot \frac{5}{5}$$

$$= \frac{6}{10} - \frac{5}{10}$$

$$= \frac{1}{10}$$

**29.** The LCM of 18 and 24 is 72.

$$\frac{11}{18} - \frac{7}{24} = \frac{11}{18} \cdot \frac{4}{4} - \frac{7}{24} \cdot \frac{3}{3}$$

$$= \frac{44}{72} - \frac{21}{72}$$

$$= \frac{23}{72}$$

**31.** The LCM of 90 and 120 is 360.

$$\frac{13}{90} - \frac{17}{120} = \frac{13}{90} \cdot \frac{4}{4} - \frac{17}{120} \cdot \frac{3}{3}$$

$$= \frac{52}{360} - \frac{51}{360}$$

$$= \frac{1}{360}$$

**33.** The LCM of 3 and 9 is 9.

$$\frac{2}{3}x - \frac{4}{9}x = \frac{2}{3} \cdot \frac{3}{3} \cdot x - \frac{4}{9}x$$

$$= \frac{6}{9}x - \frac{4}{9}x$$

$$= \frac{2}{9}x$$

**35.** The LCM of 5 and 4 is 20.

$$\frac{3}{5}a - \frac{3}{4}a = \frac{3}{5} \cdot \frac{4}{4} \cdot a - \frac{3}{4} \cdot \frac{5}{5}a$$

$$= \frac{12}{20}a - \frac{15}{20}a$$

$$= \frac{-3}{20}a, \text{ or } -\frac{3}{20}a$$

**37.**
$$x - \frac{5}{9} = \frac{2}{9}$$

$$x - \frac{5}{9} + \frac{5}{9} = \frac{2}{9} + \frac{5}{9} \quad \text{Adding } \frac{5}{9} \text{ on both sides}$$

$$x + 0 = \frac{7}{9}$$

$$x = \frac{7}{9}$$

The solution is $\frac{7}{9}$.

**39.**  $a + \dfrac{2}{11} = \dfrac{8}{11}$

$a + \dfrac{2}{11} - \dfrac{2}{11} = \dfrac{8}{11} - \dfrac{2}{11}$    Subtracting $\dfrac{2}{11}$ on both sides

$a + 0 = \dfrac{6}{11}$

$a = \dfrac{6}{11}$

The solution is $\dfrac{6}{11}$.

**41.**  $x + \dfrac{2}{3} = \dfrac{7}{9}$

$x + \dfrac{2}{3} - \dfrac{2}{3} = \dfrac{7}{9} - \dfrac{2}{3}$    Subtracting $\dfrac{2}{3}$ on both sides

$x + 0 = \dfrac{7}{9} - \dfrac{2}{3} \cdot \dfrac{3}{3}$    The LCD is 9. We multiply by 1 to get the LCD.

$x = \dfrac{7}{9} - \dfrac{6}{9} = \dfrac{1}{9}$

The solution is $\dfrac{1}{9}$.

**43.**  $a - \dfrac{3}{8} = \dfrac{3}{4}$

$a - \dfrac{3}{8} + \dfrac{3}{8} = \dfrac{3}{4} + \dfrac{3}{8}$    Adding $\dfrac{3}{8}$ on both sides

$a + 0 = \dfrac{3}{4} \cdot \dfrac{2}{2} + \dfrac{3}{8}$    The LCD is 8. We multiply by 1 to get the LCD.

$a = \dfrac{6}{8} + \dfrac{3}{8} = \dfrac{9}{8}$

The solution is $\dfrac{9}{8}$.

**45.**  $\dfrac{2}{3} + x = \dfrac{4}{5}$

$\dfrac{2}{3} + x - \dfrac{2}{3} = \dfrac{4}{5} - \dfrac{2}{3}$    Subtracting $\dfrac{2}{3}$ on both sides

$x + 0 = \dfrac{4}{5} \cdot \dfrac{3}{3} - \dfrac{2}{3} \cdot \dfrac{5}{5}$    The LCD is 15. We multiply by 1 to get the LCD.

$x = \dfrac{12}{15} - \dfrac{10}{15} = \dfrac{2}{15}$

The solution is $\dfrac{2}{15}$.

**47.**  $\dfrac{3}{8} + a = \dfrac{1}{12}$

$\dfrac{3}{8} + a - \dfrac{3}{8} = \dfrac{1}{12} - \dfrac{3}{8}$    Subtracting $\dfrac{3}{8}$ on both sides

$a + 0 = \dfrac{1}{12} \cdot \dfrac{2}{2} - \dfrac{3}{8} \cdot \dfrac{3}{3}$    The LCD is 24. We multiply by 1 to get the LCD.

$a = \dfrac{2}{24} - \dfrac{9}{24} = \dfrac{2-9}{24}$

$a = \dfrac{-7}{24}$, or $-\dfrac{7}{24}$

The solution is $-\dfrac{7}{24}$.

**49.**  $n - \dfrac{1}{10} = -\dfrac{1}{30}$

$n - \dfrac{1}{10} + \dfrac{1}{10} = -\dfrac{1}{30} + \dfrac{1}{10}$    Adding $\dfrac{1}{10}$ on both sides

$n + 0 = -\dfrac{1}{30} + \dfrac{1}{10} \cdot \dfrac{3}{3}$    The LCD is 30. We multiply by 1 to get the LCD.

$n = -\dfrac{1}{30} + \dfrac{3}{30}$

$n = \dfrac{-1}{30} + \dfrac{3}{30} = \dfrac{2}{30}$

$n = \dfrac{2 \cdot 1}{2 \cdot 15} = \dfrac{2}{2} \cdot \dfrac{1}{15}$

$n = \dfrac{1}{15}$

The solution is $\dfrac{1}{15}$.

**51.**  $x + \dfrac{3}{4} = -\dfrac{1}{2}$

$x + \dfrac{3}{4} - \dfrac{3}{4} = -\dfrac{1}{2} - \dfrac{3}{4}$    Subtracting $\dfrac{3}{4}$ on both sides

$x + 0 = -\dfrac{1}{2} \cdot \dfrac{2}{2} - \dfrac{3}{4}$    The LCD is 4. We multiply by 1 to get the LCD.

$x = -\dfrac{2}{4} - \dfrac{3}{4} = \dfrac{-2}{4} - \dfrac{3}{4}$

$x = \dfrac{-2-3}{4}$

$x = \dfrac{-5}{4}$, or $-\dfrac{5}{4}$

The solution is $-\dfrac{5}{4}$.

**53. *Familiarize*.** We visualize the situation. Let $t =$ the number of hours Monica listened to Brahms.

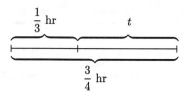

**Translate**. This is a "how much more" situation that can be translated as follows:

| Time spent listening to Beethoven | plus | Time spent listening to Brahms | is | Total listening time |
|---|---|---|---|---|
| $\downarrow$ | $\downarrow$ | $\downarrow$ | $\downarrow$ | $\downarrow$ |
| $\frac{1}{3}$ | $+$ | $t$ | $=$ | $\frac{3}{4}$ |

**Solve**. We subtract $\frac{1}{3}$ on both sides of the equation.

$$\frac{1}{3} + t - \frac{1}{3} = \frac{3}{4} - \frac{1}{3}$$

$$t + 0 = \frac{3}{4} \cdot \frac{3}{3} - \frac{1}{3} \cdot \frac{4}{4} \qquad \text{The LCD is 12. We multiply by 1 to get the LCD.}$$

$$t = \frac{9}{12} - \frac{4}{12} = \frac{5}{12}$$

**Check**. We return to the original problem and add.

$$\frac{1}{3} + \frac{5}{12} = \frac{1}{3} \cdot \frac{4}{4} + \frac{5}{12} = \frac{4}{12} + \frac{5}{12} = \frac{9}{12} = \frac{3}{3} \cdot \frac{3}{4} = \frac{3}{4}$$

**State**. Monica spent $\frac{5}{12}$ hr listening to Brahms.

**55. *Familiarize*.** We visualize the situation. Let $d =$ the distance that remains to be walked.

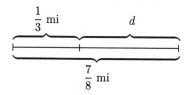

**Translate**. This is a "how much more" situation that can be translate as follows:

| Distance already walked | plus | Distance remaining to be walked | is | Total distance |
|---|---|---|---|---|
| $\downarrow$ | $\downarrow$ | $\downarrow$ | $\downarrow$ | $\downarrow$ |
| $\frac{1}{3}$ | $+$ | $d$ | $=$ | $\frac{7}{8}$ |

**Solve**. We subtract $\frac{1}{3}$ on both sides of the equation.

$$\frac{1}{3} + d - \frac{1}{3} = \frac{7}{8} - \frac{1}{3}$$

$$d + 0 = \frac{7}{8} \cdot \frac{3}{3} - \frac{1}{3} \cdot \frac{8}{8} \qquad \text{The LCD is 24. We multiply by 1 to get the LCD.}$$

$$d = \frac{21}{24} - \frac{8}{24} = \frac{13}{24}$$

**Check**. We return to the original problem and add.

$$\frac{1}{3} + \frac{13}{24} = \frac{1}{3} \cdot \frac{8}{8} + \frac{13}{14} = \frac{8}{24} + \frac{13}{24} = \frac{21}{24} = \frac{3}{3} \cdot \frac{7}{8} = \frac{7}{8}$$

**State**. Hugo should walk $\frac{13}{24}$ mi farther.

**57. *Familiarize*.** We visualize the situation. Let $d =$ the number of inches by which the new tread depth exceeds the more typical depth.

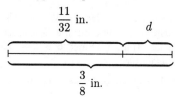

**Translate**. This is a "how much more" situation.

| Typical tread depth | plus | Excess depth | is | New tread depth |
|---|---|---|---|---|
| $\downarrow$ | $\downarrow$ | $\downarrow$ | $\downarrow$ | $\downarrow$ |
| $\frac{11}{32}$ | $+$ | $d$ | $=$ | $\frac{3}{8}$ |

**Solve**. We subtract $\frac{11}{32}$ on both sides of the equation.

$$\frac{11}{32} + d - \frac{11}{32} = \frac{3}{8} - \frac{11}{32}$$

$$d + 0 = \frac{3}{8} \cdot \frac{4}{4} - \frac{11}{32} \qquad \text{The LCD is 32. We multiply by 1 to get the LCD.}$$

$$d = \frac{12}{32} - \frac{11}{32}$$

$$d = \frac{1}{32}$$

**Check**. We return to the original problem and add.

$$\frac{11}{32} + \frac{1}{32} = \frac{12}{32} = \frac{4 \cdot 3}{4 \cdot 8} = \frac{4}{4} \cdot \frac{3}{8} = \frac{3}{8}$$

The answer checks.

**State**. The new tread depth is $\frac{1}{32}$ in. deeper than the more typical depth of $\frac{11}{32}$ in.

**59.** $\dfrac{9}{10} \div \dfrac{3}{5} = \dfrac{9}{10} \cdot \dfrac{5}{3} = \dfrac{9 \cdot 5}{10 \cdot 3} = \dfrac{3 \cdot 3 \cdot 5}{2 \cdot 5 \cdot 3} = \dfrac{3 \cdot 5}{3 \cdot 5} \cdot \dfrac{3}{2} = \dfrac{3}{2}$

**61.** $(-7) \div \dfrac{1}{3} = -7 \cdot \dfrac{3}{1} = \dfrac{-7 \cdot 3}{1} = \dfrac{-21}{1} = -21$

**63. *Familiarize*.** We visualize the situation. Repeated addition will work here.

We let $w =$ the weight of 8 boxes.

**Translate**. The problem translates to the following equation:

$$w = 8 \cdot \frac{3}{4}$$

*Solve*. We carry out the multiplication.

$$w = 8 \cdot \frac{3}{4}$$

$$w = \frac{8 \cdot 3}{4} = \frac{2 \cdot 4 \cdot 3}{4 \cdot 1}$$

$$w = \frac{4}{4} \cdot \frac{2 \cdot 3}{1}$$

$$w = 6$$

*Check*. We repeat the calculation. We can also observe that we are multiplying 8 by a number less than 1, so the product will be less than 8. Since 6 is less than 8, our answer seems reasonable.

*State*. The weight of 8 small boxes of cornflakes is 6 lb.

**65.**

**67.** ◈

**69.** Use a calculator.

$$x + \frac{7}{253} = \frac{12}{299}$$

$$x + \frac{7}{253} - \frac{7}{253} = \frac{12}{299} - \frac{7}{253}$$

$$x + 0 = \frac{12}{13 \cdot 23} - \frac{7}{11 \cdot 23} \quad \text{The LCD is } 11 \cdot 13 \cdot 23.$$

$$x = \frac{12}{13 \cdot 23} \cdot \frac{11}{11} - \frac{7}{11 \cdot 23} \cdot \frac{13}{13}$$

$$x = \frac{132}{13 \cdot 23 \cdot 11} - \frac{91}{11 \cdot 23 \cdot 13}$$

$$x = \frac{41}{3289}$$

**71.** $\frac{2}{5} - \frac{1}{6}(-3)^2 = \frac{2}{5} - \frac{1}{6} \cdot 9$

$$= \frac{2}{5} - \frac{9}{6}$$

$$= \frac{2}{5} \cdot \frac{6}{6} - \frac{9}{6} \cdot \frac{5}{5} \quad \text{The LCD is 30.}$$

$$= \frac{12}{30} - \frac{45}{30}$$

$$= \frac{-33}{30} = \frac{3}{3} \cdot \frac{-11}{10}$$

$$= \frac{-11}{10}, \text{ or } -\frac{11}{10}$$

**73.** $-4 \times \frac{3}{7} - \frac{1}{7} \times \frac{4}{5} = \frac{-12}{7} - \frac{4}{35}$

$$= \frac{-12}{7} \cdot \frac{5}{5} - \frac{4}{35} \quad \text{The LCD is 35.}$$

$$= \frac{-60}{35} - \frac{4}{35}$$

$$= \frac{-64}{35}, \text{ or } -\frac{64}{35}$$

**75. Familiarize**. Let $h$ = the height, in inches, of the pile of cold cuts. The height of the 8 turkey slices is $8 \cdot \frac{1}{16}$ in., and the height of the 3 cheese slices is $3 \cdot \frac{5}{32}$ in.

*Translate*.

| Height of turkey slices | plus | height of cheese slices | is | total height. |
|---|---|---|---|---|
| ↓ | ↓ | ↓ | ↓ | ↓ |
| $8 \cdot \frac{1}{16}$ | $+$ | $3 \cdot \frac{5}{32}$ | $=$ | $h$ |

*Solve*. We carry out the calculations on the left side of the equation.

$$8 \cdot \frac{1}{16} + 3 \cdot \frac{5}{32} = h$$

$$\frac{8}{16} + \frac{15}{32} = h$$

$$\frac{8}{16} \cdot \frac{2}{2} + \frac{15}{32} = h \quad \text{The LCD is 32.}$$

$$\frac{16}{32} + \frac{15}{32} = h$$

$$\frac{31}{32} = h$$

*Check*. We repeat the calculations on the left side of the equation. The answer checks.

*State*. The pile of cold cuts was $\frac{31}{32}$ in. high.

**77. Familiarize**. Let $t$ = the fractional piece of the estate that each twin received. Then the twins received a total of $2t$ of the estate.

*Translate*.

| First child's share | plus | second child's share | plus | twins' share | is | 1 entire estate. |
|---|---|---|---|---|---|---|
| ↓ | ↓ | ↓ | ↓ | ↓ | ↓ | ↓ |
| $\frac{1}{4}$ | $+$ | $\frac{3}{8}$ | $+$ | $2t$ | $=$ | $1$ |

*Solve*. We solve the equation.

$$\frac{1}{4} + \frac{3}{8} = 2t + 1$$

$$\frac{2}{8} + \frac{3}{8} + 2t = 1$$

$$\frac{5}{8} + 2t = 1 \quad \text{Adding on the left}$$

$$\frac{5}{8} + 2t - \frac{5}{8} = 1 - \frac{5}{8}$$

$$2t + 0 = \frac{8}{8} - \frac{5}{8}$$

$$2t = \frac{3}{8}$$

$$\frac{1}{2} \cdot 2t = \frac{1}{2} \cdot \frac{3}{8}$$

$$t = \frac{3}{16}$$

*Check*. We add the portions.

$$\frac{1}{4} + \frac{3}{8} + \frac{3}{16} + \frac{3}{16} = \frac{4}{16} + \frac{6}{16} + \frac{3}{16} + \frac{3}{16} = \frac{16}{16} = 1$$

Since the portions add up to 1 entire estate, the answer checks.

*State*. Each twin received $\frac{3}{16}$ of the estate.

## Exercise Set 4.4

**1.** $6x - 4 = 14$
$6x - 4 + 4 = 14 + 4$   Using the addition principle
$6x + 0 = 18$
$6x = 18$
$\frac{1}{6} \cdot 6x = \frac{1}{6} \cdot 18$   Using the multiplication principle
$1x = \frac{18}{6}$
$x = 3$

Check:   $\underline{6x - 4 = 14}$
$6 \cdot 3 - 4 \ ? \ 14$
$18 - 4 \ \big| $
$14 \ \big| \ 14$   TRUE

The solution is 3.

**3.** $3a + 8 = 23$
$3a + 8 - 8 = 23 - 8$   Using the addition principle
$3a + 0 = 15$
$3a = 15$
$\frac{1}{3} \cdot 3a = \frac{1}{3} \cdot 15$   Using the multiplication principle
$1a = \frac{15}{3}$
$a = 5$

Check:   $\underline{3a + 8 = 23}$
$3 \cdot 5 + 8 \ ? \ 23$
$15 + 8 \ \big| $
$23 \ \big| \ 23$   TRUE

The solution is 5.

**5.** $4a + 9 = 37$
$4a + 9 - 9 = 37 - 9$   Using the addition principle
$4a + 0 = 28$
$4a = 28$
$\frac{1}{4} \cdot 4a = \frac{1}{4} \cdot 28$   Using the multiplication principle
$1a = \frac{28}{4}$
$a = 7$

Check:   $\underline{4a + 9 = 37}$
$4 \cdot 7 + 9 \ ? \ 37$
$28 + 9 \ \big| $
$37 \ \big| \ 37$   TRUE

The solution is 7.

**7.** $31 = 3x - 5$
$31 + 5 = 3x - 5 + 5$   Using the addition principle
$36 = 3x + 0$
$36 = 3x$
$\frac{1}{3} \cdot 36 = \frac{1}{3} \cdot 3x$   Using the multiplication principle
$\frac{36}{3} = 1x$
$12 = x$

Check:   $\underline{31 = 3x - 5}$
$31 \ ? \ 3 \cdot 12 - 5$
$\big| \ 36 - 5$
$31 \ \big| \ 31$   TRUE

The solution is 12.

**9.** $-5t + 4 = 39$
$-5t + 4 - 4 = 39 - 4$   Using the addition principle
$-5t + 0 = 35$
$-5t = 35$
$-\frac{1}{5}(-5t) = -\frac{1}{5} \cdot 35$   Using the multiplication principle
$1t = -\frac{35}{5}$
$t = -7$

Check:   $\underline{-5t + 4 = 39}$
$-5(-7) + 4 \ ? \ 39$
$35 + 4 \ \big| $
$39 \ \big| \ 39$   TRUE

The solution is $-7$.

**11.** $3x + 4 = -11$
$3x + 4 - 4 = -11 - 4$   Using the addition principle
$3x + 0 = -15$
$3x = -15$
$\frac{1}{3} \cdot 3x = \frac{1}{3}(-15)$   Using the multiplication principle
$1x = -\frac{15}{3}$
$x = -5$

The number $-5$ checks and is the solution.

**13.** $\frac{4}{5}x = 20$
$\frac{5}{4} \cdot \frac{4}{5}x = \frac{5}{4} \cdot 20$   Using the multiplication principle
$1x = \frac{100}{4}$
$x = 25$

The number 25 checks and is the solution.

**15.**  $\dfrac{3}{2}x - 3 = 12$

$\dfrac{3}{2}x - 3 + 3 = 12 + 3$  Using the addition principle

$\dfrac{3}{2}x + 0 = 15$

$\dfrac{3}{2}x = 15$

$\dfrac{2}{3} \cdot \dfrac{3}{2}x = \dfrac{2}{3} \cdot 15$  Using the multiplication principle

$1x = \dfrac{30}{3}$

$x = 10$

The number 10 checks and is the solution.

**17.**  $\dfrac{3}{5}t - 4 = 8$

$\dfrac{3}{5}t - 4 + 4 = 8 + 4$  Using the addition principle

$\dfrac{3}{5}t + 0 = 12$

$\dfrac{3}{5}t = 12$

$\dfrac{5}{3} \cdot \dfrac{3}{5}t = \dfrac{5}{3} \cdot 12$  Using the multiplication principle

$1t = \dfrac{60}{3}$

$t = 20$

The number 20 checks and is the solution.

**19.**  $x + \dfrac{7}{3} = \dfrac{19}{6}$

$x + \dfrac{7}{3} - \dfrac{7}{3} = \dfrac{19}{6} - \dfrac{7}{3}$  Using the addition principle

$x + 0 = \dfrac{19}{6} - \dfrac{7}{3} \cdot \dfrac{2}{2}$

$x = \dfrac{19}{6} - \dfrac{14}{6}$

$x = \dfrac{5}{6}$

The number $\dfrac{5}{6}$ checks and is the solution.

**21.**  $7 = a + \dfrac{14}{5}$

$7 - \dfrac{14}{5} = a + \dfrac{14}{5} - \dfrac{14}{5}$  Using the addition principle

$7 \cdot \dfrac{5}{5} - \dfrac{14}{5} = a + 0$

$\dfrac{35}{5} - \dfrac{14}{5} = a$

$\dfrac{21}{5} = a$

The number $\dfrac{21}{5}$ checks and is the solution.

**23.**  $\dfrac{2}{5}t - 1 = \dfrac{7}{5}$

$\dfrac{2}{5}t - 1 + 1 = \dfrac{7}{5} + 1$  Using the addition principle

$\dfrac{2}{5}t + 0 = \dfrac{7}{5} + 1 \cdot \dfrac{5}{5}$

$\dfrac{2}{5}t = \dfrac{7}{5} + \dfrac{5}{5}$

$\dfrac{2}{5}t = \dfrac{12}{5}$

$\dfrac{5}{2} \cdot \dfrac{2}{5}t = \dfrac{5}{2} \cdot \dfrac{12}{5}$  Using the multiplication principle

$1t = \dfrac{5 \cdot 12}{2 \cdot 5}$

$t = \dfrac{5 \cdot 2 \cdot 6}{2 \cdot 5 \cdot 1} = \dfrac{5 \cdot 2}{5 \cdot 2} \cdot \dfrac{6}{1}$

$t = 6$

The number 6 checks and is the solution.

**25.**  $\dfrac{39}{8} = \dfrac{11}{4} + \dfrac{1}{2}x$

$\dfrac{39}{8} - \dfrac{11}{4} = \dfrac{11}{4} + \dfrac{1}{2}x - \dfrac{11}{4}$  Using the addition principle

$\dfrac{39}{8} - \dfrac{11}{4} \cdot \dfrac{2}{2} = \dfrac{1}{2}x + 0$

$\dfrac{39}{8} - \dfrac{22}{8} = \dfrac{1}{2}x$

$\dfrac{17}{8} = \dfrac{1}{2}x$

$2 \cdot \dfrac{17}{8} = 2 \cdot \dfrac{1}{2}x$  Using the multiplication principle

$\dfrac{2 \cdot 17}{8} = 1x$

$\dfrac{2 \cdot 17}{2 \cdot 4} = x$

$\dfrac{17}{4} = x$

The number $\dfrac{17}{4}$ checks and is the solution.

**27.**  $\dfrac{13}{3}x + \dfrac{11}{2} = \dfrac{35}{4}$

$\dfrac{13}{3}x + \dfrac{11}{2} - \dfrac{11}{2} = \dfrac{35}{4} - \dfrac{11}{2}$  Using the addition principle

$\dfrac{13}{3}x + 0 = \dfrac{35}{4} - \dfrac{11}{2} \cdot \dfrac{2}{2}$

$\dfrac{13}{3}x = \dfrac{35}{4} - \dfrac{22}{4}$

$\dfrac{13}{3}x = \dfrac{13}{4}$

$\dfrac{3}{13} \cdot \dfrac{13}{3}x = \dfrac{3}{13} \cdot \dfrac{13}{4}$  Using the multiplication principle

$1x = \dfrac{3 \cdot 13}{13 \cdot 4}$

$x = \dfrac{3}{4}$

The number $\dfrac{3}{4}$ checks and is the solution.

**29.** $-200$ represents the \$200 withdrawal, 90 represents the \$90 deposit, and $-40$ represents the \$40 withdrawal. We add these numbers to find the change in the balance.

$$-200 + 90 + (-40) = -110 + (-40) = -150$$

The account balance decreased by \$150.

**31.** $\dfrac{10}{7} \div 2m = \dfrac{10}{7} \cdot \dfrac{1}{2m} = \dfrac{10 \cdot 1}{7 \cdot 2m} = \dfrac{2 \cdot 5 \cdot 1}{7 \cdot 2 \cdot m} =$

$\dfrac{2}{2} \cdot \dfrac{5 \cdot 1}{7 \cdot m} = \dfrac{5}{7m}$

**33.** $3(a + b) = 3 \cdot a + 3 \cdot b = 3a + 3b$

**35.** ◈

**37.** ◈

**39.** Use a calculator.

$$\dfrac{553}{2451}a - \dfrac{13}{57} = \dfrac{29}{43}$$

$$\dfrac{553}{2451}a - \dfrac{13}{57} + \dfrac{13}{57} = \dfrac{29}{43} + \dfrac{13}{57}$$

$$\dfrac{553}{2451}a = \dfrac{29}{43} + \dfrac{13}{57}$$

$$\dfrac{553}{2451}a = \dfrac{29}{43} \cdot \dfrac{57}{57} + \dfrac{13}{57} \cdot \dfrac{43}{43}$$

$$\dfrac{553}{2451}a = \dfrac{1653}{43 \cdot 57} + \dfrac{559}{57 \cdot 43}$$

$$\dfrac{553}{2451}a = \dfrac{2212}{2451}$$

$$\dfrac{2451}{553} \cdot \dfrac{553}{2451}a = \dfrac{2451}{553} \cdot \dfrac{2212}{2451}$$

$$a = 4$$

**41.**

$$\dfrac{47}{5} - \dfrac{a}{4} = \dfrac{44}{7}$$

$$\dfrac{47}{5} - \dfrac{a}{4} - \dfrac{47}{5} = \dfrac{44}{7} - \dfrac{47}{5}$$

$$-\dfrac{a}{4} = \dfrac{44}{7} \cdot \dfrac{5}{5} - \dfrac{47}{5} \cdot \dfrac{7}{7}$$

$$-\dfrac{a}{4} = \dfrac{220}{35} - \dfrac{329}{35}$$

$$-\dfrac{a}{4} = -\dfrac{109}{35}$$

$$-4\left(-\dfrac{a}{4}\right) = -4\left(-\dfrac{109}{35}\right)$$

$$a = \dfrac{436}{35}$$

**43.** *Familiarize.* The perimeter $P$ is the sum of the lengths of the sides, so we have $P = \dfrac{5}{4}x + x + \dfrac{5}{2} + 6 + 2$.

*Translate.* We substitute 15 for $P$.

$$\dfrac{5}{4}x + x + \dfrac{5}{2} + 6 + 2 = 15$$

*Solve.* We solve the equation. We begin by collecting like terms on the left side.

$$\dfrac{5}{4}x + x + \dfrac{5}{2} + 6 + 2 = 15$$

$$\left(\dfrac{5}{4} + 1\right)x + \dfrac{5}{2} + 6 \cdot \dfrac{2}{2} + 2 \cdot \dfrac{2}{2} = 15$$

$$\left(\dfrac{5}{4} + \dfrac{4}{4}\right)x + \dfrac{5}{2} + \dfrac{12}{2} + \dfrac{4}{2} = 15$$

$$\dfrac{9}{4}x + \dfrac{21}{2} = 15$$

$$\dfrac{9}{4}x + \dfrac{21}{2} - \dfrac{21}{2} = 15 - \dfrac{21}{2}$$

$$\dfrac{9}{4}x + 0 = 15 \cdot \dfrac{2}{2} - \dfrac{21}{2}$$

$$\dfrac{9}{4}x = \dfrac{30}{2} - \dfrac{21}{2}$$

$$\dfrac{9}{4}x = \dfrac{9}{2}$$

$$\dfrac{4}{9} \cdot \dfrac{9}{4}x = \dfrac{4}{9} \cdot \dfrac{9}{2}$$

$$1x = \dfrac{2 \cdot 2 \cdot 9}{9 \cdot 2}$$

$$x = 2$$

*Check.* $\dfrac{5}{4} \cdot 2 + 2 + \dfrac{5}{2} + 6 + 2 = \dfrac{5}{2} + 2 + \dfrac{5}{2} + 6 + 2 = 15$, so the result checks.

*State.* $x$ is 2 cm.

---

## Exercise Set 4.5

**1.** [b]    [a] Multiply: $3 \cdot 5 = 15$.

$3\dfrac{2}{5} = \dfrac{17}{5}$  [b] Add: $15 + 2 = 17$.

[a]  [c] Keep the denominator.

**3.** [b]    [a] Multiply: $6 \cdot 4 = 24$.

$6\dfrac{1}{4} = \dfrac{25}{4}$  [b] Add: $24 + 1 = 25$.

[a]  [c] Keep the denominator.

**5.** $-20\dfrac{1}{8} = -\dfrac{161}{8}$   ($20 \cdot 8 = 160; 160 + 1 = 161$; include the negative sign)

**7.** $5\dfrac{1}{10} = \dfrac{51}{10}$   ($5 \cdot 10 = 50; 50 + 1 = 51$)

**9.** $20\dfrac{3}{5} = \dfrac{103}{5}$   ($20 \cdot 5 = 100; 100 + 3 = 103$)

**11.** $-9\dfrac{5}{6} = -\dfrac{59}{6}$   ($9 \cdot 6 = 54; 54 + 5 = 59$; include the negative sign)

**13.** $6\dfrac{9}{10} = \dfrac{69}{10}$   ($6 \cdot 10 = 60; 60 + 9 = 69$)

**15.** $-12\frac{3}{4} = -\frac{51}{4}$   $(12 \cdot 4 = 48; \; 48 + 3 = 51;$ include the negative sign)

**17.** $5\frac{7}{10} = \frac{57}{10}$   $(5 \cdot 10 = 50; \; 50 + 7 = 57)$

**19.** $-5\frac{7}{100} = -\frac{507}{100}$   $(5 \cdot 100 = 500; \; 500 + 7 = 507;$ include the negative sign)

**21.** To convert $\frac{14}{3}$ to a mixed numeral, we divide.

$$\begin{array}{r} 4 \\ 3\overline{)1\,4} \\ 1\,2 \\ \hline 2 \end{array} \qquad \frac{14}{3} = 4\frac{2}{3}$$

**23.** To convert $\frac{27}{6}$ to a mixed numeral, we divide.

$$\begin{array}{r} 4 \\ 6\overline{)2\,7} \\ 2\,4 \\ \hline 3 \end{array} \qquad \frac{27}{6} = 4\frac{3}{6} = 4\frac{1}{2}$$

Since $\frac{27}{6} = 4\frac{1}{2}$, we have $-\frac{27}{6} = -4\frac{1}{2}$.

**25.**
$$\begin{array}{r} 5 \\ 1\,0\overline{)5\,7} \\ 5\,0 \\ \hline 7 \end{array} \qquad \frac{57}{10} = 5\frac{7}{10}$$

**27.**
$$\begin{array}{r} 7 \\ 7\overline{)5\,3} \\ 4\,9 \\ \hline 4 \end{array} \qquad \frac{53}{7} = 7\frac{4}{7}$$

**29.**
$$\begin{array}{r} 7 \\ 6\overline{)4\,5} \\ 4\,2 \\ \hline 3 \end{array} \qquad \frac{45}{6} = 7\frac{3}{6} = 7\frac{1}{2}$$

**31.**
$$\begin{array}{r} 1\,1 \\ 4\overline{)4\,6} \\ 4\,0 \\ \hline 6 \\ 4 \\ \hline 2 \end{array} \qquad \frac{46}{4} = 11\frac{2}{4} = 11\frac{1}{2}$$

**33.**
$$\begin{array}{r} 1 \\ 8\overline{)1\,2} \\ 8 \\ \hline 4 \end{array} \qquad \frac{12}{8} = 1\frac{4}{8} = 1\frac{1}{2}$$

Since $\frac{12}{8} = 1\frac{1}{2}$, we have $-\frac{12}{8} = -1\frac{1}{2}$.

**35.**
$$\begin{array}{r} 4 \\ 6\overline{)2\,8} \\ 2\,4 \\ \hline 4 \end{array} \qquad \frac{28}{6} = 4\frac{4}{6} = 4\frac{2}{3}$$

**37.**
$$\begin{array}{r} 5\,5 \\ 4\overline{)2\,2\,3} \\ 2\,0\,0 \\ \hline 2\,3 \\ 2\,0 \\ \hline 3 \end{array} \qquad \frac{223}{4} = 55\frac{3}{4}$$

Since $\frac{223}{4} = 55\frac{3}{4}$, we have $-\frac{223}{4} = -55\frac{3}{4}$.

**39.** We first divide as usual.

$$\begin{array}{r} 1\,0\,8 \\ 8\overline{)8\,6\,9} \\ 8\,0\,0 \\ \hline 6\,9 \\ 6\,4 \\ \hline 5 \end{array}$$

The answer is 108 R 5. We write a mixed numeral for the quotient as follows: $108\frac{5}{8}$.

**41.** We first divide as usual.

$$\begin{array}{r} 9\,0\,6 \\ 7\overline{)6\,3\,4\,5} \\ 6\,3\,0\,0 \\ \hline 4\,5 \\ 4\,2 \\ \hline 3 \end{array}$$

The answer is 906 R 3. We write a mixed numeral for the quotient as follows: $906\frac{3}{7}$.

**43.**
$$\begin{array}{r} 4\,0 \\ 2\,1\overline{)8\,5\,2} \\ 8\,4\,0 \\ \hline 1\,2 \end{array}$$

We get $40\frac{12}{21}$. This simplifies as $40\frac{4}{7}$.

**45.** First we find $302 \div 15$.

$$\begin{array}{r} 2\,0 \\ 1\,5\overline{)3\,0\,2} \\ 3\,0\,0 \\ \hline 2 \\ 0 \\ \hline 2 \end{array} \qquad \frac{302}{15} = 20\frac{2}{15}$$

Since $302 \div 15 = 20\frac{2}{15}$, we have $-302 \div 15 = -20\frac{2}{15}$.

**47.** First we find $471 \div 21$.

$$\begin{array}{r} 2\,2 \\ 2\,1\overline{)4\,7\,1} \\ 4\,2\,0 \\ \hline 5\,1 \\ 4\,2 \\ \hline 9 \end{array} \qquad \frac{471}{21} = 22\frac{9}{21} = 22\frac{3}{7}$$

Since $471 \div 21 = 22\frac{3}{7}$, we have $471 \div (-21) = -22\frac{3}{7}$.

**49.** There are 5 items from Boston Chicken in the list. We add the grams of fat in these items and then divide by 5.

$$\frac{0+0+3+4+4}{5} = \frac{11}{5} = 2\frac{1}{5}$$

On average, the foods from Boston Chicken contain $2\frac{1}{5}$ g of fat.

**51.** There are 20 items on the list. When we add the grams of fat in these items we get 46 g. Then we divide by 20.

$$\frac{46}{20} = 2\frac{6}{20} = 2\frac{3}{10}$$

The average number of grams of fat for the entire list is $2\frac{3}{10}$ g.

**53.** $\frac{7}{9} \cdot \frac{24}{21} = \frac{7 \cdot 24}{9 \cdot 21}$

$$= \frac{7 \cdot 3 \cdot 8}{3 \cdot 3 \cdot 3 \cdot 7}$$

$$= \frac{3 \cdot 7}{3 \cdot 7} \cdot \frac{8}{3 \cdot 3}$$

$$= \frac{8}{9}$$

**55.** $\frac{5}{12}(-6) = \frac{5(-6)}{12}$

$$= \frac{5(-1)(6)}{2 \cdot 6}$$

$$= \frac{5(-1)}{2} \cdot \frac{6}{6}$$

$$= \frac{-5}{2}, \text{ or } -\frac{5}{2}$$

**57.** ◈

**59.** ◈

**61.** Use a calculator.

$$\frac{103,676}{349} = 297\frac{23}{349}$$

**63.** $\frac{72}{12} + \frac{5}{6} = 6 + \frac{5}{6} \qquad (72 \div 12 = 6)$

$$= 6\frac{5}{6}$$

**65.**
```
     5 2
7 ⌐ 3 6 6
    3 5 0
    ─────
      1 6
      1 4
     ────
        2
```
$\frac{366}{7} = 52\frac{2}{7}$

**67.** Find the average wire length first. Recall that 1 ft = 12 in.

3 ft $10\frac{1}{4}$ in. $= 3 \times 12$ in. $+ 10\frac{1}{4}$ in. $= 36$ in. $+ 10\frac{1}{4}$ in. $=$ $46\frac{1}{4}$ in. $= \frac{185}{4}$ in.

3 ft $11\frac{3}{4}$ in. $= 3 \times 12$ in. $+ 11\frac{3}{4}$ in. $= 36$ in. $+ 11\frac{3}{4}$ in. $=$ $47\frac{3}{4}$ in. $= \frac{191}{4}$ in.

We add $\frac{185}{4}$ and $\frac{191}{4}$ and then divide by 2:

$$\frac{185}{4} + \frac{191}{4} = \frac{376}{4} = 94$$

$$\frac{94}{2} = 47$$

The average wire length is 47 in. We divide by 12 to convert to feet.

$$\frac{47}{12} = 3\frac{11}{12} \text{ ft} = 3 \text{ ft } 11 \text{ in.}$$

Now find the average diameter. We add $4\frac{3}{8}$ in. and $5\frac{1}{8}$ in. and then divide by 2.

$$4\frac{3}{8} + 5\frac{1}{8} = 9\frac{4}{8} = 9\frac{1}{2}$$

$$9\frac{1}{2} \div 2 = \frac{19}{2} \div 2 = \frac{19}{2} \cdot \frac{1}{2} = \frac{19}{4} = 4\frac{3}{4}$$

The average diameter is $4\frac{3}{4}$ in.

---

## Exercise Set 4.6

**1.** $\quad 5\frac{7}{8}$

$\quad +3\frac{5}{8}$

$\overline{\quad 8\frac{12}{8}} = 8 + \frac{12}{8}$

$\qquad = 8 + 1\frac{1}{2}$

$\qquad = 9\frac{1}{2}$

To find a mixed numeral for $\frac{12}{8}$ we divide:

```
      1
8 ⌐ 1 2
    8
   ──
    4
```
$\frac{12}{8} = 1\frac{4}{8} = 1\frac{1}{2}$

**3.** The LCD is 12.

$1 \boxed{\frac{1}{4} \cdot \frac{3}{3}} = 1\frac{3}{12}$

$+1 \boxed{\frac{2}{3} \cdot \frac{4}{4}} = +1\frac{8}{12}$

$\overline{\qquad\qquad 2\frac{11}{12}}$

**5.** The LCD is 12.

$7 \boxed{\frac{3}{4} \cdot \frac{3}{3}} = 7\frac{9}{12}$

$+5 \boxed{\frac{5}{6} \cdot \frac{2}{2}} = +5\frac{10}{12}$

$\overline{\qquad\qquad 12\frac{19}{12}} = 12 + \frac{19}{12}$

$\qquad\qquad\qquad = 12 + 1\frac{7}{12}$

$\qquad\qquad\qquad = 13\frac{7}{12}$

**7.** The LCD is 10.

$$3 \;\boxed{\frac{2}{5} \cdot \frac{2}{2}} = \; 3\,\frac{4}{10}$$
$$+8\,\frac{7}{10} \; = +8\,\frac{7}{10}$$
$$\rule{3cm}{0.4pt}$$
$$11\,\frac{11}{10} = 11 + \frac{11}{10}$$
$$= 11 + 1\frac{1}{10}$$
$$= 12\frac{1}{10}$$

**9.** The LCD is 24.

$$6 \;\boxed{\frac{3}{8} \cdot \frac{3}{3}} = \; 6\,\frac{9}{24}$$
$$+10 \;\boxed{\frac{5}{6} \cdot \frac{4}{4}} = +10\,\frac{20}{24}$$
$$\rule{3cm}{0.4pt}$$
$$16\,\frac{29}{24} = 16 + \frac{29}{24}$$
$$= 16 + 1\frac{5}{24}$$
$$= 17\frac{5}{24}$$

**11.** The LCD is 10.

$$12 \;\boxed{\frac{4}{5} \cdot \frac{2}{2}} = 12\,\frac{8}{10}$$
$$+8\,\frac{7}{10} \; = +8\,\frac{7}{10}$$
$$\rule{3cm}{0.4pt}$$
$$20\,\frac{15}{10} = 20 + \frac{15}{10}$$
$$= 20 + 1\frac{5}{10}$$
$$= 21\frac{5}{10}$$
$$= 21\frac{1}{2}$$

**13.** The LCD is 8.

$$14\,\frac{5}{8} \; = \; 14\,\frac{5}{8}$$
$$+13 \;\boxed{\frac{1}{4} \cdot \frac{2}{2}} = +13\,\frac{2}{8}$$
$$\rule{3cm}{0.4pt}$$
$$27\,\frac{7}{8}$$

**15.**
$$4\frac{1}{5} = \; 3\frac{6}{5}$$
$$-2\frac{3}{5} = -2\frac{3}{5}$$
$$\rule{2cm}{0.4pt}$$
$$1\frac{3}{5}$$

> Since $\frac{1}{5}$ is smaller than $\frac{3}{5}$, we cannot subtract until we borrow:
>
> $4\frac{1}{5} = 3 + \frac{5}{5} + \frac{1}{5} = 3 + \frac{6}{5} = 3\frac{6}{5}$

**17.** The LCD is 10.

$$6 \;\boxed{\frac{3}{5} \cdot \frac{2}{2}} = \; 6\,\frac{6}{10}$$
$$-2 \;\boxed{\frac{1}{2} \cdot \frac{5}{5}} = -2\,\frac{5}{10}$$
$$\rule{3cm}{0.4pt}$$
$$4\,\frac{1}{10}$$

**19.** The LCD is 24.

$$34 \;\boxed{\frac{1}{3} \cdot \frac{8}{8}} = \; 34\,\frac{8}{24} = \; 33\,\frac{32}{24}$$
$$-12 \;\boxed{\frac{5}{8} \cdot \frac{3}{3}} = -12\,\frac{15}{24} = -12\,\frac{15}{24}$$
$$\rule{6cm}{0.4pt}$$
$$21\,\frac{17}{24}$$

$\left(\text{Since } \frac{8}{24} \text{ is smaller than } \frac{15}{24}, \text{ we cannot subtract until we borrow: } 34\frac{8}{24} = 33 + \frac{24}{24} + \frac{8}{24} = 33 + \frac{32}{24} = 33\frac{32}{24}.\right)$

**21.**
$$21 \; = \; 20\frac{4}{4} \quad \left(21 = 20 + 1 = 20 + \frac{4}{4} = 20\frac{4}{4}\right)$$
$$-\; 8\frac{3}{4} = -\; 8\frac{3}{4}$$
$$\rule{2cm}{0.4pt}$$
$$12\frac{1}{4}$$

**23.**
$$34 \; = \; 33\frac{8}{8} \quad \left(34 = 33 + 1 = 33 + \frac{8}{8} = 33\frac{8}{8}\right)$$
$$-\; 18\frac{5}{8} = -\; 18\frac{5}{8}$$
$$\rule{2cm}{0.4pt}$$
$$15\frac{3}{8}$$

**25.** The LCD is 12.

$$21 \;\boxed{\frac{1}{6} \cdot \frac{2}{2}} = \; 21\,\frac{2}{12} = \; 20\,\frac{14}{12}$$
$$-13 \;\boxed{\frac{3}{4} \cdot \frac{3}{3}} = -13\,\frac{9}{12} = -13\,\frac{9}{12}$$
$$\rule{6cm}{0.4pt}$$
$$7\,\frac{5}{12}$$

$\left(\text{Since } \frac{2}{12} \text{ is smaller than } \frac{9}{12}, \text{ we cannot subtract until we borrow: } 21\frac{2}{12} = 20 + \frac{12}{12} + \frac{2}{12} = 20 + \frac{14}{12} = 20\frac{14}{12}.\right)$

**27.** The LCD is 18.

$$25 \boxed{\frac{1}{9} \cdot \frac{2}{2}} = 25 \frac{2}{18} = 24 \frac{20}{18}$$

$$-13 \boxed{\frac{5}{6} \cdot \frac{3}{3}} = -13 \frac{15}{18} = -13 \frac{15}{18}$$

$$11 \frac{5}{18}$$

$\left(\text{Since } \dfrac{2}{18} \text{ is smaller than } \dfrac{15}{18}, \text{ we cannot subtract until we}\right.$

$\left. \text{borrow: } 25\dfrac{2}{18} = 24 + \dfrac{18}{18} + \dfrac{2}{18} = 24 + \dfrac{20}{18} = 24\dfrac{20}{18}.\right)$

**29.** $\quad 5\dfrac{3}{14}t + 3\dfrac{2}{21}t$

$= \left(5\dfrac{3}{14} + 3\dfrac{2}{21}\right)t \quad$ Using the distributive law

$= \left(5\dfrac{9}{42} + 3\dfrac{4}{42}\right)t \quad$ The LCD is 42.

$= 8\dfrac{13}{42}t \qquad\qquad$ Adding

**31.** $\quad 9\dfrac{1}{2}x - 7\dfrac{3}{8}x$

$= \left(9\dfrac{1}{2} - 7\dfrac{3}{8}\right)x \quad$ Using the distributive law

$= \left(9\dfrac{4}{8} - 7\dfrac{3}{8}\right)x \quad$ The LCD is 8.

$= 2\dfrac{1}{8}x \qquad\qquad$ Subtracting

**33.** $\quad 3\dfrac{7}{8}t + 4\dfrac{9}{10}t$

$= \left(3\dfrac{7}{8} + 4\dfrac{9}{10}\right)t \quad$ Using the distributive law

$= \left(3\dfrac{35}{40} + 4\dfrac{36}{40}\right)t \quad$ The LCD is 40.

$= 7\dfrac{71}{40}t = 8\dfrac{31}{40}t$

**35.** $\quad 37\dfrac{5}{9}t - 25\dfrac{4}{5}t$

$= \left(37\dfrac{5}{9} - 25\dfrac{4}{5}\right)t \quad$ Using the distributive law

$= \left(37\dfrac{25}{45} - 25\dfrac{36}{45}\right)t \quad$ The LCD is 45.

$= \left(36\dfrac{70}{45} - 25\dfrac{36}{45}\right)t$

$= 11\dfrac{34}{45}t$

**37.** $\quad 2\dfrac{5}{6}x + 3\dfrac{1}{3}x$

$= \left(2\dfrac{5}{6} + 3\dfrac{1}{3}\right)x \quad$ Using the distributive law

$= \left(2\dfrac{5}{6} + 3\dfrac{2}{6}\right)x \quad$ The LCD is 6.

$= 5\dfrac{7}{6}x = 6\dfrac{1}{6}x$

**39.** $\quad 4\dfrac{3}{11}x + 5\dfrac{2}{3}x$

$= \left(4\dfrac{3}{11} + 5\dfrac{2}{3}\right)x \quad$ Using the distributive law

$= \left(4\dfrac{9}{33} + 5\dfrac{22}{33}\right)x \quad$ The LCD is 33.

$= 9\dfrac{31}{33}x$

**41.** *Familiarize*. We let $w = $ the total weight of the fish.

*Translate*. We write an equation.

| Weight of one fish | plus | Weight of second fish | is | Total weight |
|:---:|:---:|:---:|:---:|:---:|
| ↓ | ↓ | ↓ | ↓ | ↓ |
| $1\frac{1}{2}$ | $+$ | $2\frac{3}{4}$ | $=$ | $w$ |

*Solve*. We carry out the addition. The LCD is 12.

$$1 \boxed{\frac{1}{2} \cdot \frac{2}{2}} = 1 \frac{2}{4}$$

$$+2 \frac{3}{4} = +2 \frac{3}{4}$$

$$3\frac{5}{4} = 3 + \frac{5}{4}$$

$$= 3 + 1\frac{1}{4}$$

$$= 4\frac{1}{4}$$

Thus, $w = 4\dfrac{1}{4}$.

*Check*. We repeat the calculation. We also note that the answer is larger than either of the individual weights, so the answer seems reasonable.

*State*. The total weight of the fish was $4\dfrac{1}{4}$ lb.

**43.** *Familiarize*. We let $h = $ Rocky's excess height.

*Translate*. We have a "how much more" situation.

| Height of daughter | plus | How much more height | is | Rocky's height |
|:---:|:---:|:---:|:---:|:---:|
| ↓ | ↓ | ↓ | ↓ | ↓ |
| $180\frac{3}{4}$ | $+$ | $h$ | $=$ | $187\frac{1}{10}$ |

*Solve*. We solve the equation as follows:

$$h = 187\frac{1}{10} - 180\frac{3}{4}$$

$$187 \boxed{\frac{1}{10} \cdot \frac{2}{2}} = 187 \frac{2}{20}$$

$$180 \boxed{\frac{3}{4} \cdot \frac{5}{5}} = 180 \frac{15}{20}$$

$$187\frac{1}{10} = \quad 187\frac{2}{20} = \quad 186\frac{22}{20}$$
$$-\,180\frac{3}{4} = -\,180\frac{15}{20} = -\,180\frac{15}{20}$$
$$\overline{\qquad\qquad\qquad\qquad\qquad 6\frac{7}{20}}$$

Thus, $h = 6\frac{7}{20}$.

**Check.** We add Rocky's excess height to his daughter's height:

$$180\frac{3}{4} + 6\frac{7}{20} = 180\frac{15}{20} + 6\frac{7}{20} = 186\frac{22}{20} = 187\frac{2}{20} = 187\frac{1}{10}$$

The answer checks.

**State.** Rocky is $6\frac{7}{20}$ cm taller.

**45. Familiarize.** We draw a picture, letting $x =$ the amount of pipe that was used.

$$\vdash\!\!-\!\!-\!\!- 10\frac{5}{16} \text{ ft} \!\!-\!\!-\!\!- \dashv \!\!-\!\!-\!\!- 8\frac{3}{4} \text{ ft} \!\!-\!\!-\!\!-\dashv$$
$$\vdash\!\!-\!\!-\!\!-\!\!-\!\!-\!\!-\!\!-\!\!- x \!\!-\!\!-\!\!-\!\!-\!\!-\!\!-\!\!-\!\!-\dashv$$

**Translate.** We write an addition sentence.

First length plus Second length is Total length

$$10\frac{5}{16} \quad + \quad 8\frac{3}{4} \quad = \quad x$$

**Solve.** We carry out the addition. The LCD is 16.

$$10\frac{5}{16} \quad = \quad 10\frac{5}{16}$$
$$+\,8\boxed{\frac{3}{4}\cdot\frac{4}{4}} = +\,8\frac{12}{16}$$
$$\overline{\qquad\qquad\qquad 18\frac{17}{16}} = 18 + \frac{17}{16}$$
$$= 18 + 1\frac{1}{16}$$
$$= 19\frac{1}{16}$$

Thus, $x = 19\frac{1}{16}$.

**Check.** We repeat the calculation. We also note that the total length is larger than either of the individual lengths, so the answer seems reasonable.

**State.** Janet used $19\frac{1}{16}$ ft of pipe.

**47. Familiarize.** We draw a picture.

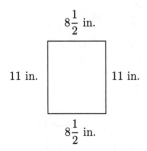

$8\frac{1}{2}$ in.

11 in.      11 in.

$8\frac{1}{2}$ in.

**Translate.** We let $D =$ the distance around the sheet of paper.

| Top distance | plus | Right-side distance | plus | Bottom distance | plus | Left-side distance | is | Total distance |
|---|---|---|---|---|---|---|---|---|
| $8\frac{1}{2}$ | + | 11 | + | $8\frac{1}{2}$ | + | 11 | = | $D$ |

**Solve.** To solve we carry out the addition. The LCD is 4.

$$8\frac{1}{2}$$
$$11$$
$$8\frac{1}{2}$$
$$+\,11$$
$$\overline{38\frac{2}{2}} = 38 + \frac{2}{2} = 38 + 1 = 39$$

Thus, $D = 39$.

**Check.** We repeat the calculation.

**State.** The distance around the sheet of paper is 39 in.

**49. Familiarize.** Let $d =$ the difference between the high and low prices of the stock.

**Translate.** We write an equation.

| High price | minus | Low price | is | Difference in prices |
|---|---|---|---|---|
| $37\frac{5}{8}$ | − | $20\frac{1}{2}$ | = | $d$ |

**Solve.** To solve we carry out the subtraction. The LCD is 8.

$$37\frac{5}{8} \quad = \quad 37\frac{5}{8}$$
$$-\,20\boxed{\frac{1}{2}\cdot\frac{4}{4}} = -\,20\frac{4}{8}$$
$$\overline{\qquad\qquad\qquad 17\frac{1}{8}}$$

**Check.** We add the difference in price to the low price:

$$17\frac{1}{8} + 20\frac{1}{2} = 17\frac{1}{8} + 20\frac{4}{8} = 37\frac{5}{8}$$

This checks.

**State**. The difference between the high and low prices of the stock was $17\frac{1}{8}$.

**51. Familiarize**. We make a drawing. Let $l$ = the length, in inches, of the wood that remains.

**Translate**.

| Length cut off | + | Thickness of blade | + | Length left over | is | Original length |
|---|---|---|---|---|---|---|
| $15\frac{3}{4}$ | + | $\frac{1}{8}$ | + | $l$ | = | 36 |

**Solve**. This is a two-step problem. First we add $15\frac{3}{4}$ and $\frac{1}{8}$. The LCD is 8.

$$15\boxed{\frac{3}{4}\cdot\frac{2}{2}} = 15\frac{6}{8}$$
$$+ \frac{1}{8} = +\frac{1}{8}$$
$$\overline{\qquad\qquad 15\frac{7}{8}}$$

Now we have $15\frac{7}{8} + l = 36$. We subtract $15\frac{7}{8}$ on both sides of the equation.

$$15\frac{7}{8} + l = 36$$
$$l = 36 - 15\frac{7}{8}$$

$$36 = 35\frac{8}{8}$$
$$-15\frac{7}{8} = -15\frac{7}{8}$$
$$\overline{\qquad\qquad 20\frac{1}{8}}$$

**Check**. We repeat the calculations.

**State**. The piece of wood that remains is $20\frac{1}{8}$ in. long.

**53. Familiarize**. We make a drawing. We let $t$ = the number of hours Sue worked on the third day.

**Translate**. We write an addition sentence.

$$2\frac{1}{2} + 4\frac{1}{5} + t = 10\frac{1}{2}$$

**Solve**. This is a two-step problem.

First we add $2\frac{1}{2} + 4\frac{1}{5}$ to find the time worked on the first two days. The LCD is 10.

$$2\boxed{\frac{1}{2}\cdot\frac{5}{5}} = 2\frac{5}{10}$$
$$+4\boxed{\frac{1}{5}\cdot\frac{2}{2}} = +4\frac{2}{10}$$
$$\overline{\qquad\qquad 6\frac{7}{10}}$$

Then we subtract $6\frac{7}{10}$ from $10\frac{1}{2}$ to find the time worked on the third day. The LCD is 10.

$$6\frac{7}{10} + t = 10\frac{1}{2}$$
$$t = 10\frac{1}{2} - 6\frac{7}{10}$$

$$10\boxed{\frac{1}{2}\cdot\frac{5}{5}} = 10\frac{5}{10} = 9\frac{15}{10}$$
$$-6\frac{7}{10} = -6\frac{7}{10} = -6\frac{7}{10}$$
$$\overline{\qquad\qquad 3\frac{8}{10} = 3\frac{4}{5}}$$

**Check**. We repeat the calculations.

**State**. Sue worked $3\frac{4}{5}$ hr the third day.

**55.** The length of each of the five sides is $5\frac{3}{4}$ yd. We add to find the distance around the figure.

$$5\frac{3}{4} + 5\frac{3}{4} + 5\frac{3}{4} + 5\frac{3}{4} + 5\frac{3}{4} = 25\frac{15}{4} = 25 + 3\frac{3}{4} = 28\frac{3}{4}$$

The distance is $28\frac{3}{4}$ yd.

**57.** We see that $d$ and the two smallest distances combined are the same as the largest distance. We translate and solve.

$$2\frac{3}{4} + d + 2\frac{3}{4} = 12\frac{7}{8}$$
$$d = 12\frac{7}{8} - 2\frac{3}{4} - 2\frac{3}{4}$$
$$= 10\frac{1}{8} - 2\frac{3}{4} \quad \text{Subtracting } 2\frac{3}{4} \text{ from } 12\frac{7}{8}$$
$$= 7\frac{3}{8} \quad \text{Subtracting } 2\frac{3}{4} \text{ from } 10\frac{1}{8}$$

The length of $d$ is $7\frac{3}{8}$ ft.

**59. Familiarize**. We let $b$ = the length of the bolt.

**Translate**. From the drawing we see that the length of the small bolt is the sum of the diameters of the two tubes and the thicknesses of the two washers and the nut. Thus, we have

$$b = \frac{1}{2} + \frac{1}{16} + \frac{3}{4} + \frac{1}{16} + \frac{3}{16}.$$

*Solve.* We carry out the addition. The LCD is 16.

$$b = \frac{1}{2} + \frac{1}{16} + \frac{3}{4} + \frac{1}{16} + \frac{3}{16} =$$

$$\frac{1}{2} \cdot \frac{8}{8} + \frac{1}{16} + \frac{3}{4} \cdot \frac{4}{4} + \frac{1}{16} + \frac{3}{16} =$$

$$\frac{8}{16} + \frac{1}{16} + \frac{12}{16} + \frac{1}{16} + \frac{3}{16} = \frac{25}{16} = 1\frac{9}{16}$$

*Check.* We repeat the calculation.

*State.* The smallest bolt is $1\frac{9}{16}$ in. long.

**61.** $8\frac{3}{5} - 9\frac{2}{5} = 8\frac{3}{5} + \left(-9\frac{2}{5}\right)$

Since $9\frac{2}{5}$ is greater than $8\frac{3}{5}$, the answer will be negative.

The difference in absolute values is

$$\begin{array}{r} 9\frac{2}{5} = \phantom{-}8\frac{7}{5} \\ -8\frac{3}{5} = -8\frac{3}{5} \\ \hline \frac{4}{5} \end{array}$$

so $8\frac{3}{5} - 9\frac{2}{5} = -\frac{4}{5}$.

**63.** $3\frac{1}{2} - 6\frac{3}{4} = 3\frac{1}{2} + \left(-6\frac{3}{4}\right)$

Since $6\frac{3}{4}$ is greater than $3\frac{1}{2}$, the answer will be negative.

The difference in absolute values is

$$\begin{array}{r} 6\frac{3}{4} = \phantom{-}6\frac{3}{4} = \phantom{-}6\frac{3}{4} \\ -3\frac{1}{2} = -3\boxed{\frac{1}{2}\cdot\frac{1}{2}} = -3\frac{2}{4} \\ \hline 3\frac{1}{4} \end{array}$$

so $3\frac{1}{2} - 6\frac{3}{4} = -3\frac{1}{4}$.

**65.** $3\frac{4}{5} - 7\frac{2}{3} = 3\frac{4}{5} + \left(-7\frac{2}{3}\right)$

Since $7\frac{2}{3}$ is greater than $3\frac{4}{5}$, the answer will be negative.

The difference in absolute values is

$$\begin{array}{r} 7\frac{2}{3} = \phantom{-}7\boxed{\frac{2}{3}\cdot\frac{5}{5}} = \phantom{-}7\frac{10}{15} = 6\frac{25}{15} \\ -3\frac{4}{5} = -3\boxed{\frac{4}{5}\cdot\frac{3}{3}} = -3\frac{12}{15} = -3\frac{12}{15} \\ \hline 3\frac{13}{15} \end{array}$$

so $3\frac{4}{5} - 7\frac{2}{3} = -3\frac{13}{15}$.

**67.** $-3\frac{1}{5} - 4\frac{2}{5} = -3\frac{1}{5} + \left(-4\frac{2}{5}\right)$

We add the absolute values and make the answer negative.

$$\begin{array}{r} 3\frac{1}{5} \\ +4\frac{2}{5} \\ \hline 7\frac{3}{5} \end{array}$$

Thus, $-3\frac{1}{5} - 4\frac{2}{5} = -7\frac{3}{5}$.

**69.** $-4\frac{2}{5} - 6\frac{3}{7} = -4\frac{2}{5} + \left(-6\frac{3}{7}\right)$

We add the absolute values and make the answer negative.

$$\begin{array}{r} 4\frac{2}{5} = \phantom{+}4\boxed{\frac{2}{5}\cdot\frac{7}{7}} = \phantom{+}4\frac{14}{35} \\ +6\frac{3}{7} = +6\boxed{\frac{3}{7}\cdot\frac{5}{5}} = +\phantom{0}6\frac{15}{35} \\ \hline 10\frac{29}{35} \end{array}$$

Thus, $-4\frac{2}{5} - 6\frac{3}{7} = -10\frac{29}{35}$.

**71.** $-6\frac{1}{9} - \left(-4\frac{2}{9}\right) = -6\frac{1}{9} + 4\frac{2}{9}$

Since $-6\frac{1}{9}$ has the greater absolute value, the answer will be negative. The difference in absolute values is

$$\begin{array}{r} 6\frac{1}{9} = \phantom{-}5\frac{10}{9} \\ -4\frac{2}{9} = -4\frac{2}{9} \\ \hline 1\frac{8}{9} \end{array}$$

so $-6\frac{1}{9} - \left(-4\frac{2}{9}\right) = -1\frac{8}{9}$.

**73.** $\frac{12}{25} \div \frac{24}{5} = \frac{12}{25} \cdot \frac{5}{24} = \frac{12 \cdot 5}{25 \cdot 24} = \frac{12 \cdot 5 \cdot 1}{5 \cdot 5 \cdot 12 \cdot 2} =$

$\frac{12 \cdot 5}{12 \cdot 5} \cdot \frac{1}{5 \cdot 2} = \frac{1}{10}$

**75.** $\left(-\frac{3}{4}\right)\left(-\frac{32}{33}\right) = \frac{3 \cdot 32}{4 \cdot 33} = \frac{3 \cdot 4 \cdot 8}{4 \cdot 3 \cdot 11} = \frac{3 \cdot 4}{3 \cdot 4} \cdot \frac{8}{11} = \frac{8}{11}$

**77.** ◈

**79.** ◈

**81.** Use a calculator.

$$\begin{array}{r} 5798\boxed{\frac{17}{53}\cdot\frac{43}{43}} = \phantom{-}5798\frac{731}{2279} = \phantom{-}5797\frac{3010}{2279} \\ -3909\frac{1957}{2279} = -3909\frac{1957}{2279} = -3909\frac{1957}{2279} \\ \hline 1888\frac{1053}{2279} \end{array}$$

**83.** *Familiarize*. We visualize the situation.

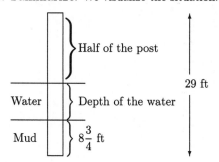

*Translate and Solve*. First we find the length of the post that extends above the water's surface. We let $p =$ this length.

$$\underbrace{\text{Half}}_{\downarrow} \underbrace{\text{of the post}}_{\downarrow} \text{ is } \underbrace{\text{above the water}}_{\downarrow}$$

$$\frac{1}{2} \quad \cdot \quad 29 \quad = \quad p$$

To solve we carry out the multiplication.

$$\frac{1}{2} \cdot 29 = p$$

$$\frac{29}{2} = p$$

$$14\frac{1}{2} = p$$

Now we find the depth of the water. We let $d =$ the depth.

| Length of post above water | plus | Depth of water | plus | Length of post in mud | is | Total length of post |
|---|---|---|---|---|---|---|
| $14\frac{1}{2}$ | $+$ | $d$ | $+$ | $8\frac{3}{4}$ | $=$ | $29$ |

We solve the equation.

$$14\frac{1}{2} + d + 8\frac{3}{4} = 29$$

$$23\frac{1}{4} + d = 29 \qquad \text{Adding on the left side}$$

$$d = 29 - 23\frac{1}{4} \qquad \text{Subtracting } 23\frac{1}{4} \text{ on both sides}$$

$$d = 5\frac{3}{4} \qquad \text{Simplifying}$$

*Check*. We repeat the calculations.

*State*. The water is $5\frac{3}{4}$ ft deep at that location.

**85.**
$$42\frac{7}{9} = x - 13\frac{2}{5}$$

$$42\frac{7}{9} + 13\frac{2}{5} = x - 13\frac{2}{5} + 13\frac{2}{5}$$

$$42\frac{35}{45} + 13\frac{18}{45} = x$$

$$55\frac{53}{45} = x$$

$$55 + 1\frac{8}{45} = x$$

$$56\frac{8}{45} = x$$

The solution is $56\frac{8}{45}$.

---

## Exercise Set 4.7

**1.** $10 \cdot 2\frac{5}{6}$

$= \frac{10}{1} \cdot \frac{17}{6}$    Writing fractional notation

$= \frac{10 \cdot 17}{1 \cdot 6} = \frac{2 \cdot 5 \cdot 17}{1 \cdot 2 \cdot 3} = \frac{2}{2} \cdot \frac{5 \cdot 17}{1 \cdot 3} = \frac{85}{3} = 28\frac{1}{3}$

**3.** $6\frac{2}{3} \cdot \frac{1}{4}$

$= \frac{20}{3} \cdot \frac{1}{4}$    Writing fractional notation

$= \frac{20 \cdot 1}{3 \cdot 4} = \frac{4 \cdot 5 \cdot 1}{3 \cdot 4} = \frac{4}{4} \cdot \frac{5 \cdot 1}{3} = \frac{5}{3} = 1\frac{2}{3}$

**5.** $-10 \cdot 7\frac{1}{3} = -\frac{10}{1} \cdot \frac{22}{3} = -\frac{220}{3} = -73\frac{1}{3}$

**7.** $3\frac{1}{2} \cdot 4\frac{2}{3} = \frac{7}{2} \cdot \frac{14}{3} = \frac{7 \cdot 14}{2 \cdot 3} = \frac{7 \cdot 2 \cdot 7}{2 \cdot 3} = \frac{2}{2} \cdot \frac{7 \cdot 7}{3} = \frac{49}{3} = 16\frac{1}{3}$

**9.** $-2\frac{3}{10} \cdot 4\frac{2}{5} = -\frac{23}{10} \cdot \frac{22}{5} = -\frac{23 \cdot 22}{10 \cdot 5} = -\frac{23 \cdot 2 \cdot 11}{2 \cdot 5 \cdot 5} = \frac{2}{2}\left(-\frac{23 \cdot 11}{5 \cdot 5}\right) = -\frac{253}{25} = -10\frac{3}{25}$

**11.** $6\frac{3}{10} \cdot 5\frac{7}{10} = \frac{63}{10} \cdot \frac{57}{10} = \frac{3591}{100} = 35\frac{91}{100}$

**13.** $20 \div 2\frac{3}{5}$

$= 20 \div \frac{13}{5}$    Writing fractional notation

$= 20 \cdot \frac{5}{13}$    Multiplying by the reciprocal

$= \frac{20 \cdot 5}{13} = \frac{100}{13} = 7\frac{9}{13}$

**15.** $8\frac{2}{5} \div 7$

$= \frac{42}{5} \div 7$    Writing fractional notation

$= \frac{42}{5} \cdot \frac{1}{7}$    Multiplying by the reciprocal

$= \frac{42 \cdot 1}{5 \cdot 7} = \frac{6 \cdot 7}{5 \cdot 7} = \frac{7}{7} \cdot \frac{6}{5} = \frac{6}{5} = 1\frac{1}{5}$

**17.** $4\frac{3}{4} \div 1\frac{1}{3} = \frac{19}{4} \div \frac{4}{3} = \frac{19}{4} \cdot \frac{3}{4} = \frac{19 \cdot 3}{4 \cdot 4} = \frac{57}{16} = 3\frac{9}{16}$

**19.** $-1\frac{7}{8} \div 1\frac{2}{3} = -\frac{15}{8} \div \frac{5}{3} = -\frac{15}{8} \cdot \frac{3}{5} = -\frac{15 \cdot 3}{8 \cdot 5} = -\frac{5 \cdot 3 \cdot 3}{8 \cdot 5}$

$= \frac{5}{5}\left(-\frac{3 \cdot 3}{8}\right) = -\frac{3 \cdot 3}{8} = -\frac{9}{8} = -1\frac{1}{8}$

**21.** $5\frac{1}{10} \div 4\frac{3}{10} = \frac{51}{10} \div \frac{43}{10} = \frac{51}{10} \cdot \frac{10}{43} = \frac{51 \cdot 10}{10 \cdot 43}$

$= \frac{10}{10} \cdot \frac{51}{43} = \frac{51}{43} = 1\frac{8}{43}$

**23.** $20\frac{1}{4} \div (-90) = \frac{81}{4} \div (-90) = \frac{81}{4}\left(-\frac{1}{90}\right) = -\frac{81 \cdot 1}{4 \cdot 90} =$

$-\frac{9 \cdot 9 \cdot 1}{4 \cdot 9 \cdot 10} = \frac{9}{9} \cdot \left(-\frac{9 \cdot 1}{4 \cdot 10}\right) = -\frac{9}{40}$

**25.** $lw = 2\frac{3}{5} \cdot 9$

$= \frac{13}{5} \cdot 9$

$= \frac{117}{5} = 23\frac{2}{5}$

**27.** $rs = 5 \cdot 3\frac{1}{7}$

$= 5 \cdot \frac{22}{7}$

$= \frac{110}{7} = 15\frac{5}{7}$

**29.** $mt = 6\frac{2}{9}\left(-4\frac{3}{5}\right)$

$= \frac{56}{9}\left(-\frac{23}{5}\right)$

$= -\frac{1288}{45} = -28\frac{28}{45}$

**31.** $R \cdot S \div T = 4\frac{2}{3} \cdot 1\frac{3}{7} \div (-5)$

$= \frac{14}{3} \cdot \frac{10}{7} \div (-5)$

$= \frac{14 \cdot 10}{3 \cdot 7} \div (-5)$

$= \frac{2 \cdot 7 \cdot 10}{3 \cdot 7} \div (-5)$

$= \frac{7}{7} \cdot \frac{2 \cdot 10}{3} \div (-5)$

$= \frac{20}{3} \div (-5)$

$= \frac{20}{3} \cdot \left(-\frac{1}{5}\right)$

$= -\frac{20 \cdot 1}{3 \cdot 5}$

$= -\frac{4 \cdot 5}{3 \cdot 5}$

$= -\frac{4}{3}$

$= -1\frac{1}{3}$

**33.** $r + ps = 5\frac{1}{2} + 3 \cdot 2\frac{1}{4}$

$= 5\frac{1}{2} + \frac{3}{1} \cdot \frac{9}{4}$

$= 5\frac{1}{2} + \frac{27}{4}$

$= 5\frac{1}{2} + 6\frac{3}{4}$

$= 5\frac{2}{4} + 6\frac{3}{4}$

$= 11\frac{5}{4}$

$= 12\frac{1}{4}$

**35.** $m + n \div p = 7\frac{2}{5} + 4\frac{1}{2} \div 6$

$= 7\frac{2}{5} + \frac{9}{2} \div 6$

$= 7\frac{2}{5} + \frac{9}{2} \cdot \frac{1}{6}$

$= 7\frac{2}{5} + \frac{9 \cdot 1}{2 \cdot 6} = 7\frac{2}{5} + \frac{3 \cdot 3 \cdot 1}{2 \cdot 2 \cdot 3}$

$= 7\frac{2}{5} + \frac{3 \cdot 1}{2 \cdot 2} = 7\frac{2}{5} + \frac{3}{4}$

$= 7\frac{8}{20} + \frac{15}{20} = 7\frac{23}{20}$

$= 8\frac{3}{20}$

**37. Familiarize.** Let $c$ = the number of cassettes that can be placed on each shelf.

**Translate.** A division corresponds to this situation.

$$c = 27 \div 1\frac{1}{8}$$

**Solve.** We carry out the division.

$$c = 27 \div 1\frac{1}{8} = 27 \div \frac{9}{8} = 27 \cdot \frac{8}{9} = \frac{27 \cdot 8}{9} =$$

$$\frac{3 \cdot 3 \cdot 3 \cdot 8}{3 \cdot 3 \cdot 1} = \frac{3 \cdot 3}{3 \cdot 3} \cdot \frac{3 \cdot 8}{1} = \frac{24}{1} = 24$$

**Check.** We can check by multiplying the number of cassettes by the width of a cassette.

$$24 \cdot 1\frac{1}{8} = 24 \cdot \frac{9}{8} = \frac{24 \cdot 9}{8} = \frac{3 \cdot 8 \cdot 3 \cdot 3}{8 \cdot 1} =$$

$$\frac{8}{8} \cdot \frac{3 \cdot 3 \cdot 3}{1} = 27$$

The answer checks.

**State.** June can place 24 cassettes on the shelf.

**39. Familiarize.** Let $s$ = the number of teaspoons 10 average American women consume in one day.

**Translate.** A multiplication corresponds to this situation.

$$s = 10 \cdot 1\frac{1}{3}$$

**Solve.** We carry out the multiplication.

$$s = 10 \cdot 1\frac{1}{3} = 10 \cdot \frac{4}{3} = \frac{10 \cdot 4}{3} = \frac{40}{3} = 13\frac{1}{3}$$

*Check*. We repeat the calculation. The answer checks.

*State*. In one day 10 average American women consume $13\frac{1}{3}$ tsp of sodium.

**41.** *Familiarize*. We draw a picture.

| $\frac{1}{3}$ lb | $\frac{1}{3}$ lb | $\cdots$ | $\frac{1}{3}$ lb |
|---|---|---|---|

$$\longleftarrow 5\frac{1}{2} \text{ lb} \longrightarrow$$

We let $s$ = the number of servings that can be prepared from $5\frac{1}{2}$ lb of flounder fillet.

*Translate*. The situation corresponds to a division sentence.

$$s = 5\frac{1}{2} \div \frac{1}{3}$$

*Solve*. We carry out the division.

$$s = 5\frac{1}{2} \div \frac{1}{3} = \frac{11}{2} \div \frac{1}{3}$$

$$= \frac{11}{2} \cdot \frac{3}{1} = \frac{33}{2}$$

$$= 16\frac{1}{2}$$

*Check*. We check by multiplying. If $16\frac{1}{2}$ servings are prepared, then

$$16\frac{1}{2} \cdot \frac{1}{3} = \frac{33}{2} \cdot \frac{1}{3} = \frac{3 \cdot 11 \cdot 1}{2 \cdot 3} = \frac{3}{3} \cdot \frac{11 \cdot 1}{2} = \frac{11}{2} = 5\frac{1}{2} \text{ lb}$$

of flounder is used. Our answer checks.

*State*. $16\frac{1}{2}$ servings can be prepared from $5\frac{1}{2}$ lb of flounder fillet.

**43.** *Familiarize*. We let $w$ = the weight of $5\frac{1}{2}$ cubic feet of water.

*Translate*. We write an equation.

$$\underbrace{\text{Weight per}}_{\text{cubic foot}} \cdot \underbrace{\text{Number of}}_{\text{cubic feet}} = \underbrace{\text{Total}}_{\text{weight}}$$

$$\downarrow \qquad \downarrow \qquad \downarrow \qquad \downarrow \qquad \downarrow$$

$$62\frac{1}{2} \quad \cdot \quad 5\frac{1}{2} \quad = \quad w$$

*Solve*. To solve the equation we carry out the multiplication.

$$w = 62\frac{1}{2} \cdot 5\frac{1}{2}$$

$$= \frac{125}{2} \cdot \frac{11}{2} = \frac{125 \cdot 11}{2 \cdot 2}$$

$$= \frac{1375}{4} = 343\frac{3}{4}$$

*Check*. We repeat the calculation. We also note that $62\frac{1}{2} \approx 60$ and $5\frac{1}{2} \approx 5$. Then the product is about 300. Our answer seems reasonable.

*State*. The weight of $5\frac{1}{2}$ cubic feet of water is $343\frac{3}{4}$ lb.

**45.** *Familiarize*. We let $t$ = the number of inches of tape used in 60 sec of recording.

*Translate*. We write an equation.

$$\underbrace{\text{Inches per}}_{\text{second}} \cdot \underbrace{\text{Number of}}_{\text{seconds}} = \underbrace{\text{Tape}}_{\text{used}}$$

$$\downarrow \qquad \downarrow \qquad \downarrow \qquad \downarrow \qquad \downarrow$$

$$1\frac{3}{8} \quad \cdot \quad 60 \quad = \quad t$$

*Solve*. We carry out the multiplication.

$$t = 1\frac{3}{8} \cdot 60 = \frac{11}{8} \cdot 60$$

$$= \frac{11 \cdot 4 \cdot 15}{2 \cdot 4} = \frac{11 \cdot 15}{2} \cdot \frac{4}{4}$$

$$= \frac{165}{2} = 82\frac{1}{2}$$

*Check*. We repeat the calculation.

*State*. $82\frac{1}{2}$ in. of tape are used in 60 sec of recording in short-play mode.

**47.** *Familiarize*. We let $t$ = the Fahrenheit temperature.

*Translate*.

$$\underbrace{\text{Celsius}}_{\text{temperature}} \text{ times } 1\frac{4}{5} \text{ plus } 32° \text{ is } \underbrace{\text{Fahrenheit}}_{\text{temperature}}$$

$$\downarrow \qquad \downarrow \quad \downarrow \quad \downarrow \quad \downarrow \quad \downarrow \qquad \downarrow$$

$$20 \quad \cdot \quad 1\frac{4}{5} \quad + \quad 32 \quad = \qquad t$$

*Solve*. We multiply and then add, according to the rules for order of operations.

$$t = 20 \cdot 1\frac{4}{5} + 32 = \frac{20}{1} \cdot \frac{9}{5} + 32 = \frac{20 \cdot 9}{1 \cdot 5} + 32 =$$

$$\frac{4 \cdot 5 \cdot 9}{1 \cdot 5} + 32 = \frac{5}{5} \cdot \frac{4 \cdot 9}{1} + 32 = 36 + 32 = 68$$

*Check*. We repeat the calculation.

*State*. 68° Fahrenheit corresponds to 20° Celsius.

**49.** *Familiarize*. Let $w$ = Shihong's body weight.

*Translate*.

$$\underbrace{103\frac{1}{2} \text{ kg}} \text{ is } 1\frac{1}{2} \text{ times } \underbrace{\text{Shihong's body weight.}}$$

$$\downarrow \qquad \downarrow \quad \downarrow \quad \downarrow \qquad \downarrow$$

$$103\frac{1}{2} \quad = \quad 1\frac{1}{2} \quad \cdot \qquad w$$

*Solve*. We solve the equation.

$$103\frac{1}{2} = 1\frac{1}{2} \cdot w$$

$$\frac{207}{2} = \frac{3}{2}w$$

$$\frac{2}{3} \cdot \frac{207}{2} = \frac{2}{3} \cdot \frac{3}{2}w$$

$$\frac{2 \cdot 207}{3 \cdot 2} = w$$

$$\frac{2 \cdot 3 \cdot 69}{3 \cdot 2 \cdot 1} = w$$

$$\frac{2 \cdot 3}{2 \cdot 3} \cdot \frac{69}{1} = w$$

$$69 = w$$

*Check*. We find $1\frac{1}{2}$ times 69: $1\frac{1}{2} \cdot 69 = \frac{3}{2} \cdot 69 = \frac{207}{2} = 103\frac{1}{2}$. Since $1\frac{1}{2}$ times 69 kg is $103\frac{1}{2}$ kg, the answer checks.

*State*. Shihong weighed 69 kg.

**51.** *Familiarize*. To compute an average, we add the values and then divide the sum by the number of values. Let $w =$ the average birthweight, in pounds.

*Translate*. We have

$$w = \frac{2\frac{9}{16} + 2\frac{9}{32} + 2\frac{1}{8} + 2\frac{5}{16}}{4}.$$

*Solve*. First we add.

$$2\frac{9}{16} + 2\frac{9}{32} + 2\frac{1}{8} + 2\frac{5}{16}$$

$$= 2\frac{18}{32} + 2\frac{9}{32} + 2\frac{4}{32} + 2\frac{10}{32}$$

$$= 8\frac{41}{32} = 9\frac{9}{32}$$

Then we divide:

$$9\frac{9}{32} \div 4 = \frac{297}{32} \div 4$$

$$= \frac{297}{32} \cdot \frac{1}{4}$$

$$= \frac{297}{128} = 2\frac{41}{128}$$

*Check*. As a partial check we note that the average weight is larger than the smallest weight and smaller than the largest weight. We could also repeat the calculations.

*State*. The average birthweight of the quadruplets was $2\frac{41}{128}$ lb.

**53.** *Familiarize*. We will add the values and then divide the sum by the number of values. Let $d =$ the number of days the bulbs burned, on average.

*Translate*. We have

$$d = \frac{17 + 16\frac{1}{2} + 20 + 18\frac{1}{2} + 21}{5}.$$

*Solve*. First we add.

$$17 + 16\frac{1}{2} + 20 + 18\frac{1}{2} + 21$$

$$= 92\frac{2}{2} = 93$$

Then we divide.

$$\frac{93}{5} = 18\frac{3}{5}$$

*Check*. As a partial check we note that the average is larger than the smallest value and smaller than the largest value. We could also repeat the calculations.

*State*. On average, the bulbs burned $18\frac{3}{5}$ days.

**55.** *Familiarize*. The figure contains a square with sides of $10\frac{1}{2}$ ft and a rectangle with dimensions of $8\frac{1}{2}$ ft by 4 ft. The area of the shaded region consists of the area of the square less the area of the rectangle. Let $A =$ the area of the shaded region, in square feet.

*Translate*. We write an equation.

$$A = 10\frac{1}{2} \cdot 10\frac{1}{2} - 8\frac{1}{2} \cdot 4$$

*Solve*. We multiply and then subtract.

$$A = 10\frac{1}{2} \cdot 10\frac{1}{2} - 8\frac{1}{2} \cdot 4$$

$$A = \frac{21}{2} \cdot \frac{21}{2} - \frac{17}{2} \cdot 4$$

$$A = \frac{441}{4} - \frac{68}{2}$$

$$A = \frac{441}{4} - \frac{68}{2} \cdot \frac{2}{2}$$

$$A = \frac{441}{4} - \frac{136}{4}$$

$$A = \frac{305}{4} = 76\frac{1}{4}$$

*Check*. We repeat the calculation.

*State*. The area of the shaded region is $76\frac{1}{4}$ ft$^2$.

**57.** *Familiarize*. First we will find the total width of the columns and determine if this is less than or greater than $8\frac{1}{2}$ in. Then, if the total is less than $8\frac{1}{2}$ in., we will find the difference between $8\frac{1}{2}$ in. and the total width and then divide by 2 to find the width of each margin. Let $w =$ the total width of the columns.

*Translate*. First we write an equation for finding the total width of the columns.

$$w = 2 \cdot 1\frac{1}{2} + 5 \cdot \frac{3}{4}$$

*Solve*. We solve the equation.

$$w = 2 \cdot 1\frac{1}{2} + 5 \cdot \frac{3}{4}$$

$$w = 2 \cdot \frac{3}{2} + 5 \cdot \frac{3}{4}$$

$$w = \frac{2 \cdot 3}{2} + \frac{5 \cdot 3}{4}$$

$$w = 3 + \frac{15}{4} = 3 + 3\frac{3}{4}$$

$$w = 6\frac{3}{4}$$

Since the total width of the columns is less than $8\frac{1}{2}$ in., the table will fit on a piece of standard paper. Let $l =$ the number of inches by which the width of the paper exceeds the width of the table. Then we have:

$$l = 8\frac{1}{2} - 6\frac{3}{4} = 8\frac{2}{4} - 6\frac{3}{4}$$

$$l = 7\frac{6}{4} - 6\frac{3}{4}$$

$$l = 1\frac{3}{4}$$

This tells us that the total width of the margins will be $1\frac{3}{4}$ in. Since the margins are of equal width, we divide by 2 to find the width of each margin. Let $m =$ this width, in inches. We have:

$$m = 1\frac{3}{4} \div 2$$

$$m = \frac{7}{4} \div 2$$

$$m = \frac{7}{4} \cdot \frac{1}{2} = \frac{7}{8}$$

**Check**. We repeat the calculations.

**State**. The table will fit on a standard piece of paper. Each margin will be $\frac{7}{8}$ in. wide.

**59.** $-8(x-3) = -8 \cdot x - (-8)(3) = -8x - (-24) = -8x + 24$

**61.** $-9x = 189$

$$\frac{-9x}{-9} = \frac{189}{-9}$$

$$x = -21$$

The solution is $-21$.

**63.** $-198 \div (-6) = \frac{-198}{-6} = 33$

**65.** ◈

**67.** ◈

**69.** $17\frac{23}{31} \cdot 19\frac{13}{15} = \frac{550}{31} \cdot \frac{298}{15} = \frac{163,900}{465} =$

$$352\frac{220}{465} = 352\frac{44}{93}$$

**71.**

$$\left(\frac{5}{9} - \frac{1}{4}\right)(-12) + \left(-4 - \frac{3}{4}\right)^2$$

$$= \left(\frac{20}{36} - \frac{9}{36}\right)(-12) + \left(-\frac{16}{4} - \frac{3}{4}\right)^2$$

$$= \frac{11}{36}(-12) + \left(-\frac{19}{4}\right)^2$$

$$= \frac{11}{36}(-12) + \frac{361}{16} = \frac{11(-12)}{36} + \frac{361}{16}$$

$$= \frac{12}{12} \cdot \frac{11(-1)}{3} + \frac{361}{16} = \frac{-11}{3} + \frac{361}{16}$$

$$= \frac{-176}{48} + \frac{1083}{48} = \frac{907}{48} = 18\frac{43}{48}$$

**73.**

$$\frac{7}{8} - 1\frac{1}{8} \times \frac{2}{3} + \frac{9}{10} \div \frac{3}{5}$$

$$= \frac{7}{8} - \frac{9}{8} \times \frac{2}{3} + \frac{9}{10} \times \frac{5}{3}$$

$$= \frac{7}{8} - \frac{9 \cdot 2}{8 \cdot 3} + \frac{9 \cdot 5}{10 \cdot 3} = \frac{7}{8} - \frac{2 \cdot 3}{2 \cdot 3} \cdot \frac{3}{4} + \frac{3 \cdot 5}{3 \cdot 5} \cdot \frac{3}{2}$$

$$= \frac{7}{8} - \frac{3}{4} + \frac{3}{2} = \frac{7}{8} - \frac{6}{8} + \frac{12}{8}$$

$$= \frac{13}{8}, \text{ or } 1\frac{5}{8}$$

**75.** $a^3 + a^2 = \left(-3\frac{1}{2}\right)^3 + \left(-3\frac{1}{2}\right)^2$

$$= \left(-\frac{7}{2}\right)^3 + \left(-\frac{7}{2}\right)^2$$

$$= -\frac{343}{8} + \frac{49}{4}$$

$$= -\frac{343}{8} + \frac{98}{8}$$

$$= -\frac{245}{8} = -30\frac{5}{8}$$

$$a^2(a+1) = \left(-3\frac{1}{2}\right)^2\left(-3\frac{1}{2} + 1\right)$$

$$= \left(-\frac{7}{2}\right)^2\left(-3\frac{1}{2} + 1\right)$$

$$= \frac{49}{4}\left(-3\frac{1}{2} + 1\right)$$

$$= \frac{49}{4}\left(-2\frac{1}{2}\right) = \frac{49}{4}\left(-\frac{5}{2}\right)$$

$$= -\frac{245}{8} = -30\frac{5}{8}$$

**77. Familiarize**. We will express each height in inches. (Recall that 1 ft = 12 in.) Then we will add the heights and divide the sum by the number of heights, 5.

6 ft, 10 in. = $6 \cdot 12 + 10 = 72 + 10 = 82$ in.;

6 ft, 7 in. = $6 \cdot 12 + 7 = 72 + 7 = 79$ in.;

5 ft, 10 in. = $5 \cdot 12 + 10 = 60 + 10 = 70$ in.;

6 ft, 6 in. = $6 \cdot 12 + 6 = 72 + 6 = 78$ in.;

7 ft, 1 in. = $7 \cdot 12 + 1 = 84 + 1 = 85$ in.;

Let $h =$ the average height.

**Translate**. We write an equation.

$$h = \frac{82 + 79 + 70 + 78 + 85}{5}$$

**Solve**. We carry out the calculation.

$$h = \frac{82 + 79 + 70 + 78 + 85}{5}$$

$$h = \frac{394}{5} = 78\frac{4}{5}$$

Now we convert $78\frac{4}{5}$ in. to feet and inches.

$78\frac{4}{5} = 72 + 6\frac{4}{5}$ and 72 in. = 6 ft, so $78\frac{4}{5}$ in. = 6 ft, $6\frac{4}{5}$ in.

**Check**. We repeat the calculations.

**State**. The average height of the given NBA stars is 6 ft, $6\frac{4}{5}$ in.

**79.** *Familiarize.* First we will find the number $n$ by which we multiply 4 to get 10. Then we will multiply the amount of each ingredient by this number.

*Translate.* We write an equation.

4  times  what number  is  10?

$$4 \quad \cdot \quad n \quad = \quad 10$$

*Solve.* We solve the equation.

$$4n = 10$$

$$\frac{4n}{4} = \frac{10}{4}$$

$$n = \frac{10}{4} = \frac{2 \cdot 5}{2 \cdot 2}$$

$$n = \frac{5}{2}$$

Now we multiply the amount of each ingredient by $\frac{5}{2}$.

$$2 \cdot \frac{5}{2} = \frac{2 \cdot 5}{2} = 5;$$

$$1\frac{1}{2} \cdot \frac{5}{2} = \frac{3}{2} \cdot \frac{5}{2} = \frac{15}{4} = 3\frac{3}{4};$$

$$3 \cdot \frac{5}{2} = \frac{15}{2} = 7\frac{1}{2};$$

$$2\frac{1}{2} \cdot \frac{5}{2} = \frac{5}{2} \cdot \frac{5}{2} = \frac{25}{4} = 6\frac{1}{4};$$

$$1 \cdot \frac{5}{2} = \frac{5}{2} = 2\frac{1}{2};$$

$$\frac{1}{3} \cdot \frac{5}{2} = \frac{5}{6};$$

$$\frac{1}{4} \cdot \frac{5}{2} = \frac{5}{8}$$

*Check.* We repeat the calculations.

*State.* The ingredients for 10 servings are 5 chicken bouillon cubes, $3\frac{3}{4}$ cups hot water, $7\frac{1}{2}$ tablespoons margarine, $7\frac{1}{2}$ tablespoons flour, $6\frac{1}{4}$ cups diced cooked chicken, $2\frac{1}{2}$ cups cooked peas, $2\frac{1}{2}$ 4-oz cans sliced mushrooms (drained), $\frac{5}{6}$ cup sliced cooked carrots, $\frac{5}{8}$ cup chopped onions, 5 tablespoons chopped pimiento, $2\frac{1}{2}$ teaspoons salt.

# Chapter 5

# Decimal Notation

## Exercise Set 5.1

**1.** 481.27

    a) Write a word name for the whole number.      | Four hundred eighty-one |

    b) Write "and" for the decimal point.

    Four hundred eighty-one | and |

    c) Write a word name for the number to the right of the decimal point, followed by the place value of the last digit.

    Four hundred eighty-one and | twenty-seven hundredths |

A word name for 481.27 is four hundred eighty-one and twenty-seven hundredths.

**3.** $1.5599

    a) Write a word name for the whole number. | One |

    b) Write "and" for the decimal point.

    One | and |

    c) Write a word name for the number to the right of the decimal point, followed by the place value of the last digit.

    One and | five thousand five hundred ninety-nine ten thousandths |

A word name for 1.5599 is one and five thousand five hundred ninety-nine ten thousandths.

**5.**

Thirty-four — and — eight hundred ninety-one thousandths — $\underbrace{34}$ . $\underbrace{891}$

**7.** Write "and 48 cents" as "and $\dfrac{48}{100}$ dollars." A word name for $326.48 is three hundred twenty-six and $\dfrac{48}{100}$ dollars.

**9.** Write "and 72 cents" as "and $\dfrac{72}{100}$ dollars." A word name for $36.72 is thirty-six and $\dfrac{72}{100}$ dollars.

**11.**    8.$\underline{3}$      8.3.      $\dfrac{83}{10}$

   1 place   Move 1 place.   1 zero

$8.3 = \dfrac{83}{10}$

**13.**    203.$\underline{6}$      203.6.      $\dfrac{2036}{10}$

   1 place   Move 1 place.   1 zero

$203.6 = \dfrac{2036}{10}$

**15.**    $-2.\underline{703}$      $-2.703.$      $\dfrac{-2073}{1000}$

   3 places   Move 3 places.   3 zeros

$-2.703 = \dfrac{-2073}{1000}, \text{ or } -\dfrac{2073}{1000}$

**17.**    0.$\underline{0109}$      0.0109.      $\dfrac{109}{10,000}$

   4 places   Move 4 places.   4 zeros

$0.0109 = \dfrac{109}{10,000}$

**19.**    $-6.\underline{004}$      $-6.004.$      $\dfrac{-6004}{1000}$

   3 places   Move 3 places.   3 zeros

$-6.004 = \dfrac{-6004}{1000}, \text{ or } -\dfrac{6004}{1000}$

**21.**    $\dfrac{8}{10}$      0.8.

   1 zero   Move 1 place.

$\dfrac{8}{10} = 0.8$

**23.**    $-\dfrac{59}{100}$      $-0.59.$

   2 zeros   Move 2 places.

$-\dfrac{59}{100} = -0.59$

**25.**    $\dfrac{3798}{1000}$      3.798.

   3 zeros   Move 3 places.

$\dfrac{3798}{1000} = 3.798$

**79. _Familiarize._** First we will find the number $n$ by which we multiply 4 to get 10. Then we will multiply the amount of each ingredient by this number.

_**Translate.**_ We write an equation.

$$4 \text{ times } \underbrace{\text{what number}} \text{ is } 10?$$

$$
\begin{array}{ccccc}
\downarrow & \downarrow & \downarrow & \downarrow & \downarrow \\
4 & \cdot & n & = & 10
\end{array}
$$

_**Solve.**_ We solve the equation.

$$4n = 10$$

$$\frac{4n}{4} = \frac{10}{4}$$

$$n = \frac{10}{4} = \frac{2 \cdot 5}{2 \cdot 2}$$

$$n = \frac{5}{2}$$

Now we multiply the amount of each ingredient by $\frac{5}{2}$.

$$2 \cdot \frac{5}{2} = \frac{2 \cdot 5}{2} = 5;$$

$$1\frac{1}{2} \cdot \frac{5}{2} = \frac{3}{2} \cdot \frac{5}{2} = \frac{15}{4} = 3\frac{3}{4};$$

$$3 \cdot \frac{5}{2} = \frac{15}{2} = 7\frac{1}{2};$$

$$2\frac{1}{2} \cdot \frac{5}{2} = \frac{5}{2} \cdot \frac{5}{2} = \frac{25}{4} = 6\frac{1}{4};$$

$$1 \cdot \frac{5}{2} = \frac{5}{2} = 2\frac{1}{2};$$

$$\frac{1}{3} \cdot \frac{5}{2} = \frac{5}{6};$$

$$\frac{1}{4} \cdot \frac{5}{2} = \frac{5}{8}$$

_**Check.**_ We repeat the calculations.

_**State.**_ The ingredients for 10 servings are 5 chicken bouillon cubes, $3\frac{3}{4}$ cups hot water, $7\frac{1}{2}$ tablespoons margarine, $7\frac{1}{2}$ tablespoons flour, $6\frac{1}{4}$ cups diced cooked chicken, $2\frac{1}{2}$ cups cooked peas, $2\frac{1}{2}$ 4-oz cans sliced mushrooms (drained), $\frac{5}{6}$ cup sliced cooked carrots, $\frac{5}{8}$ cup chopped onions, 5 tablespoons chopped pimiento, $2\frac{1}{2}$ teaspoons salt.

**67.**

$-0.03\boxed{4}8$   Thousandths digit is 4 or lower.
    ↓      Round from $-0.0348$ to $-0.03$.
$-0.03$

**69.**

$0.324\boxed{6}$   Ten-thousandths digit is 5 or higher.
    ↓      Round up.
$0.325$

**71.**

$17.001\boxed{5}$   Ten-thousandths digit is 5 or higher.
    ↓      Round up.
$17.002$

**73.**

$-20.202\boxed{0}2$   Ten-thousandths digit is 4 or lower.
    ↓      Round $-20.20202$ to $-20.202$.
$-20.202$

**75.**

$9.984\boxed{8}$   Ten-thousandths digit is 5 or higher.
    ↓      Round up.
$9.985$

**77.**

$809.4\boxed{7}32$   Hundredths digit is 5 or higher.
    ↓      Round up.
$809.5$

**79.**

$809.47\boxed{3}2$   Thousandths digit is 4 or lower.
    ↓      Round down.
$809.47$

**81.** $\dfrac{0}{n} = 0$, for any integer $n$ that is not 0.

Thus, $\dfrac{0}{-19} = 0$.

**83.** $\dfrac{4}{9} - \dfrac{2}{3} = \dfrac{4}{9} - \dfrac{2}{3} \cdot \dfrac{3}{3}$

$= \dfrac{4}{9} - \dfrac{6}{9}$

$= \dfrac{-2}{9}$, or $-\dfrac{2}{9}$

**85.** $3x - 8 = 21$

$3x - 8 + 8 = 21 + 8$

$3x = 29$

$\dfrac{3x}{3} = \dfrac{29}{3}$

$x = \dfrac{29}{3}$

The solution is $\dfrac{29}{3}$.

**87.** ◈

**89.** ◈

**91.** From the graph we see that the last year for which the corresponding vertical bar falls below $-0.6$ is 1979.

**93.** From the graph we see that the only year for which the corresponding vertical bar goes above 1.0 is 1998.

**95.** $0.07070\boxed{707}$ ← Drop all decimal places
    ↓       past the fifth place.
$0.07070$

The answer is $0.07070$.

---

## Exercise Set 5.2

**1.**

$$\begin{array}{r} \overset{1}{\phantom{0}}\phantom{0} \\ 3\,1\,6.2\,5 \\ +\ \ 1\,8.1\,2 \\ \hline 3\,3\,4.3\,7 \end{array}$$

Add hundredths.
Add tenths.
Write a decimal point in the answer.
Add ones.
Add tens.
Add hundreds.

**3.**

$$\begin{array}{r} \overset{1\ \ 1}{\phantom{0}} \\ 6\,5\,9.4\,0\,3 \\ +\ \ 9\,1\,6.8\,1\,2 \\ \hline 1\,5\,7\,6.2\,1\,5 \end{array}$$

Add thousandths.
Add hundredths.
Add tenths.
Write a decimal point in the answer.
Add ones.
Add tens.
Add hundreds.

**5.**

$$\begin{array}{r} \overset{1\ \ \ \ \ 1}{\phantom{0}} \\ 9.1\,0\,4 \\ +\ 1\,2\,3.4\,5\,6 \\ \hline 1\,3\,2.5\,6\,0 \end{array}$$

**7.**

$$\begin{array}{r} \overset{\ \ \ \ \ 1}{\phantom{0}} \\ 6\,1.0\,0\,6 \\ +\ \ \ 3.4\,0\,7 \\ \hline 6\,4.4\,1\,3 \end{array}$$

**9.** Line up the decimal points.

$$\begin{array}{r} 2\,0.0\,1\,2\,4 \\ +\ 3\,0.0\,1\,2\,4 \\ \hline 5\,0.0\,2\,4\,8 \end{array}$$

**11.** Line up the decimal points.

$$\begin{array}{r} 0.8\,3\,0 \\ +\ 0.0\,0\,5 \\ \hline 0.8\,3\,5 \end{array}$$

Writing an extra zero

**13.** Line up the decimal points.

$$\begin{array}{r} \overset{1}{\phantom{0}}\phantom{0} \\ 0.3\,4\,0 \\ 3.5\,0\,0 \\ 0.1\,2\,7 \\ +\ 7\,6\,8.0\,0\,0 \\ \hline 7\,7\,1.9\,6\,7 \end{array}$$

Writing an extra zero
Writing 2 extra zeros

Writing in the decimal point
and 3 extra zeros
Adding

**15.**

$$\begin{array}{r} \overset{1\ \ \ 1\ 1}{\phantom{0}} \\ 1\,7.0\,0\,0\,0 \\ 3.2\,4\,0\,0 \\ 0.2\,5\,6\,0 \\ +\ \ \ 0.3\,6\,8\,9 \\ \hline 2\,0.8\,6\,4\,9 \end{array}$$

Writing in the decimal point.
You may find it helpful to
write extra zeros.

**17.**
```
    1 2 1    1
      2.7 0 3 0
    7 8.3 3 0 0
   ,2 8.0 0 0 9
 + 1 1 8.4 3 4 1
 ─────────────────
   2 2 7.4 6 8 0
```

**19.**
```
    5.2      Subtract tenths.
  − 3.1      Write a decimal point in the answer.
  ─────
    2.1      Subtract ones.
```

**21.**
```
    4 11 2 11     Borrow tenths to subtract hundredths.
    5 1.3 1       Subtract hundredths.
  −    2.2 9      Subtract tenths.
  ──────────      Write a decimal point in the answer.
    4 9.0 2       Borrow tens to subtract ones.
                  Subtract ones.
                  Subtract tens.
```

**23.**
```
    4 9 9 10
    2.5 0 0 0      Writing 3 extra zeros
  − 0.0 0 2 5
  ────────────
    2.4 9 7 5
```

**25.**
```
        11
     8  1 13
     9 2.3 4 1
  −    6.4 2
  ──────────
    8 5.9 2 1
```

**27.**
```
    2 9 10 6 14
    3.0 0 7 4
  − 1.3 4 0 8
  ────────────
    1.6 6 6 6
```

**29.**
```
          6 9 10
    6.0 7 0 0      Writing 2 extra zeros
  − 2.0 0 7 8
  ────────────
    4.0 6 2 2
```

**31.** Line up the decimal points. Write an extra zero if desired.
```
      17 11
    1 7 1 10
    2 8.2 0
  − 1 9.3 5
  ──────────
      8.8 5
```

**33.**
```
      3 10
    3 4.0 7
  − 3 0.7
  ──────────
      3.3 7
```

**35.**
```
        4 10
    8.4 5 0
  − 7.4 0 5
  ──────────
    1.0 4 5
```

**37.**
```
      5 10
    6.0 0 3
  − 2.3
  ──────────
    3.7 0 3
```

**39.**
```
    1 9 9 9 10
    2.0 0 0 0      Writing in the decimal point
  − 1.0 9 0 8      and 4 extra zeros
  ────────────
    0.9 0 9 2      Subtracting
```

**41.**
```
          13
    1 3 9 10
    6 2 4.0 0
  −    1 8.7 9
  ────────────
    6 0 5.2 1
```

**43.**
```
    2 9 9 10
    3.0 0 0
  − 2.0 0 6
  ──────────
    0.9 9 4
```

**45.**
```
          6 10
    2 6 3.7 0
  − 1 0 2.0 8
  ────────────
    1 6 1.6 2
```

**47.**
```
      4 9 9 10
    4 5.0 0 0
  −    0.9 9 9
  ────────────
    4 4.0 0 1
```

**49.** −8.02 + 9.73   A positive and a negative number

a) |−8.02| = 8.02, |9.73| = 9.73, and |9.73| > |−8.02|, so the answer is positive.

b)
```
      9 .73      Finding the difference in
    − 8 .02      the absolute values
    ──────
      1 .71
```

c) −8.02 + 9.73 = 1.71

**51.** 12.9 − 15.4 = 12.9 + (−15.4)
We add the opposite of 15.4. We have a positive and a negative number.

a) |12.9| = 12.9, |−15.4| = 15.4, and |15.4| > |12.9|, so the answer is negative.

b)
```
      4 14
    1 5.4      Finding the difference in
  − 1 2.9      the absolute values
  ──────
      2.5
```

c) 12.9 − 15.4 = −2.5

**53.** −2.9 + (−4.3)   Two negative numbers

a)
```
      1
    2 .9
  + 4 .3      Adding the absolute values
  ──────
    7 .2
```

b) −2.9 + (−4.3) = −7.2  The sum of two negative numbers is negative.

**55.** −4.301 + 7.68   A negative and a positive number

a) |−4.301| = 4.301, |7.68| = 7.68, and |7.68| > |−4.301|, so the answer is positive.

b)
```
          7 10
    7.6 8 0      Finding the difference in
  − 4.3 0 1      the absolute values
  ──────────
    3.3 7 9
```

c) −4.301 + 7.68 = 3.379

**57.**
```
      −13.4 − 9.2
   = −13.4 + (−9.2)   Adding the opposite of 9.2
   = −22.6            The sum of two negatives is
                      negative.
```

**59.** $-2.1 - (-4.6)$
$= -2.1 + 4.6$  Adding the opposite of $-4.6$
$= 2.5$  Subtracting absolute values.
Since 4.6 has the larger absolute value, the answer is positive.

**61.** $14.301 + (-17.82)$
$= -3.519$  Subtracting absolute values. Since $-17.82$ has the larger absolute value, the answer is negative.

**63.** $7.201 - (-2.4)$
$= 7.201 + 2.4$  Adding the opposite of $-2.4$
$= 9.601$  Adding

**65.** $23.9 + (-9.4)$
$= 14.5$  Subtracting absolute values. Since 23.9 has the larger absolute value, the answer is positive.

**67.** $-8.9 - (-12.7)$
$= -8.9 + 12.7$  Adding the opposite of $-12.7$
$= 3.8$  Subtracting absolute values. Since 12.7 has the larger absolute value, the answer is positive.

**69.** $-4.9 - 5.392$
$= -4.9 + (-5.392)$  Adding the opposite of 5.392
$= -10.292$  The sum of two negatives is negative.

**71.** $14.7 - 23.5$
$= 14.7 + (-23.5)$  Adding the opposite of 23.5
$= -8.8$  Subtracting absolute values. Since $-23.5$ has the larger absolute value, the answer is negative.

**73.** $5.1x + 3.6x$
$= (5.1 + 3.6)x$  Using the distributive law
$= 8.7x$  Adding

**75.** $17.59a - 12.73a$
$= (17.59 - 12.73)a$
$= 4.86a$

**77.** $15.2t + 7.9 + 5.9t$
$= 15.2t + 5.9t + 7.9$  Using the commutative law
$= (15.2 + 5.9)t + 7.9$  Using the distributive law
$= 21.1t + 7.9$

**79.** $9.208t - 14.519t$
$= (9.208 - 14.519)t$  Using the distributive law
$= (9.208 + (-14.519))t$  Adding the opposite of 14.519
$= -5.311t$  Subtracting absolute values. The coefficient is negative since $-14.519$ has the larger absolute value.

**81.** $4.906y - 7.1 + 3.2y$
$= 4.906y + 3.2y - 7.1$
$= (4.906 + 3.2)y - 7.1$
$= 8.106y - 7.1$

**83.** $4.8x + 1.9y - 5.7x + 1.2y$
$= 4.8x + 1.9y + (-5.7x) + 1.2y$  Rewriting as addition
$= 4.8x + (-5.7x) + 1.9y + 1.2y$  Using the commutative law
$= (4.8 + (-5.7))x + (1.9 + 1.2)y$
$= -0.9x + 3.1y$

**85.** $4.9 - 3.9t + 2.3 - 4.5t$
$= 4.9 + (-3.9t) + 2.3 + (-4.5t)$
$= 4.9 + 2.3 + (-3.9t) + (-4.5t)$
$= (4.9 + 2.3) + (-3.9 + (-4.5))t$
$= 7.2 + (-8.4t)$
$= 7.2 - 8.4t$

**87.** $\dfrac{0}{n} = 0$, for any integer $n$ that is not 0.
Thus, $\dfrac{0}{-92} = 0$.

**89.** $\dfrac{3}{5} - \dfrac{7}{10} = \dfrac{3}{5} \cdot \dfrac{2}{2} - \dfrac{7}{10}$  The LCM is 10.
$= \dfrac{6}{10} - \dfrac{7}{10}$
$= \dfrac{-1}{10}$, or $-\dfrac{1}{10}$

**91.** $7x + 19 = 40$
$7x + 19 - 19 = 40 - 19$
$7x = 21$
$\dfrac{7x}{7} = \dfrac{21}{7}$
$x = 3$
The solution is 3.

**93.** ◈

**95.** ◈

**97.** $79.02x + 0.0093y - 53.14z - 0.02001y - 37.987z - 97.203x - 0.00987y$
$= 79.02x - 97.203x + 0.0093y - 0.02001y - 0.00987y - 53.14z - 37.987z$
$= -18.183x - 0.02058y - 91.127z$

**99.** First, "undo" the incorrect addition by subtracting 235.7 from the incorrect answer:

$$\begin{array}{r} 8\ 1\ 7.\ 2 \\ -\ 2\ 3\ 5.\ 7 \\ \hline 5\ 8\ 1.\ 5 \end{array}$$

The original minuend was 581.5. Now subtract 235.7 from this as the student originally intended:

$$\begin{array}{r} 5\ 8\ 1.\ 5 \\ -\ 2\ 3\ 5.\ 7 \\ \hline 3\ 4\ 5.\ 8 \end{array}$$

The correct answer is 345.8.

**101.** $8767.73 - 23.56 = 8744.17$, so 8744.16 should be 8744.17;
$8744.17 - 20.49 = 8723.68$, so 8764.65 should be 8723.68;
$8723.68 + 85.00 = 8808.68$, so 8848.65 should be 8808.68;
$8808.68 - 48.60 = 8760.08$, so 8801.05 should be 8760.08;
$8760.08 - 267.95 = 8492.13$, so 8533.09 should be 8492.13.

## Exercise Set 5.3

**1.**     6. 8    (1 decimal place)
   × ___7___    (0 decimal places)
     4 7. 6    (1 decimal place)

**3.**     0. 8 4    (2 decimal places)
   × ____8___    (0 decimal places)
     6. 7 2    (2 decimal places)

**5.**     6. 3    (1 decimal place)
   × 0. 0 4    (2 decimal places)
     0. 2 5 2    (3 decimal places)

**7.**     1 7. 2    (1 decimal place)
   × 0. 0 0 6    (3 decimal places)
     0. 1 0 3 2    (4 decimal places)

**9.** 1$\underline{0}$ × 42.63          42.6.3

1 zero          Move 1 place to the right.

10 × 42.63 = 426.3

**11.** −1000 × 783.686852 = −(1000 × 783.686852)

−(1$\underline{000}$ × 783.686852)          −783.686.852

3 zeros                Move 3 places to the right.

−1000 × 783.686852 = −783,686.852

**13.** −7.8 × 1$\underline{00}$          −7.80.

2 zeros          Move 2 places to the right.

−7.8 × 100 = −780

**15.** 0.$\underline{1}$ × 79.18          7.9.18

1 decimal place      Move 1 place to the left.

0.1 × 79.18 = 7.918

**17.** 0.$\underline{001}$ × 97.68          0.097.68

3 decimal places      Move 3 places to the left.

0.001 × 97.68 = 0.09768

**19.** 28.7 × (−0.01) = −(28.7 × 0.01)

−(28.7 × 0.$\underline{01}$)          −0.28.7

2 decimal places      Move 2 places to the left.

28.7 × (−0.01) = −0.287

**21.**     2. 7 3    (2 decimal places)
   × ___1 6___    (0 decimal places)
     1 6 3 8
     2 7 3 0
     4 3. 6 8    (2 decimal places)

**23.**     0. 9 8 4    (3 decimal places)
   × ____3. 3___    (1 decimal place)
       2 9 5 2
     2 9 5 2 0
     3. 2 4 7 2    (4 decimal places)

**25.** We multiply the absolute values.

       3 7. 4    (1 decimal place)
   × ____2. 4___    (1 decimal place)
     1 4 9 6
     7 4 8 0
     8 9. 7 6    (2 decimal places)

Since the product of two negative numbers is positive, the answer is 89.76.

**27.** We multiply the absolute values.

       7 4 9    (0 decimal places)
   × ___0. 4 3___    (2 decimal places)
       2 2 4 7
     2 9 9 6 0
     3 2 2. 0 7    (2 decimal places)

Since the product of a positive number and a negative number is negative, the answer is −322.07.

**29.**     0. 8 7    (2 decimal places)
   × ____6 4___    (0 decimal places)
       3 4 8
     5 2 2 0
     5 5. 6 8    (2 decimal places)

**31.**     4 6. 5 0    (2 decimal places)
   × _____7 5___    (0 decimal places)
     2 3 2 5 0
     3 2 5 5 0 0
     3 4 8 7. 5 0    (2 decimal places)

Since the last decimal place is 0, we could also write this answer as 3487.5.

**33.** We multiply the absolute values.

       0. 2 3 1    (3 decimal places)
   × ____0. 5___    (1 decimal place)
     0. 1 1 5 5    (4 decimal places)

Since the product of two negative numbers is positive, the answer is 0.1155.

**35.** 9.42 × (−1000) = −(9.42 × 1000)

−(9.42 × 1$\underline{000}$)          −9.420.

3 zeros          Move 3 places to the right.

9.42 × (−1000) = −9420

**37.** −95.3 × (−0.0001) = 95.3 × 0.0001

95.3 × 0.$\underline{0001}$          0.0095.3

4 decimal places      Move 4 places to the left.

−95.3 × (−0.0001) = 0.00953

**39.** Move 2 places to the right.

$28.88.¢

Change from \$ sign in front to ¢ sign at end.

$28.88 = 2888¢

**41.** Move 2 places to the right.

$0.66.¢

Change from \$ sign in front to ¢ sign at end.

$0.66 = 66¢

**43.** Move 2 places to the left.

$0.34.¢

Change from ¢ sign at end to \$ sign in front.

34¢ = \$0.34

**45.** Move 2 places to the left.

$34.45.¢

Change from ¢ sign at end to \$ sign in front.

3345¢ = \$34.45

**47.** $32.279 \text{ billion} = \$32.279 \times 1,\underbrace{000,000,000}_{9 \text{ zeros}}$

$32.279000000.

Move 9 places to the right.

$32.279 billion = \$32,279,000,000

**49.** $1.03 \text{ million} = 1.03 \times 1,\underbrace{000,000}_{6 \text{ zeros}}$

1.030000.

Move 6 places to the right.

1.03 million = 1,030,000

**51.**
$$
\begin{aligned}
&P + Prt \\
&= 10,000 + 10,000(0.04)(2.5) && \text{Substituting} \\
&= 10,000 + 400(2.5) && \text{Multiplying and dividing} \\
&= 10,000 + 1000 && \text{in order from left to right} \\
&= 11,000 && \text{Adding}
\end{aligned}
$$

**53.**
$$
\begin{aligned}
&vt + at^2 \\
&= 10(1.5) + 4.9(1.5)^2 \\
&= 10(1.5) + 4.9(1.5)(1.5) \\
&= 10(1.5) + 4.9(2.25) && \text{Squaring first} \\
&= 15 + 11.025 \\
&= 26.025
\end{aligned}
$$

**55. a)**
$$
\begin{aligned}
P &= 2l + 2w \\
&= 2(12.5) + 2(9.5) \\
&= 25 + 19 \\
&= 44
\end{aligned}
$$

The perimeter is 44 ft.

**b)**
$$
\begin{aligned}
A &= l \cdot w \\
&= (12.5)(9.5) \\
&= 118.75
\end{aligned}
$$

The area is 118.75 ft$^2$.

**57.** $\dfrac{n}{n} = 1$, for any integer $n$ that is not 0.

Thus, $\dfrac{-109}{-109} = 1$.

**59.** $\dfrac{2}{9} - \dfrac{5}{18} = \dfrac{2}{9} \cdot \dfrac{2}{2} - \dfrac{5}{18}$    The LCM is 18.

$$
\begin{aligned}
&= \dfrac{4}{18} - \dfrac{5}{18} \\
&= \dfrac{-1}{18}, \text{ or } -\dfrac{1}{18}
\end{aligned}
$$

**61.** $-\dfrac{3}{20} + \dfrac{3}{4} = -\dfrac{3}{20} + \dfrac{3}{4} \cdot \dfrac{5}{5}$    The LCM is 20.

$$
\begin{aligned}
&= -\dfrac{3}{20} + \dfrac{15}{20} \\
&= \dfrac{12}{20} \\
&= \dfrac{4 \cdot 3}{4 \cdot 5} = \dfrac{4}{4} \cdot \dfrac{3}{5} \\
&= \dfrac{3}{5}
\end{aligned}
$$

**63.** ◈

**65.** ◈

**67.** Use a calculator.
$$
\begin{aligned}
&3.14r^2 + 6.28rh \\
&= 3.14(5.756)^2 + 6.28(5.756)(9.047) \\
&= 3.14(33.131536) + 6.28(5.756)(9.047) \\
&= 104.033023 + 36.14768(9.047) \\
&= 104.033023 + 327.028061 \\
&= 431.061084
\end{aligned}
$$

**69.** $(1 \text{ million}) \cdot (1 \text{ billion})$
$$
\begin{aligned}
&= 1,\underbrace{000,000}_{6 \text{ zeros}} \times 1,\underbrace{000,000,000}_{9 \text{ zeros}} \\
&= 1,\underbrace{000,000,000,000,000}_{15 \text{ zeros}} \\
&= 10^{15}
\end{aligned}
$$

**71.** The period from June 20 to July 20 consists of 30 days, so the "customer charge" is $30 \times \$0.35$, or \$10.50. The "energy charge" for the first 250 kilowatt-hours is $250 \times \$0.10470$, or \$26.18 (rounding to the nearest cent).

We subtract to find the number of kilowatt-hours in excess of 250: $430 - 250 = 180$. Then the "energy charge" for the 180 kilowatt-hours in excess of 250 kilowatt-hours is $180 \times \$0.09079$, or \$16.34 (rounding to the nearest cent).

Finally, we add to find the total bill:

$10.50 + \$26.18 + \$16.34 = \$53.02.

# Exercise Set 5.4

**1.**
```
      16.4
   5 )82.0
      5 0
      ───
      3 2
      3 0
      ───
        20    Write an extra 0.
        20
      ───
         0
```

**3.**
```
      23.78
   4 )95.12
      80 00
      ─────
      15 12
      12 00
      ─────
       3 12
       2 80
      ─────
         32
         32
      ─────
          0
```
Divide as though dividing whole numbers. Place the decimal point directly above the decimal point in the dividend.

**5.**
```
       7.48
   12 )89.76
       84 00
       ─────
        5 76
        4 80
       ─────
          96
          96
       ─────
           0
```

**7.**
```
       7.2
   33 )237.6
       231 0
       ─────
          66
          66
       ─────
           0
```

**9.** We first consider $9.144 \div 8$.
```
      1.143
   8 )9.144
      8 000
      ─────
      1 144
        800
      ─────
        344
        320
      ─────
         24
         24
      ─────
          0
```
Since a positive number divided by a negative number is negative, the answer is $-1.143$.

**11.** We first consider $5.4 \div 6$.
```
      0.9
   6 )5.4
      5 4
      ───
        0
```
Since a negative number divided by a positive number is negative, the answer is $-0.9$.

**13.**
```
          70.
   0.1 2∧)8.4 0∧
          8 40
          ────
             0
```
Multiply the divisor by 100 (move the decimal point 2 places). Multiply the same way in the dividend (move 2 places). Then divide.

**15.**
```
           40.
   2. 6∧)104.0∧
          104 0
          ─────
              0
```
Put a decimal point at the end of the whole number. Multiply the divisor by 10 (move the decimal point 1 place). Multiply the same way in the dividend (move 1 place), adding an extra 0. Then divide.

**17.** We first consider $1.8 \div 12$.
```
       0.15
   12 )1.80
       1 2
       ───
         60
         60
       ───
          0
```
Divide as though dividing whole numbers. Place the decimal point directly above the decimal point in the dividend.

Since a positive number divided by a negative number is negative, the answer is $-0.15$.

**19.**
```
           48.
   2. 7∧)129.6∧
         108 0
         ─────
           2 16
           2 16
         ─────
              0
```

**21.**
```
          3.2
   8.5∧)27.2∧0
         25 5
         ────
          1 70    Write an extra zero.
          1 70
         ────
             0
```

**23.** We first consider $5 \div 8$.
```
      0.625
   8 )5.000
      4 8
      ───
        20    Write an extra 0.
        16
      ───
         40    Write an extra 0.
         40
      ───
          0
```
Since a negative number divided by a negative number is positive, the answer is $0.625$.

**25.**
```
            0.26
   0.4 7∧)0.1 2∧22
            9 40
           ─────
            2 82
            2 82
           ─────
               0
```

**27.**
```
              2.34
   0.0 3 2∧)0.0 7 4∧88
              6 400
             ──────
             1 088
               960
             ──────
               1 28
               1 28
             ──────
                  0
```

**29.** We first consider $24.969 \div 82$.

$$
\begin{array}{r}
0.3\,0\,4\,5 \\
8\,2\,\overline{\smash{\big)}\,2\,4.9\,6\,9\,0} \\
2\,4\,6\,0\,0 \\
\hline
3\,6\,9 \\
3\,2\,8 \\
\hline
4\,1\,0 \\
4\,1\,0 \\
\hline
0
\end{array}
$$
Write an extra 0.

Since a negative number divided by a positive number is negative, the answer is $-0.3045$.

**31.** $\dfrac{-213.4567}{100}$ $\qquad$ $-2.13.4567$

2 zeros $\qquad$ Move 2 places to the left.

$\dfrac{-213.4567}{100} = -2.134567$

**33.** $\dfrac{1.0237}{0.001}$ $\qquad$ $1.023.7$

3 decimal places $\qquad$ Move 3 places to the right.

$\dfrac{1.0237}{0.001} = 1023.7$

**35.** $\dfrac{56.78}{-0.001} = -\dfrac{56.78}{0.001}$

$-\dfrac{56.78}{0.001}$ $\qquad$ $-56.780.$

3 decimal places $\qquad$ Move 3 places to the right.

$\dfrac{56.78}{-0.001} = -56,780$

**37.** $\dfrac{0.97}{0.1}$ $\qquad$ $0.9.7$

1 decimal place $\qquad$ Move 1 place to the right.

$\dfrac{0.97}{0.1} = 9.7$

**39.** $\dfrac{75.3}{-0.001} = -\dfrac{75.3}{0.001}$

$-\dfrac{75.3}{0.001}$ $\qquad$ $-75.300.$

3 decimal places $\qquad$ Move 3 places to the right.

$\dfrac{75.3}{-0.001} = -75,300$

**41.** $\dfrac{23,001}{100}$ $\qquad$ $230.01.$

2 zeros $\qquad$ Move 2 places to the left.

$\dfrac{23,001}{100} = 230.01$

**43.** $\quad 14 \times (82.6 + 67.9)$
$= 14 \times (150.5)$ $\quad$ Doing the calculation inside the parentheses
$= 2107$ $\quad$ Multiplying

**45.** $0.003 + 3.03 \div (-0.01) = 0.003 - 303$ $\quad$ Dividing first
$\qquad\qquad\qquad\qquad\quad = -302.997$ $\quad$ Subtracting

**47.** $\quad (4.9 - 18.6) \times 13$
$= -13.7 \times 13$ $\quad$ Doing the calculation inside the parentheses
$= -178.1$ $\quad$ Multiplying

**49.** $\quad 123.3 - 4.24 \times 1.01$
$= 123.3 - 4.2824$ $\quad$ Multiplying
$= 119.0176$ $\quad$ Subtracting

**51.** $\quad 12 \div (-0.03) - 12 \times 0.03^2$
$= 12 \div (-0.03) - 12 \times 0.0009$ $\quad$ Evaluating the exponential expression
$= -400 - 0.0108$ $\quad$ Dividing and multiplying in order from left to right
$= -400.0108$ $\quad$ Subtracting

**53.** $\quad (4 - 2.5)^2 \div 100 + 0.1 \times 6.5$
$= (1.5)^2 \div 100 + 0.1 \times 6.5$ $\quad$ Doing the calculation inside the parentheses
$= 2.25 \div 100 + 0.1 \times 6.5$ $\quad$ Evaluating the exponential expression
$= 0.0225 + 0.65$ $\quad$ Dividing and multiplying in order from left to right
$= 0.6725$ $\quad$ Adding

**55.** $\quad 6 \times 0.9 - 0.1 \div 4 + 0.2^3$
$= 6 \times 0.9 - 0.1 \div 4 + 0.008$ $\quad$ Evaluating the exponential expression
$= 5.4 - 0.025 + 0.008$ $\quad$ Multiplying and dividing in order from left to right
$= 5.383$ $\quad$ Subtracting and adding in order from left to right

**57.** $\quad 12^2 \div (12 + 2.4) - [(2 - 2.4) \div 0.8]$
$= 12^2 \div (12 + 2.4) - [-0.4 \div 0.8]$ $\quad$ Doing the calculations in the innermost parentheses first
$= 12^2 \div 14.4 - [-0.5]$ $\quad$ Doing the calculations inside the parentheses
$= 12^2 \div 14.4 + 0.5$ $\quad$ Simplifying
$= 144 \div 14.4 + 0.5$ $\quad$ Evaluating the exponential expression
$= 10 + 0.5$ $\quad$ Dividing
$= 10.5$ $\quad$ Adding

**59.** We add the populations and divide by the number of addends, 6. Note that the populations are in billions.

$(2.565 + 3.050 + 3.721 + 4.477 + 5.320 + 6.241) \div 6 =$
$25.374 \div 6 = 4.229$

The average population was 4.229 billion.

**61.** We add the temperature for the years 1992 through 1996 and then divide by the number of addends, 5:

$(59.23 + 59.36 + 59.56 + 59.72 + 59.58) \div 5 = 297.45 \div 5 = 59.49$

The average temperature for the years 1992 through 1996 was 59.49°F.

**63.** $-4\dfrac{1}{3} + 7\dfrac{5}{6}$

Since $7\dfrac{5}{6}$ is greater than $4\dfrac{1}{3}$, the answer will be negative. Find the difference in absolute values.

$$7\dfrac{5}{6} = \quad 7\dfrac{5}{6} \quad = \quad 7\dfrac{5}{6}$$

$$-4\dfrac{1}{3} = -4\,\boxed{\dfrac{1}{3}\cdot\dfrac{2}{2}} = -4\dfrac{2}{6}$$

$$3\dfrac{3}{6} = 3\dfrac{1}{2}$$

Thus, $-4\dfrac{1}{3} + 7\dfrac{5}{6} = 3\dfrac{1}{2}$.

**65.**
$$-3x + 7 = 31$$
$$-3x + 7 - 7 = 31 - 7$$
$$-3x = 24$$
$$\dfrac{-3x}{-3} = \dfrac{24}{-3}$$
$$x = -8$$

The solution is $-8$.

**67.**
$$-28 = 5 - 3x$$
$$-28 - 5 = 5 - 3x - 5$$
$$-33 = -3x$$
$$\dfrac{-33}{-3} = \dfrac{-3x}{-3}$$
$$11 = x$$

The solution is 11.

**69.** ◈

**71.** ◈

**73.**
$$23.042(7 - 4.037 \times 1.46 - 0.932^2)$$
$$= 23.042(7 - 4.037 \times 1.46 - 0.868624)$$
$$= 23.042(7 - 5.89402 - 0.868624)$$
$$= 23.042(0.237356)$$
$$= 5.469156952$$

**75.**  $5.2738 \div 0.01 \times 1000 \div \underline{\quad} = 52.738$
$$527,380 \div \underline{\quad} = 52.738$$

We need to divide by a number that will move the decimal point 4 places to the left. Thus, the missing value is 10,000.

**77.** We divide. Note that 18.5 million $= 18.5 \times 1,000,000 = 18,500,000$.

```
                    1 8. 8 7
   9 8 0,0 0 0 | 1 8,5 0 0,0 0 0.0 0
                 9 8 0 0 0 0 0
                 ───────────
                   8 7 0 0 0 0 0
                   7 8 4 0 0 0 0
                   ───────────
                     8 6 0 0 0 0 0
                     7 8 4 0 0 0 0
                     ───────────
                       7 6 0 0 0 0 0
                       6 8 6 0 0 0 0
                       ───────────
                         7 4 0 0 0 0
```

Rounding to the nearest tenth, we get 18.9 rating points.

**79.** The period from July 20 to August 20 consists of 31 days.

The "customer charge" is $31 \times \$0.35 = \$10.85$. The "energy charge" for the first 250 kilowatt-hours is $250 \times \$0.10470 = \$26.18$ (rounded to the nearest cent).

Subtract to find the "energy charge" for the kilowatt-hours in excess of 250:

$\$67.89 - \$10.85 - \$26.18 = \$30.86$

Divide to find the number of kilowatt-hours in excess of 250:

$30.86 \div \$0.09079 = 340$ (rounded to the nearest hour)

The total number of kilowatt-hours of electricity used is $250 + 340 = 590$.

## Exercise Set 5.5

**1.** $\dfrac{5}{16} = 5 \div 16$

```
        0.3 1 2 5
   1 6 | 5.0 0 0 0
         4 8
         ───
           2 0
           1 6
           ───
             4 0
             3 2
             ───
               8 0
               8 0
               ───
                 0
```

$\dfrac{5}{16} = 0.3125$

(Note that we could also have multiplied $\dfrac{5}{16}$ by $\dfrac{625}{625}$ to get a denominator of 10,000.)

**3.**  $\dfrac{19}{40} = \dfrac{19}{40} \cdot \dfrac{25}{25}$    We use $\dfrac{25}{25}$ for 1 to get a denominator of 1000.

$$= \dfrac{475}{1000} = 0.475$$

(Note that we could also have performed the division $19 \div 40$.)

**5.** $-\dfrac{1}{5} = -\dfrac{1}{5} \cdot \dfrac{2}{2} = -\dfrac{2}{10} = -0.2$

**7.** $\dfrac{13}{20} = \dfrac{13}{20} \cdot \dfrac{5}{5} = \dfrac{65}{100} = 0.65$

**9.** $\dfrac{17}{40} = \dfrac{17}{40} \cdot \dfrac{25}{25} = \dfrac{425}{1000} = 0.425$

**11.** $\dfrac{49}{40} = \dfrac{49}{40} \cdot \dfrac{25}{25} = \dfrac{1225}{1000} = 1.225$

**13.** $-\dfrac{13}{25} = -\dfrac{13}{25} \cdot \dfrac{4}{4} = -\dfrac{52}{100} = -0.52$

**15.** $\dfrac{2502}{125} = \dfrac{2502}{125} \cdot \dfrac{8}{8} = \dfrac{20,016}{1000} = 20.016$

**17.** $\dfrac{-1}{4} = \dfrac{-1}{4} \cdot \dfrac{25}{25} = \dfrac{-25}{100} = -0.25$

**19.** $\dfrac{23}{40} = \dfrac{23}{40} \cdot \dfrac{25}{25} = \dfrac{575}{1000} = 0.575$

**21.** $-\dfrac{5}{8} = -\dfrac{5}{8} \cdot \dfrac{125}{125} = -\dfrac{625}{1000} = -0.625$

**23.** $\dfrac{37}{25} = \dfrac{37}{25} \cdot \dfrac{4}{4} = \dfrac{148}{100} = 1.48$

**25.** $\dfrac{8}{15} = 8 \div 15$

```
     0. 5 3 3
1 5 )8. 0 0 0
     7 5
     ___
       5 0
       4 5
       ___
         5 0
         4 5
         ___
           5
```

Since 5 keeps reappearing as a remainder, the digits repeat and
$\dfrac{8}{15} = 0.533\ldots$ or $0.5\overline{3}$.

**27.** $\dfrac{1}{3} = 1 \div 3$

```
    0. 3 3 3
3 )1. 0 0 0
    9
    __
    1 0
      9
    ___
      1 0
         9
      ___
         1
```

Since 1 keeps reappearing as a remainder, the digits repeat and
$\dfrac{1}{3} = 0.333\ldots$ or $0.\overline{3}$.

**29.** First consider $\dfrac{4}{3}$.

$\dfrac{4}{3} = 4 \div 3$

```
    1. 3 3
3 )4. 0 0
    3
    __
    1 0
      9
    ___
      1 0
         9
      ___
         1
```

Since 1 keeps reappearing as a remainder, the digits repeat and
$\dfrac{4}{3} = 1.333\ldots$ or $1.\overline{3}$.

Thus, $\dfrac{-4}{3} = -1.\overline{3}$.

**31.** $\dfrac{7}{6} = 7 \div 6$

```
    1. 1 6 6
6 )7. 0 0 0
    6
    __
    1 0
      6
    ___
      4 0
      3 6
      ___
        4 0
        3 6
        ___
          4
```

Since 4 keeps reappearing as a remainder, the digits repeat and
$\dfrac{7}{6} = 1.166\ldots$ or $1.1\overline{6}$.

**33.** $\dfrac{4}{7} = 4 \div 7$

```
    0. 5 7 1 4 2 8 5
7 )4. 0 0 0 0 0 0 0
    3 5
    ___
      5 0
      4 9
      ___
        1 0
           7
        ___
           3 0
           2 8
           ___
             2 0
             1 4
             ___
               6 0
               5 6
               ___
                 4 0
                 3 5
                 ___
                   5
```

Since 5 reappears as a remainder, the sequence repeats and
$\dfrac{4}{7} = 0.571428571428\ldots$ or $0.\overline{571428}$.

**35.** First consider $\dfrac{11}{12}$.

$$\dfrac{11}{12} = 11 \div 12$$

```
        0. 9 1 6 6
  1 2 ⟌ 1 1. 0 0 0 0
        1 0 8
        ———
            2 0
            1 2
            ———
              8 0
              7 2
              ———
                8 0
                7 2
                ———
                  8
```

Since 8 keeps reappearing as a remainder, the digits repeat and $\dfrac{11}{12} = 0.91666\ldots$ or $0.91\overline{6}$.

Thus, $-\dfrac{11}{12} = -0.91\overline{6}$.

**37.** Round $0.\underline{5}\boxed{3}3\,3\ldots$ to the nearest tenth.
Hundredths digit is 4 or less.

$0.\,5$            Round down.

Round $0.\,5\,\underline{3}\boxed{3}\,3\ldots$ to the nearest hundredth.
Thousandths digit is 4 or less.

$0.\,5\,3$          Round down.

Round $0.\,5\,3\,\underline{3}\boxed{3}\ldots$ to the nearest thousandth.
Ten-thousandths digit is 4 or less.

$0.\,5\,3\,3$        Round down.

**39.** Round $0.\underline{3}\boxed{3}3\,3\ldots$ to the nearest tenth.
Hundredths digit is 4 or less.

$0.\,3$            Round down.

Round $0.\,3\,\underline{3}\boxed{3}\,3\ldots$ to the nearest hundredth.
Thousandths digit is 4 or less.

$0.\,3\,3$          Round down.

Round $0.\,3\,3\,\underline{3}\boxed{3}\ldots$ to the nearest thousandth.
Ten-thousandths digit is 4 or less.

$0.\,3\,3\,3$        Round down.

**41.** Round $-1.\,\underline{3}\boxed{3}3\,3\ldots$ to the nearest tenth.
Hundredths digit is 4 or less.
Round $-1.3333\ldots$ to $-1.3$.

Round $-1.\,3\,\underline{3}\boxed{3}3\ldots$ to the nearest hundredth.
Thousandths digit is 4 or less.
Round $-1.3333\ldots$ to $-1.33$.

Round $-1.\,3\,3\,\underline{3}\boxed{3}\ldots$ to the nearest thousandth.
Ten-thousandths digit is 4 or less.

Round $-1.3333\cdots$ to $-1.333$.

**43.** Round $1.\,\underline{1}\boxed{6}6\,6\ldots$ to the nearest tenth.
Hundredths digit is 5 or more.

$1.\,2$            Round up.

Round $1.\,1\,\underline{6}\boxed{6}6\ldots$ to the nearest hundredth.
Thousandths digit is 5 or more.

$1.\,1\,7$          Round up.

Round $1.\,1\,6\,\underline{6}\boxed{6}\ldots$ to the nearest thousandth.
Ten-thousandths digit is 5 or more.

$1.\,1\,6\,7$        Round up.

**45.** $0.\overline{571428}$

Round to the nearest tenth.

$0.\underline{5}\boxed{7}1428571428\ldots$
Hundredths digit is 5 or more.

$0.6$            Round up.

Round to the nearest hundredth.

$0.5\underline{7}\boxed{1}428571428\ldots$
Thousandths digit is 4 or less.

$0.57$            Round down.

Round to the nearest thousandth.

$0.571\boxed{4}28571428\ldots$
Ten-thousandths digit is 4 or less.

$0.571$            Round down.

**47.** Round $-0.\,\underline{9}\boxed{1}6\,6\ldots$ to the nearest tenth.
Hundredths digit is 4 or less.
Round $-0.9166\cdots$ to $-0.9$.

Round $-0.\,9\,\underline{1}\boxed{6}6\ldots$ to the nearest hundredth.
Thousandths digit is 5 or more.
Round $-0.9166\cdots$ to $-0.92$.

Round $-0.\,9\,1\,\underline{6}\boxed{6}\ldots$ to the nearest thousandth.
Ten-thousandths digit is 5 or more.

Round $-0.9166\cdots$ to $-0.917$.

**49.** Round $0.\,\underline{7}\boxed{4}7\,4\ldots$ to the nearest tenth.
Hundredths digit is 4 or less.

$0.\,7$            Round down.

Round $0.\,7\,\underline{4}\boxed{7}4\ldots$ to the nearest hundredth.
Thousandths digit is 5 or more.

$0.\,7\,5$          Round up.

Round $0.\,7\,4\,\underline{7}\boxed{4}\ldots$ to the nearest thousandth.
Ten-thousandths digit is 4 or less.

$0.\,7\,4\,7$        Round down.

**51.** Round $-7.\,\underline{9}\,\boxed{6}\,6\,6\,\ldots$ to the nearest tenth.

       ⌐⎯⎯⎯ Hundredths digit is 5 or more.

Round $-7.9666\cdots$ to $-8.0$.

Round $-7.\,9\,\underline{6}\,\boxed{6}\,6\,\ldots$ to the nearest hundredth.

        ⌐⎯⎯⎯ Thousandths digit is 5 or more.

Round $-7.9666\cdots$ to $-7.97$.

Round $-7.\,9\,6\,\underline{6}\,\boxed{6}\,\ldots$ to the nearest thousandth.

       ⌐⎯⎯⎯ Ten-thousandths digit is 5 or

              more.

Round $-7.9666\cdots$ to $-7.967$.

**53.** We will use the first method discussed in the text.

$$\frac{7}{8}(10.84) = \frac{7}{8} \times \frac{10.84}{1} = \frac{7 \times 10.84}{8} = \frac{75.88}{8} = 9.485$$

**55.** We will use the second method discussed in the text.

$$\frac{47}{9}(-79.95) = 5.\overline{2}(-79.95)$$
$$\approx 5.222(-79.95)$$
$$= -417.4989$$

Note that this answer is not as accurate as those found using either of the other methods, due to rounding.

**57.** We will use the first method discussed in the text.

$$\left(\frac{1}{6}\right)0.0765 + \left(\frac{3}{4}\right)0.1124 = \frac{1}{6} \times \frac{0.0765}{1} + \frac{3}{4} \times \frac{0.1124}{1}$$
$$= \frac{0.0765}{6} + \frac{3 \times 0.1124}{4}$$
$$= \frac{0.0765}{6} + \frac{0.3372}{4}$$
$$= 0.01275 + 0.0843$$
$$= 0.09705$$

**59.** We will use the third method discussed in the text.

$$\frac{3}{4} \times 2.56 - \frac{7}{8} \times 3.94$$
$$= \frac{3}{4} \times \frac{256}{100} - \frac{7}{8} \times \frac{394}{100}$$
$$= \frac{768}{400} - \frac{2758}{800}$$
$$= \frac{768}{400} \cdot \frac{2}{2} - \frac{2758}{800}$$
$$= \frac{1536}{800} - \frac{2758}{800}$$
$$= \frac{-1222}{800} = -\frac{1222}{800}$$
$$= -\frac{2 \cdot 611}{2 \cdot 400} = \frac{2}{2} \cdot \left(-\frac{611}{400}\right)$$
$$= -\frac{611}{400}, \text{ or } -1.5275$$

**61.** We will use the second method discussed in the text.

$$5.2 \times 1\frac{7}{8} \div 0.4 = 5.2 \times 1.875 \div 0.4$$
$$= 9.75 \div 0.4$$
$$= 24.375$$

**63.** *Familiarize.* We draw a picture and recall that the formula for the area $A$ of a triangle with base $b$ and height $h$ is $A = \frac{1}{2} \times b \times h$.

1.8 m

1.2 m

*Translate.* We substitute 1.2 for $b$ and 1.8 for $h$.

$$A = \frac{1}{2} \times b \times h = \frac{1}{2} \times 1.2 \times 1.8$$

*Solve.* We carry out the computation.

$$A = \frac{1}{2} \times 1.2 \times 1.8$$
$$= \frac{1.2}{2} \times 1.8 \quad \text{Multiplying } \frac{1}{2} \text{ and } 1.2$$
$$= 0.6 \times 1.8 \quad \text{Dividing}$$
$$= 1.08 \quad \text{Multiplying}$$

*Check.* We repeat the calculations using a different method.

$$\frac{1}{2} \times 1.2 \times 1.8 = 0.5 \times (1.2 \times 1.8) = 0.5 \times 2.16 = 1.08$$

Our answer checks.

*State.* The area of the shawl is 1.08 m$^2$.

**65.** *Familiarize.* We draw a picture and recall that the formula for the area $A$ of a triangle with base $b$ and height $h$ is $A = \frac{1}{2} \times b \times h$.

3.4 cm

3.4 cm

*Translate.* We substitute 3.4 for $b$ and 3.4 for $h$.

$$A = \frac{1}{2} \times b \times h = \frac{1}{2} \times 3.4 \times 3.4$$

*Solve.* We carry out the computation.

$$A = \frac{1}{2} \times 3.4 \times 3.4$$
$$= \frac{3.4}{2} \times 3.4 \quad \text{Multiplying } \frac{1}{2} \text{ and } 3.4$$
$$= 1.7 \times 3.4 \quad \text{Dividing}$$
$$= 5.78 \quad \text{Multiplying}$$

*Check.* We repeat the calculations using a different method.

$$\frac{1}{2} \times 3.4 \times 3.4 = 0.5 \times (3.4 \times 3.4) = 0.5 \times 11.56 = 5.78$$

Our answer checks.

*State.* The area of the stamp is 5.78 cm$^2$.

**67.**
$$\begin{array}{r} 20 = 19\frac{5}{5} \\ -16\frac{3}{5} = -16\frac{3}{5} \\ \hline 3\frac{2}{5} \end{array}$$

**69.** $\frac{n}{1} = n$, for any integer $n$.

Thus, $\frac{95}{-1} = \frac{-95}{1} = -95$.

**71.**     $9 - 4 + 2 \div (-1) \cdot 6$

$= 9 - 4 - 2 \cdot 6$      Multiplying and dividing in

$= 9 - 4 - 12$      order from left to right

$= 5 - 12$      Adding and subtracting in

$= -7$      order from left to right

**73.** ◈

**75.** ◈

**77.** Using a calculator we find that $\frac{2}{7} = 2 \div 7 = 0.\overline{285714}$.

**79.** Using a calculator we find that $\frac{4}{7} = 4 \div 7 = 0.\overline{571428}$.

**81.** From the pattern $0.\overline{142857}$, $0.\overline{285714}$, $0.\overline{428571}$, $0.\overline{571428}$, $0.\overline{714285}$ we would guess that decimal notation for $\frac{6}{7}$ is $0.\overline{857142}$. This is confirmed by finding $6 \div 7$ on a calculator.

**83.** Using a calculator we find that $\frac{1}{99} = 1 \div 99 = 0.\overline{01}$.

**85.** From the pattern $0.\overline{1}$, $0.\overline{01}$, $0.\overline{001}$ we would guess that decimal notation for $\frac{1}{9999}$ is $0.\overline{0001}$. This is confirmed by finding $1 \div 9999$ on a calculator.

**87.** Substitute $\frac{22}{7}$ for $\pi$ and 1.4 for $r$.

$A = \pi r^2 \approx \frac{22}{7}(1.4)^2 \approx \frac{22}{7}(1.96) \approx \frac{22}{7} \times \frac{1.96}{1} \approx$

$\frac{22 \times 1.96}{7} \approx \frac{43.12}{7} \approx 6.16$ cm$^2$.

**89.** Substitute 3.14 for $\pi$ and $4\frac{1}{2}$ for $r$.

$A = \pi r^2 \approx 3.14\left(4\frac{1}{2}\right)^2 \approx 3.14\left(\frac{9}{2}\right)^2 \approx 3.14\left(\frac{81}{4}\right) \approx$

$\frac{3.14}{1} \times \frac{81}{4} \approx \frac{3.14 \times 81}{4} \approx \frac{254.34}{4} \approx 63.585$ yd$^2$.

When we do the calculation using the $\pi$ key on a calculator, we get 63.61725124 yd$^2$.

## Exercise Set 5.6

**1.** We are estimating the sum

$$\$109.95 + \$249.95.$$

We round both numbers to the nearest ten. The estimate is

$$\$110 + \$250 = \$360.$$

**3.** We are estimating the difference

$$\$299 - \$249.95.$$

We round both numbers to the nearest ten. The estimate is

$$\$300 - \$250 = \$50.$$

**5.** We are estimating the product

$$9 \times \$299.$$

We round $299 to the nearest ten. The estimate is

$$9 \times \$300 = \$2700.$$

**7.** We are estimating the quotient

$$\$1700 \div \$299.$$

Rounding $299, we get $300. Since $1700 is close to $1800, which is a multiple of $300, the estimate is

$$\$1800 \div \$300 = 6.$$

We could also have used the estimate

$$\$1500 \div \$300 = 5.$$

**9.** This is about $0.0 + 1.3 + 0.3$, so the answer is about 1.6.

**11.** This is about $6 + 0 + 0$, so the answer is about 6.

**13.** This is about $52 + 1 + 7$, so the answer is about 60.

**15.** This is about $2.7 - 0.4$, so the answer is about 2.3.

**17.** This is about $200 - 20$, so the answer is about 180.

**19.** This is about $50 \times 8$, rounding 49 to the nearest ten and 7.89 to the nearest one, so the answer is about 400. Answer (a) is correct.

**21.** This is about $100 \times 0.08$, rounding 98.4 to the nearest ten and 0.083 to the nearest hundredth, so the answer is about 8. Answer (c) is correct.

**23.** This is about $4 \div 4$, so the answer is about 1. Answer (b) is correct.

**25.** This is about $75 \div 25$, so the answer is about 3. Answer (b) is correct.

**27.** We estimate the quotient by rounding the total revenue to 56 million and the average revenue to 7000.

53.6 million $\div\ 6716 = 53,600,000 \div 6716 \approx$

$56,000,000 \div 7000 \approx 8000$

About 8000 screens were showing the movie.

**29.**

$$
\begin{array}{r}
3 \\
3\,\overline{\big)\,9} \\
3\,\overline{\big)\,2\ 7} \\
2\,\overline{\big)\,5\ 4} \\
2\,\overline{\big)\,1\ 0\ 8}
\end{array}
$$

$108 = 2 \cdot 2 \cdot 3 \cdot 3 \cdot 3$

**31.**

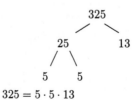

$325 = 5 \cdot 5 \cdot 13$

**33.** $\dfrac{125}{400} = \dfrac{25 \cdot 5}{25 \cdot 16} = \dfrac{25}{25} \cdot \dfrac{5}{16} = \dfrac{5}{16}$

**35.** $\dfrac{72}{81} = \dfrac{9 \cdot 8}{9 \cdot 9} = \dfrac{9}{9} \cdot \dfrac{8}{9} = \dfrac{8}{9}$

**37.** ◈

**39.** ◈

**41.** We round 14,973.35 to 15,000 and 298.75 to 300. The estimate is $15,000 \div 300 = 50$. The estimate is not close to the result given, so the decimal point was not placed correctly.

**43.** Round the first number to the nearest 10 and the other two numbers to the nearest one. The estimate is $30 \div 5 - 3 = 6 - 3 = 3$. The estimate is close to the result given, so the decimal point was placed correctly.

## Exercise Set 5.7

**1.** $5x = 27$

$\dfrac{5x}{5} = \dfrac{27}{5}$   Dividing by 5 on both sides

$x = 5.4$

Check:    $5x = 27$

$5(5.4)\ ?\ 27$

$27\ |\ 27$    TRUE

The solution is 5.4.

**3.** $100t = 52.39$

$\dfrac{100t}{100} = \dfrac{52.39}{100}$   Dividing by 100 on both sides

$t = 0.5239$

Check:    $100t = 52.39$

$100(0.5239)\ ?\ 52.39$

$52.39\ |\ 52.39$    TRUE

The solution is 0.5239.

**5.** $-23.4 = 5.2a$

$\dfrac{-23.4}{5.2} = \dfrac{5.2a}{5.2}$   Dividing by 5.2 on both sides

$-4.5 = a$

The solution is $-4.5$.

**7.** $-9.2x = -94.76$

$\dfrac{-9.2x}{-9.2} = \dfrac{-94.76}{-9.2}$

$x = 10.3$

The solution is 10.3.

**9.** $t - 19.27 = 24.51$

$t - 19.27 + 19.27 = 24.51 + 19.27$   Adding 19.27 on both sides

$t = 43.78$

To check, note that $43.78 - 19.27 = 24.51$. The solution is 43.78.

**11.** $4.1 = -3.6 + n$

$4.1 + 36 = -3.6 + n + 3.6$   Adding 3.6 on both sides

$7.7 = n$

To check, note that $-3.6 + 7.7 = 4.1$. The solution is 7.7.

**13.** $x + 13.9 = 4.2$

$x + 13.9 - 13.9 = 4.2 - 13.9$   Adding $-13.9$ (or subtracting 13.9) on both sides

$x = -9.7$

The solution is $-9.7$.

**15.** $4x - 7 = 13$

$4x - 7 + 7 = 13 + 7$   Adding 7 on both sides

$4x = 20$    Simplifying

$\dfrac{4x}{4} = \dfrac{20}{4}$    Dividing by 4 on both sides

$x = 5$

Check:    $4x - 7 = 13$

$4 \cdot 5 - 7\ ?\ 13$

$20 - 7\ |$

$13\ |\ 13$    TRUE

The solution is 5.

**17.** $7.1x - 9.3 = 8.45$

$7.1x - 9.3 + 9.3 = 8.45 + 9.3$   Adding 9.3 on both sides

$7.1x = 17.75$

$\dfrac{7.1x}{7.1} = \dfrac{17.75}{7.1}$    Dividing by 7.1 on both sides

$x = 2.5$

Check:    $7.1x - 9.3 = 8.45$

$7.1(2.5) - 9.3\ ?\ 8.45$

$17.75 - 9.3\ |$

$8.45\ |\ 8.45$    TRUE

The solution is 2.5.

**19.** $12.4 + 3.7t = 2.04$

$12.4 + 3.7t - 12.4 = 2.04 - 12.4$   Subtracting 12.04 on both sides

$3.7t = -10.36$

$\dfrac{3.7t}{3.7} = \dfrac{-10.36}{3.7}$    Dividing by 3.7 on both sides

$t = -2.8$

Check:    $12.4 + 3.7t = 2.04$

$12.4 + 3.7(-2.8)\ ?\ 2.04$

$12.4 - 10.36\ |$

$2.04\ |\ 2.04$    TRUE

The solution is $-2.8$.

**21.** $-26.05 = 7.5x + 9.2$

$-26.05 - 9.2 = 7.5x + 9.2 - 9.2$

$-35.25 = 7.5x$

$-4.7 = x$

The solution is $-4.7$.

**23.**
$$-4.2x + 3.04 = -4.1$$
$$-4.2x + 3.04 - 3.04 = -4.1 - 3.04$$
$$-4.2x = -7.14$$
$$\frac{-4.2x}{-4.2} = \frac{-7.14}{-4.2}$$
$$x = 1.7$$

The solution is 1.7.

**25.**
$$9x - 6 = 5x + 30$$
$$9x - 6 + 6 = 5x + 30 + 6 \quad \text{Adding 6 on both sides}$$
$$9x = 5x + 36$$
$$9x - 5x = 5x + 36 - 5x \quad \text{Subtracting } 5x \text{ on both sides}$$
$$4x = 36$$
$$\frac{4x}{4} = \frac{36}{4} \quad \text{Dividing by 4 on both sides}$$
$$x = 9$$

Check:
$$\begin{array}{c|c} \multicolumn{2}{c}{9x - 6 = 5x + 30} \\ \hline 9 \cdot 9 - 6 \ ? \ 5 \cdot 9 + 30 \\ 81 - 6 \ \big| \ 45 + 30 \\ 75 \ \big| \ 75 \qquad \text{TRUE} \end{array}$$

The solution is 9.

**27.**
$$3x + 15 = 11x + 5$$
$$3x + 15 - 5 = 11x + 5 - 5 \quad \text{Subtracting 5 on both sides}$$
$$3x + 10 = 11x$$
$$3x + 10 - 3x = 11x - 3x \quad \text{Subtracting } -3x \text{ on both sides}$$
$$10 = 8x$$
$$\frac{10}{8} = \frac{8x}{8} \quad \text{Dividing by 8 on both sides}$$
$$1.25 = x$$

Check:
$$\begin{array}{c|c} \multicolumn{2}{c}{3x + 15 = 11x + 5} \\ \hline 3(1.25) + 15 \ ? \ 11(1.25) + 5 \\ 3.75 + 15 \ \big| \ 13.75 + 5 \\ 18.75 \ \big| \ 18.75 \qquad \text{TRUE} \end{array}$$

The solution is 1.25.

**29.**
$$6y - 5 = 8 + 10y$$
$$6y - 5 - 6y = 8 + 10y - 6y \quad \text{Subtracting } 6y \text{ on both sides}$$
$$-5 = 8 + 4y$$
$$-5 - 8 = 8 + 4y - 8 \quad \text{Subtracting 8 on both sides}$$
$$-13 = 4y$$
$$\frac{-13}{4} = \frac{4y}{4} \quad \text{Dividing by 4 on both sides}$$
$$-3.25 = y$$

The solution is $-3.25$.

**31.**
$$5.9x + 67 = 7.6x + 16$$
$$5.9x + 67 - 16 = 7.6x + 16 - 16$$
$$5.9x + 51 = 7.6x$$
$$5.9x + 51 - 5.9x = 7.6x - 5.9x$$
$$51 = 1.7x$$
$$\frac{51}{1.7} = \frac{1.7x}{1.7}$$
$$30 = x$$

The solution is 30.

**33.**
$$7.8a + 2 = 2.4a + 19.28$$
$$7.8a + 2 - 2 = 2.4a + 19.28 - 2$$
$$7.8a = 2.4a + 17.28$$
$$7.8a - 2.4a = 2.4a + 17.28 - 2.4a$$
$$5.4a = 17.28$$
$$\frac{5.4a}{5.4} = \frac{17.28}{5.4}$$
$$a = 3.2$$

The solution is 3.2

**35.**
$$5(x + 2) = 3x + 18$$
$$5x + 10 = 3x + 18 \quad \text{Using the distributive law}$$
$$5x + 10 - 10 = 3x + 18 - 10$$
$$5x = 3x + 8$$
$$5x - 3x = 3x + 8 - 3x$$
$$2x = 8$$
$$\frac{2x}{2} = \frac{8}{2}$$
$$x = 4$$

Check:
$$\begin{array}{c|c} \multicolumn{2}{c}{5(x + 2) = 3x + 18} \\ \hline 5(4 + 2) \ ? \ 3 \cdot 4 + 18 \\ 5(6) \ \big| \ 12 + 18 \\ 30 \ \big| \ 30 \qquad \text{TRUE} \end{array}$$

The solution is 4.

**37.**
$$2(x + 3) = 4x - 11$$
$$2x + 6 = 4x - 11 \quad \text{Using the distributive law}$$
$$2x + 6 - 2x = 4x - 11 - 2x$$
$$6 = 2x - 11$$
$$6 + 11 = 2x - 11 + 11$$
$$17 = 2x$$
$$\frac{17}{2} = \frac{2x}{2}$$
$$8.5 = x$$

Check:
$$\begin{array}{c|c} \multicolumn{2}{c}{2(x + 3) = 4x - 11} \\ \hline 2(8.5 + 3) \ ? \ 4 \cdot 8.5 - 11 \\ 2(11.5) \ \big| \ 34 - 11 \\ 23 \ \big| \ 23 \qquad \text{TRUE} \end{array}$$

The solution is 8.5.

**39.**
$$2a + 17 = 12(a - 1)$$
$$2a + 17 = 12a - 12 \quad \text{Using the distributive law}$$
$$2a + 17 - 2a = 12a - 12 - 2a$$
$$17 = 10a - 12$$
$$17 + 12 = 10a - 12 + 12$$
$$29 = 10a$$
$$\frac{29}{10} = \frac{10a}{10}$$
$$2.9 = a$$

Check:
$$\begin{array}{c|c} \multicolumn{2}{c}{2a + 17 = 12(a - 1)} \\ \hline 2 \cdot 2.9 + 17 \;\; ? & 12(2.9 - 1) \\ 5.8 + 17 & 12(1.9) \\ 22.8 & 22.8 \end{array} \qquad \text{TRUE}$$

The solution is 2.9.

**41.**
$$2(x + 7.3) = 6x - 0.83$$
$$2x + 14.6 = 6x - 0.83 \quad \text{Using the distributive law}$$
$$2x + 14.6 + 0.83 = 6x - 0.83 + 0.83$$
$$2x + 15.43 = 6x$$
$$2x + 15.43 - 2x = 6x - 2x$$
$$15.43 = 4x$$
$$3.8575 = x$$

The solution is 3.8575.

**43.**
$$-7.37 - 3.2t = 4.9(t + 6.1)$$
$$-7.37 - 3.2t = 4.9t + 29.89$$
$$-7.37 - 3.2t + 3.2t = 4.9t + 29.89 + 3.2t$$
$$-7.37 = 8.1t + 29.89$$
$$-7.37 - 29.89 = 8.1t + 29.89 - 29.89$$
$$-37.26 = 8.1t$$
$$-4.6 = t$$

The solution is $-4.6$.

**45.**
$$9(x - 4) + 8 = 4x + 7$$
$$9x - 36 + 8 = 4x + 7$$
$$9x - 28 = 4x + 7 \qquad \text{Collecting like terms}$$
$$9x - 28 + 28 = 4x + 7 + 28$$
$$9x = 4x + 35$$
$$9x - 4x = 4x + 35 - 4x$$
$$5x = 35$$
$$\frac{5x}{5} = \frac{35}{5}$$
$$x = 7$$

The solution is 7.

**47.**
$$34(5 - 3.5x) = 12(3x - 8) + 653.5$$
$$170 - 119x = 36x - 96 + 653.5$$
$$170 - 119x = 36x + 557.5$$
$$170 - 119x - 170 = 36x + 557.5 - 170$$
$$-119x = 36x + 387.5$$
$$-119x - 36x = 36x + 387.5 - 36x$$
$$-155x = 387.5$$
$$\frac{-155}{-155}x = \frac{387.5}{-155}$$
$$x = -2.5$$

The solution is $-2.5$.

**49.** $\dfrac{n}{n} = 1$, for any integer $n$ that is not 0.

Thus, $\dfrac{-43}{-43} = 1$.

**51.**
$$\frac{3}{25} - \frac{7}{10} = \frac{3}{25} \cdot \frac{2}{2} - \frac{7}{10} \cdot \frac{5}{5} \qquad \text{The LCM is 50.}$$
$$= \frac{6}{50} - \frac{35}{50}$$
$$= -\frac{29}{50}$$

**53.** We add in order from left to right.
$$-17 + 24 + (-9) = 7 + (-9) = -2$$

**55.** ◇

**57.** ◇

**59.**
$$8.701(3.4 - 5.1x) - 89.321 = 5.401x + 74.65787$$
$$29.5834 - 44.3751x - 89.321 = 5.401x + 74.65787$$
$$-59.7376 - 44.3751x = 5.401x + 74.65787$$
$$-59.7376 - 44.3751x + 44.3751x = 5.401x + 74.65787 + 44.3751x$$
$$-59.7376 = 49.7761x + 74.65787$$
$$-59.7376 - 74.65787 = 49.7761x + 74.65787 - 74.65787$$
$$-134.39547 = 49.7761x$$
$$\frac{-134.39547}{49.7761} = \frac{49.7761x}{49.7761}$$
$$-2.7 = x$$

The solution is $-2.7$.

**61.**
$$14(2.5x - 3) + 9x + 5 = 4(3.25 - x) + 2[5x - 3(x + 1)]$$
$$14(2.5x - 3) + 9x + 5 = 4(3.25 - x) + 2[5x - 3x - 3]$$
$$14(2.5x - 3) + 9x + 5 = 4(3.25 - x) + 2[2x - 3]$$
$$35x - 42 + 9x + 5 = 13 - 4x + 4x - 6$$
$$44x - 37 = 7$$
$$44x - 37 + 37 = 7 + 37$$
$$44x = 44$$
$$\frac{44x}{44} = \frac{44}{44}$$
$$x = 1$$

The solution is 1.

## Exercise Set 5.8

**1. Familiarize.** Repeated addition fits this situation. We let $C$ = the cost of 7 shirts.

$$\underbrace{\boxed{\$32.98} + \boxed{\$32.98} + \cdots + \boxed{\$32.98}}_{\text{7 addends}}$$

**Translate.**

| Price per shirt | times | Number of shirts | is | Total cost |
|---|---|---|---|---|
| ↓ | ↓ | ↓ | ↓ | ↓ |
| 32.98 | × | 7 | = | $C$ |

**Solve.** We carry out the multiplication.

$$\begin{array}{r} 3\,2.9\,8 \\ \times \qquad 7 \\ \hline 2\,3\,0.8\,6 \end{array}$$

Thus, $C = 230.86$.

**Check.** We obtain a partial check by rounding and estimating:

$$32.98 \times 7 \approx 30 \times 7 = 210 \approx 230.86.$$

**State.** Seven shirts cost $230.86.

**3. Familiarize.** Repeated addition fits this situation. We let $c$ = the cost of 20.4 gal of gasoline.

**Translate.**

| Cost per gallon | times | Number of gallons | is | Total cost |
|---|---|---|---|---|
| ↓ | ↓ | ↓ | ↓ | ↓ |
| 1.299 | · | 20.4 | = | $c$ |

**Solve.** We carry out the multiplication.

$$\begin{array}{r} 1.2\,9\,9 \\ \times \quad 2\,0.4 \\ \hline 5\,1\,9\,6 \\ 2\,5\,9\,8\,0\,0 \\ \hline 2\,6.4\,9\,9\,6 \end{array}$$

Thus, $c = 26.4996$.

**Check.** We obtain a partial check by rounding and estimating:

$$1.299 \times 20.4 \approx 1.3 \times 20 = 26 \approx 26.4996.$$

**State.** We round $26.4996 to the nearest cent and find that the cost of the gasoline is $26.50.

**5. Familiarize.** We visualize the situation. We let $c$ = the amount of change.

| $50 | |
|---|---|
| $44.68 | $c$ |

**Translate.** This is a "take-away" situation.

| Amount paid | minus | Amount of purchase | is | Amount of change |
|---|---|---|---|---|
| ↓ | ↓ | ↓ | ↓ | ↓ |
| $50 | − | $44.68 | = | $c$ |

**Solve.** To solve the equation we carry out the subtraction.

$$\begin{array}{r} {\scriptstyle 4\ \ 9\ \ 9\ \ 10} \\ 5\,0.\,0\!\!\!/\;0\!\!\!/ \\ -\,4\,4.\,6\,8 \\ \hline 5.\,3\,2 \end{array}$$

Thus, $c = \$5.32$.

**Check.** We check by adding 5.32 to 44.68 to get 50. This checks.

**State.** The change was $5.32.

**7. Familiarize.** We visualize the situation. We let $b$ = the number of milligrams of blood that is left.

| 17.85 mg | |
|---|---|
| 9.68 mg | $b$ |

**Translate.** This is a "take-away" situation.

| Amount drawn | minus | Amount used | is | Amount left |
|---|---|---|---|---|
| ↓ | ↓ | ↓ | ↓ | ↓ |
| 17.85 | − | 9.68 | = | $b$ |

**Solve.** We carry out the subtraction.

$$\begin{array}{r} {\scriptstyle 1\ \ 17\ \ 7\ \ 15} \\ 1\,7.\,8\!\!\!/\;5\!\!\!/ \\ -\ \ 9.\,6\,8 \\ \hline 8.\,1\,7 \end{array}$$

Thus, $b = 8.17$.

**Check.** We add 8.17 to 9.68 to get 17.85. The answer checks.

**State.** There will be 8.17 mg of blood left.

**9. Familiarize.** Let $p$ = the amount of each payment.

**Translate.**

| Amount of payment | is | Total owed | divided by | Number of payments |
|---|---|---|---|---|
| ↓ | ↓ | ↓ | ↓ | ↓ |
| $p$ | = | 4425 | ÷ | 12 |

**Solve.** We carry out the division.

$$\begin{array}{r} 3\,6\,8.\,7\,5 \\ 1\,2\,\overline{)\,4\,4\,2\,5.\,0\,0} \\ 3\,6 \phantom{00000} \\ \hline 8\,2 \phantom{0000} \\ 7\,2 \phantom{0000} \\ \hline 1\,0\,5 \phantom{000} \\ 9\,6 \phantom{000} \\ \hline 9\,0 \phantom{00} \\ 8\,4 \phantom{00} \\ \hline 6\,0 \phantom{0} \\ 6\,0 \phantom{0} \\ \hline 0 \end{array}$$

Thus, $p = 368.75$.

**Check**. We round and estimate.

$$4425 \div 12 \approx 4000 \div 10 = 400 \approx 368.75$$

**State**. Each payment is $368.75.

**11. Familiarize**. We are combining amounts. Let $t =$ the total amount of the injections, in mg.

**Translate**.

| First amount | + | Second amount | + | Third amount | + | Fourth amount | is | Total amount |
|---|---|---|---|---|---|---|---|---|
| ↓ | ↓ | ↓ | ↓ | ↓ | ↓ | ↓ | ↓ | ↓ |
| 0.25 | + | 0.4 | + | 0.5 | + | 0.5 | = | $t$ |

**Solve**. We carry out the addition.

$$\begin{array}{r} {\scriptstyle 1} \\ 0.2\,5 \\ 0.4\,0 \\ 0.5\,0 \\ \underline{0.5\,0} \\ 1.6\,5 \end{array}$$

Thus, $t = 1.65$.

**Check**. We repeat the calculation. The answer checks.

**State**. The total amount of the injections was 1.65 mg.

**13. Familiarize**. We visualize the situation. Let $w =$ each winner's share.

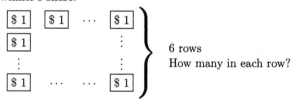

6 rows
How many in each row?

**Translate**.

| Total prize | ÷ | Number of winners | = | Each winner's share |
|---|---|---|---|---|
| ↓ | ↓ | ↓ | ↓ | ↓ |
| 127,315 | ÷ | 6 | = | $w$ |

**Solve**. We carry out the division.

$$\begin{array}{r} 2\,1,2\,1\,9.1\,6\,6 \\ 6\,\overline{)1\,2\,7,3\,1\,5.0\,0\,0} \\ \underline{1\,2\,0\,0\,0\,0} \\ 7\,3\,1\,5 \\ \underline{6\,0\,0\,0} \\ 1\,3\,1\,5 \\ \underline{1\,2\,0\,0} \\ 1\,1\,5 \\ \underline{6\,0} \\ 5\,5 \\ \underline{5\,4} \\ 1\,0 \\ \underline{6} \\ 4\,0 \\ \underline{3\,6} \\ 4\,0 \\ \underline{3\,6} \\ 4 \end{array}$$

Rounding to the nearest cent, or hundredth, we get $w = 21,219.17$.

**Check**. We can repeat the calculation. The answer checks.

**State**. Each winner's share is $21,219.17.

**15. Familiarize**. We draw a picture, letting $A =$ the area.

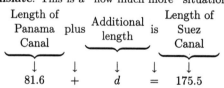

73.2 cm

61.8 cm

**Translate**. We use the formula $A = l \cdot w$.

$$A = 73.2 \times 61.8$$

**Solve**. We carry out the multiplication.

$$\begin{array}{r} 7\,3.\,2 \\ \times\ 6\,1.\,8 \\ \hline 5\,8\,5\,6 \\ 7\,3\,2\,0 \\ \underline{4\,3\,9\,2\,0\,0} \\ 4\,5\,2\,3.\,7\,6 \end{array}$$

Thus, $A = 4523.76$.

**Check**. We obtain a partial check by rounding and estimating:

$$73.2 \times 61.8 \approx 70 \times 60 = 4200 \approx 4523.76$$

**State**. The area is 4523.76 cm$^2$.

**17. Familiarize**. We visualize the situation. Let $d =$ the number of kilometers by which the length of the Suez Canal exceeds the length of the Panama Canal.

| 175.5 km | |
|---|---|
| 81.6 km | $d$ |

**Translate**. This is a "how much more" situation.

| Length of Panama Canal | plus | Additional length | is | Length of Suez Canal |
|---|---|---|---|---|
| ↓ | ↓ | ↓ | ↓ | ↓ |
| 81.6 | + | $d$ | = | 175.5 |

**Solve**. We subtract 81.6 on both sides.

$$d = 175.5 - 81.6$$
$$d = 93.9$$

$$\begin{array}{r} {\scriptstyle 17\ \ 4\ 15} \\ \cancel{1}\,\cancel{7}\,\cancel{5}.\,\cancel{5} \\ -\ 8\,1.\,6 \\ \hline 9\,3.\,9 \end{array}$$

**Check**. We check by adding 93.9 to 81.6 to get 175.5. The answer checks.

**State**. The Suez Canal is 93.9 km longer than the Panama Canal.

**19. Familiarize**. This is a two-step problem. First we find the number of thousands in the assessed value. Let $n$ represent that number.

*Translate and Solve.*

| Assessed value | divided by | 1000 | is | Number of thousands |
|---|---|---|---|---|
| ↓ | ↓ | ↓ ↓ | | ↓ |
| 124, 500 | ÷ | 1000 = | | $n$ |

To solve the equation we carry out the division. Recall that when we divide a number by 1000 we move the decimal point 3 places to the left.

$$124,500 \div 1000 = \frac{124,500}{1000} = 124.5$$

Now let $t =$ the amount of the taxes. A multiplication corresponds to the situation.

$$t = 124.5 \times 7.68$$

We carry out the multiplication.

```
      1 2 4. 5
    ×  7. 6 8
    ─────────
      9 9 6 0
    7 4 7 0 0
  8 7 1 5 0 0
  ───────────
  9 5 6. 1 6 0
```

Thus, $t = 956.16$.

**Check**. We repeat the calculations. The answer checks.

**State**. The Colavitos pay $956.16 in taxes.

**21. Familiarize**. We let $d =$ the distance around the figure.

**Translate**. We are combining lengths.

| The sum of the lengths of the 5 sides | is | the distance around the figure. |
|---|---|---|
| ↓ | ↓ | ↓ |
| $8.9 + 23.8 + 4.7 + 22.1 + 18.6$ | = | $d$ |

**Solve**. To solve we carry out the addition.

```
    2 3
    8.9
  2 3.8
    4.7
  2 2.1
+ 1 8.6
───────
  7 8.1
```

Thus, $d = 78.1$.

**Check**. To check we can repeat the addition. We can also check by rounding:

$$8.9 + 23.8 + 4.7 + 22.1 + 18.6 \approx 9 + 24 + 5 + 22 + 19 = 79 \approx 78.1$$

**State**. The distance around the figure is 78.1 cm.

**23. Familiarize**. The perimeter consists of 6 vertical segments, each of which is 2.5 cm and 6 horizontal segments, each of which is 2.25 cm. Let $P =$ the perimeter.

**Translate**.

| Total length of vertical segments | plus | Total length of horizontal segments | is | Perimeter |
|---|---|---|---|---|
| ↓ | ↓ | ↓ | ↓ | ↓ |
| $6(2.5)$ | + | $6(2.25)$ | = | $P$ |

**Solve**. We carry out the calculations.

$$6(2.5) + 6(2.25) = P$$
$$15 + 13.5 = P$$
$$28.5 = P$$

**Check**. We repeat the calculations. The answer checks.

**State**. The perimeter is 28.5 cm.

**25. Familiarize**. This is a multistep problem. First we find the sum $s$ of the two 0.8 cm segments. Then we use this length to find $d$.

**Translate and Solve.**

| Length of one small segment | plus | Length of other small segment | is | Total length. |
|---|---|---|---|---|
| ↓ | ↓ | ↓ | ↓ | ↓ |
| 0.8 | + | 0.8 | = | $s$ |

To solve we carry out the addition.

```
    1
  0. 8
+ 0. 8
──────
  1. 6
```

Thus, $s = 1.6$.

Now we find $d$.

| Total length of smaller segments | plus | length of $d$ | is | 3.91 cm |
|---|---|---|---|---|
| ↓ | ↓ | ↓ | ↓ | ↓ |
| 1.6 | + | $d$ | = | 3.91 |

To solve we subtract 1.6 on both sides of the equation.

$d = 3.91 - 1.6$
$d = 2.31$

```
  3.9 1
− 1.6 0
───────
  2.3 1
```

**Check**. We repeat the calculations.

**State**. The length $d$ is 2.31 cm.

**27. Familiarize**. This is a two-step problem. First, we find the number of miles that have been driven between fillups. This is a "how-much-more" situation. We let $n =$ the number of miles driven.

**Translate and Solve.**

| First odometer reading | plus | Number of miles driven | is | Second odometer reading |
|---|---|---|---|---|
| ↓ | ↓ | ↓ | ↓ | ↓ |
| 26, 342.8 | + | $n$ | = | 26, 736.7 |

To solve the equation we subtract 26,342.8 on both sides.

$n = 26,736.7 - 26,342.8$
$n = 393.9$

```
  2 6, 7 3 6.7
− 2 6, 3 4 2.8
─────────────
         3 9 3.9
```

Second, we divide the total number of miles driven by the number of gallons. This gives us $m =$ the number of miles per gallon.

$$393.9 \div 19.5 = m$$

To find the number $m$, we divide.

$$\begin{array}{r} 2\,0\,.2 \\ 1\,9.5_\wedge\overline{)3\,9\,3\,.\,9_\wedge 0} \\ \underline{3\,9\,0\,0} \\ 3\,9\phantom{.}\,0 \\ \underline{3\,9\phantom{.}\,0} \\ 0 \end{array}$$

Thus, $m = 20.2$.

**Check.** To check, we first multiply the number of miles per gallon times the number of gallons:

$$19.5 \times 20.2 = 393.9$$

Then we add 393.9 to 26,342.8:

$$26,342.8 + 393.9 = 26,736.7$$

The number 20.2 checks.

**State.** The van gets 20.2 miles per gallon.

**29. Familiarize.** This is a two-step problem. First, we find the total cost of the CD's. Let $c$ represent this amount. **Translate and Solve.**

$$\underbrace{\text{Price per CD}}_{\downarrow} \underbrace{\text{times}}_{\downarrow} \underbrace{\text{Number purchased}}_{\downarrow} \underbrace{\text{is}}_{\downarrow} \underbrace{\text{Total cost}}_{\downarrow}$$
$$12.99 \quad\times\quad 3 \quad=\quad c$$

To solve the equation we carry out the multiplication.

$$\begin{array}{r} 1\,2.\,9\,9 \\ \times \phantom{12.9}3 \\ \hline 3\,8.\,9\,7 \end{array}$$

Thus, $c = 38.97$.

Now we find the amount of the change. This is a "take away" situation. Let $m$ = the amount of the change. We subtract.

$$\begin{array}{r} {\scriptstyle 4\ \ 9\ \ 9\ \ 10} \\ \cancel{5\,0.\,0\,0} \\ -\,3\,8.\,9\,7 \\ \hline 1\,1.\,0\,3 \end{array}$$

Thus, $m = 11.03$.

**Check.** We round and estimate.

$$12.99 \times 3 \approx 13 \times 3 = 39 \approx 38.97$$
$$50 - 38.97 \approx 50 - 40 = 10 \approx 11.03$$

**State.** Roberto received $11.03 in change.

**31. Familiarize.** We make and label a drawing. The question deals with a rectangle and a circle, so we also list the relevant area formulas. We let $d$ = the amount of decking needed.

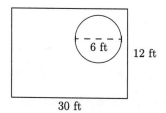

Area of a rectangle with length $l$ and width $w$:
$A = l \times w$

Area of a circle with radius $r$: $A = \pi r^2$, where $\pi \approx 3.14$

**Translate.** We subtract the area of the circle from the area of the rectangle. Recall that a circle's radius is half of its diameter.

$$\underbrace{\text{Area of rectangle}}_{\downarrow} \underbrace{\text{minus}}_{\downarrow} \underbrace{\text{Area of circle}}_{\downarrow} \underbrace{\text{is}}_{\downarrow} \underbrace{\text{Area covered by decking}}_{\downarrow}$$
$$30 \times 12 \quad-\quad 3.14\left(\frac{6}{2}\right)^2 = \quad d$$

**Solve.** We carry out the computations.

$$30 \times 12 - 3.14\left(\frac{6}{2}\right)^2 = d$$
$$30 \times 12 - 3.14(3)^2 = d$$
$$30 \times 12 - 3.14 \times 9 + d$$
$$360 - 28.26 = d$$
$$331.74 = d$$

**Check.** We can repeat the calculations. Also note that 331.74 is less than the area of the yard but more than the area of the flower garden. This agrees with the impression given by our drawing.

**State.** The amount of decking needed is 331.74 ft$^2$.

**33. Familiarize.** We make and label a drawing. The question deals with a rectangle and a circle, so we also list the relevant area formulas. We let $w$ = the amount of wood left over.

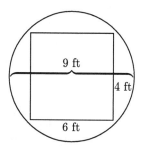

Area of rectangle with length $l$ and width $w$: $A = l \times w$

Area of circle with radius $r$: $A = \pi r^2$, where $\pi \approx 3.14$

**Translate.** We subtract the area of the rectangle from the area of the circle. Recall that a circle's radius is half of its diameter, or width.

$$\underbrace{\text{Area of circle}}_{\downarrow} \underbrace{\text{minus}}_{\downarrow} \underbrace{\text{Area of rectangle}}_{\downarrow} \underbrace{\text{is}}_{\downarrow} \underbrace{\text{Area left over}}_{\downarrow}$$
$$3.14\left(\frac{9}{2}\right)^2 \quad-\quad 6 \times 4 \quad=\quad w$$

**Solve.** We carry out the computations.

$$3.14\left(\frac{9}{2}\right)^2 - 6 \times 4 = w$$
$$3.14(4.5)^2 - 6 \cdot 4 = w$$
$$3.14(20.25) - 6 \times 4 = w$$
$$63.585 - 24 = w$$
$$39.585 = w$$

**Check.** We can repeat the calculations. Also note that 39.585 is less than the area of the rectangle which in turn is less than the area of the circle. This agrees with the impression given by our drawing.

*State*. The amount of wood left over is 39.585 ft$^2$.

35. *Familiarize*. Zachary worked $53-40$, or 13 hr of overtime. His total pay for the first 40 hr was $40 \times \$8.50$ and his total overtime pay was $13 \times \$12.75$. Let $t =$ the total amount he earned during the week.

*Translate*. We are combining amounts.

$$\underbrace{\text{Regular pay}}_{40 \times 8.50} \;\; \underbrace{\text{plus}}_{+} \;\; \underbrace{\text{Overtime pay}}_{13 \times 12.75} \;\; \underbrace{\text{is}}_{=} \;\; \underbrace{\text{Total earnings}}_{t}$$

*Solve*. We carry out the computations.

$$40 \times 8.50 + 13 \times 12.75 = t$$
$$340 + 165.75 = t$$
$$505.75 = t$$

*Check*. We repeat the calculations. The answer checks.

*State*. Zachary earned $505.75.

37. *Familiarize*. We let $m =$ the number of miles that can be driven within an $80 budget. We convert 10 cents to $0.10 so that only one unit, dollars, is used.

*Translate*. We translate to an equation.

$$\underbrace{\text{Daily rate}}_{\$34.95} \;\; \underbrace{\text{plus}}_{+} \;\; \underbrace{\text{Cost per mile}}_{\$0.10} \;\; \underbrace{\text{times}}_{\cdot} \;\; \underbrace{\text{Number of miles driven}}_{m} \;\; \underbrace{\text{is}}_{=} \;\; \underbrace{\text{Total cost}}_{80}$$

*Solve*. We solve the equation.

$$34.95 + 0.10m = 80$$
$$0.10m = 45.05 \quad \text{Subtracting 34.95 on both sides}$$
$$m = \frac{45.05}{0.10} \quad \text{Dividing by 0.10 on both sides}$$

$$m = 450.5$$

*Check*. The mileage cost is found by multiplying 450.5 by $0.10 obtaining $45.05. Then we add $45.05 to $34.95, the daily rate, and get $80. The result checks.

*State*. The businessperson can drive 450.5 mi on the car-rental allotment.

39. *Familiarize*. The total cost is the insurance charge plus the charge for the time the bike is used. The charge for the time is the cost per hour times the number of hours the bike is used. Let $t =$ the number of hours that a person can rent a bike for $25.00.

*Translate*.

$$\underbrace{\text{Insurance charge}}_{5.50} \; \underbrace{\text{plus}}_{+} \; \underbrace{\text{Cost per hour}}_{2.40} \; \underbrace{\text{times}}_{\cdot} \; \underbrace{\text{Number of hours used}}_{t} \; \underbrace{\text{is}}_{=} \; \underbrace{\text{Total cost}}_{25.00}$$

*Solve*. We solve the equation.

$$5.50 + 2.40t = 25.00$$
$$2.40t = 19.50 \quad \text{Subtracting 5.50 on both sides}$$
$$\frac{2.40t}{2.40} = \frac{19.50}{2.40} \quad \text{Dividing by 2.40 on both sides}$$

$$t = 8.125$$

*Check*. The cost for the time is found by multiplying 8.125 by $2.40 obtaining $19.50. Then we add $19.50 to $5.50, the insurance charge, and get $25.00. The result checks.

*State*. With $25.00, a person can rent a bike for 8.125 hours.

41. *Familiarize*. The total cost is the charge for a house call plus the charge for the repairperson's time. The charge for the time is the cost per hour times the number of hours the repairperson worked. Let $h =$ the number of hours the repairperson worked.

*Translate*.

$$\underbrace{\text{House call charge}}_{\$30} \; \underbrace{\text{plus}}_{+} \; \underbrace{\text{Cost per hour}}_{\$37.50} \; \underbrace{\text{times}}_{\cdot} \; \underbrace{\text{Number of hours}}_{h} \; \underbrace{\text{is}}_{=} \; \underbrace{\text{Total cost}}_{\$123.75}$$

*Solve*. We solve the equation.

$$30 + 37.50h = 123.75$$
$$37.50h = 93.75 \quad \text{Subtracting 30 on both sides}$$
$$\frac{37.50h}{37.50} = \frac{93.75}{37.50} \quad \text{Dividing by 37.50 on both sides}$$

$$h = 2.5$$

*Check*. The cost for the time worked is $37.50 \times 2.5$, or $93.75. We add $93.75 to $30, the cost of a house call, and get $123.75. Our answer checks.

*State*. The repairperson worked 2.5 hr.

43. *Familiarize*. This is a three-step problem. We will find the area $S$ of a standard soccer field and the area $F$ of a standard football field using the formula Area $= l \cdot w$. Then we will find $E$, the amount by which the area of a soccer field exceeds the area of a football field.

*Translate and Solve*.

$$S = l \cdot w = 114.9 \times 74.4 = 8548.56$$
$$F = l \cdot w = 120 \times 53.3 = 6396$$

$$\underbrace{\text{Area of football field}}_{6396} \; \underbrace{\text{plus}}_{+} \; \underbrace{\text{Excess area of soccer field}}_{E} \; \underbrace{\text{is}}_{=} \; \underbrace{\text{Area of soccer field}}_{8548.56}$$

To solve the equation we subtract 6396 on both sides.

$$E = 8548.56 - 6396$$
$$E = 2152.56$$

$$\begin{array}{r} {\scriptstyle 4\;14} \\ 8\,\cancel{5}\,\cancel{4}\,8.56 \\ -\,6\,3\,9\,6.00 \\ \hline 2\,1\,5\,2.56 \end{array}$$

**Check.** We can obtain a partial check by rounding and estimating:

$$114.9 \times 74.4 \approx 110 \times 75 = 8250 \approx 8548.56$$

$$120 \times 53.3 \approx 120 \times 50 = 6000 \approx 6396$$

$$8250 - 6000 = 2250 \approx 2152.56$$

**State.** The area of a soccer field is 2152.56 yd$^2$ greater than the area of a football field.

**45. Familiarize.** This is a multistep problem. First, we find the cost of the cheese. We let $c =$ the cost of the cheese.

**Translate and Solve.**

$$
\underbrace{\begin{array}{c}\text{Number} \\ \text{of pounds}\end{array}}_{\downarrow \atop 6} \ \underbrace{\text{times}}_{\downarrow \atop \cdot} \ \underbrace{\begin{array}{c}\text{Price per} \\ \text{pound}\end{array}}_{\downarrow \atop \$4.79} \ \underbrace{\text{is}}_{\downarrow \atop =} \ \underbrace{\begin{array}{c}\text{Cost of} \\ \text{cheese}\end{array}}_{\downarrow \atop c}
$$

To solve the equation we carry out the multiplication.

$$
\begin{array}{r}
\$4.\,7\,9 \\
\times \qquad 6 \\
\hline
\$2\,8.\,7\,4
\end{array}
$$

Thus, $c = \$28.74$.

Next, we subtract to find how much money $m$ is left to purchase seltzer.

$$
\begin{array}{ll}
m = \$40 - \$28.74 & \begin{array}{r} {\scriptstyle 3\ 9\ 9\ 10} \\ \cancel{4\,0.\,0\,0} \\ -\,2\,8.\,7\,4 \\ \hline 1\,1.\,2\,6 \end{array} \\
m = \$11.26 &
\end{array}
$$

Finally, we divide the amount of money left over by the cost of a bottle of seltzer to find how many bottles can be purchased. We let $b =$ the number of bottles of seltzer that can be purchased.

$$\$11.26 \div \$0.64 = b$$

To find $b$ we carry out the division.

$$
\begin{array}{r}
1\,7.\phantom{0} \\
0.6\,4\wedge\!\overline{\smash{)}1\,1.2\,6\wedge} \\
6\,4\,0\phantom{0} \\
\hline
4\,8\,6\phantom{0} \\
4\,4\,8\phantom{0} \\
\hline
3\,8\phantom{0}
\end{array}
$$

We stop dividing at this point, because Frank cannot purchase a fraction of a bottle. Thus, $b = 17$ (rounded to the nearest 1).

**Check.** The cost of the seltzer is $17 \cdot \$0.64$ or $\$10.88$. The cost of the cheese is $6 \cdot \$4.79$, or $\$28.74$. Frank has spent a total of $\$10.88 + \$28.74$, or $\$39.62$. Frank has $\$40 - \$39.62$, or $\$0.38$ left over. This is not enough to purchase another bottle of seltzer, so our answer checks.

**State.** Frank should buy 17 bottles of seltzer.

**47.** $\dfrac{0}{n} = 0$, for any integer $n$ that is not 0.

Thus, $\dfrac{0}{-13} = 0$.

**49.**
$$\frac{8}{11} - \frac{4}{3} = \frac{8}{11} \cdot \frac{3}{3} - \frac{4}{3} \cdot \frac{11}{11} \qquad \text{The LCM is 33.}$$
$$= \frac{24}{33} - \frac{44}{33}$$
$$= \frac{-20}{33}, \text{ or } -\frac{20}{33}$$

**51.**
$$
\begin{array}{r}
4\dfrac{1}{3} = \quad 4\dfrac{1}{3} \cdot \dfrac{2}{2} = \quad 4\dfrac{2}{6} \\[2mm]
+\,2\dfrac{1}{2} = +\,2\dfrac{1}{2} \cdot \dfrac{3}{3} = +\,2\dfrac{3}{6} \\[2mm]
\hline
6\dfrac{5}{6}
\end{array}
$$

**53.** ◈

**55.** ◈

**57. Familiarize.** This is a multistep problem. First we find the area that is to be seeded. We make a drawing.

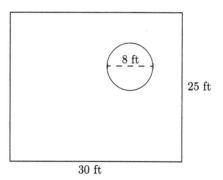

We let $a =$ the area to be seeded. Recall that the area of a rectangle with length $l$ and width $w$ is $A = l \times w$ and the area of a circle with radius $r$ is $A = \pi r^2$, where $\pi \approx 3.14$.

**Translate and Solve.** To find the area to be seeded we subtract the area of the base of the fountain from the area of the yard. Recall that a circle's radius is half of its diameter, or width.

$$
\underbrace{\begin{array}{c}\text{Area of} \\ \text{yard}\end{array}}_{\downarrow \atop 30 \times 25} \ \underbrace{\text{minus}}_{\downarrow \atop -} \ \underbrace{\begin{array}{c}\text{Area of} \\ \text{fountain}\end{array}}_{\downarrow \atop 3.14\left(\frac{8}{2}\right)^2} \ \underbrace{\text{is}}_{\downarrow \atop =} \ \underbrace{\begin{array}{c}\text{Area to} \\ \text{be seeded}\end{array}}_{\downarrow \atop a}
$$

To solve the equation we carry out the computations.

$$30 \times 25 - 3.14\left(\frac{8}{2}\right)^2 = a$$
$$30 \times 25 - 3.14(4)^2 = a$$
$$30 \times 25 - 3.14(16) = a$$
$$750 - 50.24 = a$$
$$699.76 = a$$

Now we find the number of one-pound bags $b$ of grass seed that should be purchased. We think (Number of square feet) $\div$ (Number of square feet covered by 1 pound of seed) = (Number of bags needed).

$$699.76 \div 300 = b$$

To solve the equation we carry out the division on a calculator. We find that $b \approx 2.3$.

*Check*. We recheck the calculations. Our answer checks.

*State*. Since we cannot purchase partial bags of grass seed, 3 one-pound bags should be purchased.

**59.** We must make some assumptions. First we assume that the figures are nested squares formed by connecting the midpoints of consecutive sides of the next larger square. Next assume that the shaded area is the same as the area of the innermost square. (It appears that if we folded the shaded area into the innermost square, it would exactly fill the square.) Finally assume that the length of a side of the innermost square is 5 cm. (If we project the vertices of the innermost square onto the corresponding sides of the largest square, it appears that the distance between each projection and the nearest vertex of the largest square is one-fourth the length of a side of the largest square. Thus, the distance between projections on each side of the largest square is $\frac{1}{2} \cdot 10$ cm, or 5 cm and, hence, the length of a side of the innermost square is 5 cm.) Then the area of the innermost square is 5 cm $\cdot$ 5 cm, or 25 cm$^2$, so the shaded area is 25 cm$^2$.

**61.** *Familiarize*. This is a multistep problem. First we will subtract the average costs in successive years to find the yearly increases. Let $x =$ the increase from 1995 to 1996, $y =$ the increase from 1996 to 1997, and $z =$ the increase from 1997 to 1998.

*Translate and Solve*. We have three "take away" situations.

$$x = 27.40 - 25.39 = 2.01$$
$$y = 27.00 - 27.40 = -0.40$$
$$z = 29.35 - 27.00 = 2.35$$

Now we add the increases and divide by the number of addends, 3, to find the average increase. Let $a =$ the average yearly increase.

$$a = \frac{2.01 + (-0.40) + 2.35}{3} = \frac{3.96}{3} = 1.32$$

*Check*. We repeat the calculations. The answer checks.

*State*. The average yearly increase in the cost was $1.32.

# Chapter 6

# Introduction to Graphing and Statistics

## Exercise Set 6.1

**1.** Go down the Planet column to Jupiter. Then go across to the column headed Average Distance from Sun (in miles) and read the entry, 483,612,200. The average distance from the sun to Jupiter is 483,612,200 miles.

**3.** Go down the column headed Time of Revolution in Earth Time (in years) to 164.78. Then go across the Planet column. The entry there is Neptune, so Neptune has a time of revolution of 164.78 days.

**5.** All of the entries in the column headed Average Distance from Sun (in miles) are greater than 1,000,000. Thus, all of the planets have an average distance from the sun that is greater than 1,000,000 mi.

**7.** Go down the Planet column to earth and then across to the Diameter (in miles) column to find that the diameter of earth is 7926 mi. Similarly, find that the diameter of Jupiter is 88,846 mi. Then divide:
$$\frac{88,846}{7926} \approx 11$$
It would take about 11 earth diameters to equal the diameter of Jupiter.

**9.** Find the average of all the numbers in the column headed Diameter (in miles):

$(3031 + 7520 + 7926 + 4221 + 88,846 + 74,898 + 31,763 + 31,329 + 1423)/9 = 27,884.\overline{1}$

The average of the diameters of the planets is $27,884.\overline{1}$ mi.

**11.** Look down the Average Global Temperature column to find the smallest number. It is $59.23°$. Then look across that row to see that the corresponding year is 1992.

**13.** To find the average global temperature in 1986, go down the column headed Year to 1986 and then across to find the entry $59.29°$. Similarly, find that the average global temperature in 1987 was $59.58°$.

**15.** Average for 1986 to 1988:
$$\frac{59.29° + 59.58° + 56.63°}{3} = \frac{178.50°}{3} = 59.50°$$
Average for 1989 to 1996:

$(59.45° + 59.85° + 59.74° + 59.23° + 59.36° + 59.56° +$

$59.72° + 59.58°)/8 = \frac{476.49}{8} = 59.56125°$

We subtract to find by how many degrees the latter average exceeds the former:
$$59.56125° - 59.50° = 0.06125°$$

**17.** First we find the pairs of years between which the average global temperature increased. Then we find the increases.

From 1986 to 1987: $59.58° - 59.29° = 0.29°$

From 1987 to 1988: $59.63° - 59.58° = 0.05°$

From 1989 to 1990: $59.85° - 59.45° = 0.4°$

From 1992 to 1993: $59.36° - 59.23° = 0.13°$

From 1993 to 1994: $59.56° - 59.36° = 0.2°$

From 1994 to 1995: $59.72° - 59.36° = 0.16°$

The greatest increase occurred between 1989 and 1990.

**19.** The world population in 1850 is represented by 1 symbol, so the population was 1 billion.

**21.** The 1999 (projected) population is represented by the most symbols, so the population was largest in 1999.

**23.** The smallest increase in the number of symbols is represented by $\frac{1}{2}$ symbol from 1650 to 1850 (as opposed to 1 or more symbols for each of the other pairs). Then the growth was the least between these two years.

**25.** The world population in 1975 is represented by 4 symbols so it was $4 \times 1$ billion, or 4 billion people. The population in 1999 is represented by 6 symbols so it was $6 \times 1$ billion, or 6 billion people. We subtract to find the difference:

6 billion − 4 billion = 2 billion

The world population in 1999 was 2 billion more than in 1975.

**27.** Find the year which has exactly 7 symbols above it. The year is 1995.

**29.** There are 8 symbols above 1994, so there were 8 hours per week of prime-time magazine programming in 1994.

**31.** There were 8 hours of prime-time news magazine programming in 1993 and 4 hours in 1991. Thus, the increase was 8 − 4, or 4 hours per week.

**33.** Find the pairs of years between which the hours per week increased. Then find the increases.

From 1987 to 1988: 1 hour

From 1988 to 1989: 1 hour

From 1991 to 1992: 2 hours

From 1992 to 1993: 2 hours

From 1995 to 1996: 2 hours

From 1996 to 1997: 1 hour

Thus, the number of hours of weekly prime-time news magazine programming increased the most from 1991 to 1992, from 1992 to 1993, and from 1995 to 1996.

**35.** For 1992: Note that $168,000,000 = 1.68 \times \$100,000,000$. Thus, we need 1 whole symbol and 0.68, or about $\frac{2}{3}$, of another symbol.

For 1993: Note that $312,000,000 = 3.12 \times \$100,000,000$. Thus we need 3 whole symbols and 0.12, or about $\frac{1}{10}$, of another symbol.

For 1994: Note that $577,000,000 = 5.77 \times \$100,000,000$. Thus, we need 5 whole symbols and 0.77, or about $\frac{3}{4}$, of another symbol.

For 1995: Note that $889,000,000 = 8.89 \times \$100,000,000$. Thus, we need 8 whole symbols and 0.89, or about $\frac{9}{10}$, of another symbol.

For 1996: Note that $1,100,000,000 = 11 \times \$100,000,000$. Thus, we need 11 whole symbols.

Now we draw the pictograph.

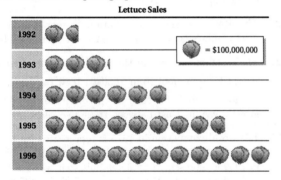

**37.**
$$3x - 2 = 7x + 10$$
$$3x - 2 + 2 = 7x + 10 + 2 \quad \text{Adding 2}$$
$$3x = 7x + 12$$
$$3x - 7x = 7x + 12 - 7x \quad \text{Subtracting } 7x$$
$$-4x = 12$$
$$\frac{-4x}{-4} = \frac{12}{-4} \quad \text{Dividing by } -4$$
$$x = -3$$

The solution is $-3$.

**39.** $\frac{3}{8} = 3 \div 8$

$$
\begin{array}{r}
0.3\,7\,5 \\
8\,\overline{)3.0\,0\,0} \\
\underline{2\,4}\phantom{000} \\
6\,0\phantom{00} \\
\underline{5\,6}\phantom{00} \\
4\,0\phantom{0} \\
\underline{4\,0}\phantom{0} \\
0
\end{array}
$$

$\frac{3}{8} = 0.375$

**41.** $\frac{29}{25} = \frac{29}{25} \cdot \frac{4}{4} = \frac{116}{100} = 1.16$

**43.** ◈

**45.** ◈

**47.** First find the increase between each pair of successive years.

1983 to 1984: 0

1984 to 1985: 0

1985 to 1986: 1

1986 to 1987: 0

1987 to 1988: 1

1988 to 1989: 1

1989 to 1990: $-1$

1990 to 1991: 0

1991 to 1992: 2

1992 to 1993: 2

1993 to 1994: 0

1994 to 1995: $-1$

1995 to 1996: 2

1996 to 1997: 1

1997 to 1998: 0

The sum of the increases is $0 + 0 + 1 + 0 + 1 + 1 + (-1) + 0 + 2 + 2 + 0 + (-1) + 2 + 1 + 0$, or 8.

We divide this number by the number of pairs of years:
$$\frac{8}{15} = 0.5\overline{3}.$$

The average yearly increase in prime-time news magazine programming for the years 1983 to 1998 was $0.5\overline{3}$ hr.

## Exercise Set 6.2

**1.** Move to the right along the bar representing 1 cup of hot cocoa with skim milk. We read that there are about 190 calories in the cup of cocoa.

**3.** The longest bar is for 1 slice of chocolate cake with fudge frosting. Thus, it has the highest caloric content.

**5.** We locate 460 calories at the bottom of the graph and then go up until we reach a bar that ends at approximately 460 calories. Now go across to the left and read the food, 1 cup of premium chocolate ice cream.

**7.** From the graph we see that 1 cup of hot cocoa made with whole milk has about 310 calories and 1 cup of hot cocoa made with skim milk has about 190 calories. We subtract to find the difference:
$$310 - 190 = 120$$

The cocoa made with whole milk has about 120 more calories than the cocoa made with skim milk.

**9.** From Exercise 5 we know that 1 cup of premium ice cream has about 460 calories. We multiply to find the caloric content of 2 cups:
$$2 \times 460 = 920$$

Kristin consumes about 920 calories.

**11.** From the graph we see that a 2-oz chocolate bar with peanuts contains about 270 calories. We multiply to find the number of extra calories Paul adds to his diet in 1 year:

$$365 \times 270 \text{ calories} = 98,550 \text{ calories}$$

Then we divide to determine the number of pounds he will gain:

$$\frac{98,550}{3500} \approx 28$$

Paul will gain about 28 pounds.

**13.** In the group of bars representing 1980 find the bar representing Latin America. Go to the top of that bar and then across to the left to read 920 on the vertical scale. Units on this scale are in thousands of hectares, so the forest area of Latin America in 1980 was about 920,000 hectares.

**15.** The heights of the pair of bars representing Latin America decrease more from 1980 to 1990 than the heights of either of the other pairs of bars. Thus, Latin America experienced the greatest loss of forest area from 1980 to 1990.

**17.** We go up the vertical scale to 600. Then we move to the right until we come to a bar in the group representing 1990 that ends at about 600. Moving down that bar we see that it represents Africa, so Africa had about 600 thousand hectares of forest area in 1990.

**19.** From Exercise 13 we know that the forest area of Latin America was about 920,000 hectares in 1980. From the graph we find that it was about 840,000 hectares in 1990. We find the average of these two numbers:

$$\frac{920,000 + 840,000}{2} = \frac{1,760,000}{2} = 880,000$$

Thus, the average forest area in Latin America for the years 1980 and 1990 was about 880,000 hectares.

**21.** On the horizontal scale in six equally spaced intervals indicate the names of the cities. Label this scale "City." Then label the vertical scale "Commuting Time (in minutes)." Note that the smallest time is 21.9 minutes and the largest is 30.6 minutes. We could start the vertical scale at 0 or we could start it at 20, using a jagged line to indicate the missing numbers. We choose the second option. Label the marks on the vertical scale by 5's. Finally, draw vertical bars above the cities to show the commuting times.

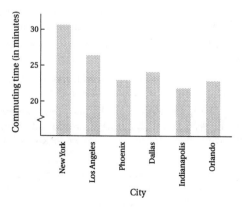

**23.** The shortest bar represents Indianapolis, so it has the least commuting time.

**25.** We add the commuting times for New York and Los Angeles and divide by the number of cities, 2.

$$\frac{30.6 + 26.4}{2} = \frac{57}{2} = 28.5 \text{ min}$$

**27.** $650 - 70 = 580$ calories per hour

**29.** Find the calories burned doing office work for 8 hours:

$$180 \cdot 8 = 1440 \text{ calories}$$

Find the calories burned sleeping for 7 hours:

$$70 \cdot 7 = 490 \text{ calories}$$

Find the sum: $1440 + 490 = 1930$ calories

**31.** The highest point on the graph lies above 1998 on the horizontal axis. Thus, the average salary was highest in 1998.

**33.** From the graph we see that the average salary was the lowest in 1991. It was about $0.85 million. From Exercise 31 we know that the average salary was the highest in 1998. It was about $1.4 million. We subtract to find the difference:

$$\$1.4 \text{ million} - \$0.85 \text{ million} = \$0.55 \text{ million}$$

The difference between the highest and lowest average salaries was about $0.55 million.

**35.** The segment connecting 1994 and 1995 drops, so the average salary decreased between 1994 and 1995.

**37.** First indicate the years on the horizontal scale and label it "Year." The smallest ozone level is 2981 parts per billion and the largest is 3148 parts per billion. We could start the vertical scale at 0, but the graph will be more compact and easier to read if we start at a higher number, say at 2980. We do this, using a jagged line to indicate the missing numbers. Mark the vertical scale appropriately, say by 20's, and label it "Ozone level (in parts per billion)." Next, at the appropriate level above each year, mark the corresponding ozone level. Finally, draw line segments connecting the points.

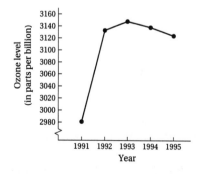

**39.** The graph falls most sharply from 1994 to 1995. (The data in the table confirms that the greatest decrease occurred between 1994 and 1995.) Thus, the decrease in the ozone level was the greatest between 1994 and 1995.

**41.** We add the ozone levels from 1992 through 1995 and divide by the number of years, 4.
$$\frac{3133 + 3148 + 3138 + 3124}{4} = \frac{12,543}{4} = 3135.75$$
The average ozone level from 1992 through 1995 is 3135.75 parts per billion.

**43.** The segment connecting 1995 and 1996 rises most steeply, so the increase was the greatest between 1995 and 1996. (The data in the table confirms this.)

**45.** $\dfrac{38.2 + 42.4 + 44.0 + 50.4 + 54.1 + 61.0}{6} = \dfrac{290.1}{6} = 48.35$
The average motion-picture expense was $48.35 million.

**47. Familiarize.** We draw a picture. We let $n =$ the number of 12-oz bottles that can be filled.

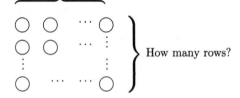

12 oz in each row

How many rows?

**Translate and Solve.** We translate to an equation and solve as follows:

$$408 \div 12 = n$$

$$\begin{array}{r} 3\,4 \\ 1\,2\,\overline{)4\,0\,8} \\ 3\,6\,0 \\ \hline 4\,8 \\ 4\,8 \\ \hline 0 \end{array}$$

**Check.** We check by multiplying the number of bottles by 12:
$$12 \times 34 = 408$$
**State.** 34 twelve-oz bottles can be filled.

**49. Familiarize.** We draw a picture. We let $n =$ the number of fluid ounces in a six-pack of Coca Cola.

12 oz in each row

6 rows
How many ounces?

**Translate and Solve.** We translate to an equation and solve as follows.
$$12 \times 6 = n$$
$$72 = n$$
**Check.** We check by dividing the total number of ounces by 6: $72 \div 6 = 12$. The answer checks.
**State.** There are 72 fluid ounces in a six-pack of Coca Cola.

**51. Familiarize.** Let $n =$ the number.
**Translate.**

$\frac{2}{3}$ of 75 is <u>what number?</u>

$\frac{2}{3} \quad \cdot \quad 75 \quad = \quad n$

**Solve.** We carry out the multiplication.
$$\frac{2}{3} \cdot 75 = n$$
$$\frac{2 \cdot 75}{3} = n$$
$$\frac{2 \cdot \cancel{3} \cdot 25}{\cancel{3} \cdot 1} = n$$
$$50 = n$$

**Check.** We repeat the calculation.
**State.** $\frac{2}{3}$ of 75 is 50.

**53.** ◈

**55.** ◈

**57.** There are 13 age groups above the age of 24. We add the incidence rates, reading from left to right across the graph beginning with the 25-29 age group.

$10 + 30 + 65 + 130 + 190 + 220 + 275 + 350 + 415 + 440 + 475 + 470 + 420 = 3490$

Average rate of incidence:
$$\frac{3490}{13} \approx 268 \text{ women per } 100,000$$

(Answers may differ slightly due to variations in the incidence rates read from the graph.)

## Exercise Set 6.3

**1.** To plot $(2,5)$, we locate 2 on the first, or horizontal, axis. Then we go up 5 units and make a dot.

To plot $(-1,3)$, we locate $-1$ on the first, or horizontal, axis. Then we go up 3 units and make a dot.

To plot $(3,-2)$, we locate 3 on the first, or horizontal, axis. Then we go down 2 units and make a dot.

To plot $(-2,-4)$, we locate $-2$ on the first, on the horizontal, axis. Then we go down 4 units and make a dot.

To plot $(0,4)$, we locate 0 on the first, or horizontal, axis. Then we go up 4 units and make a dot.

To plot $(0,-5)$, we locate 0 on the first, or horizontal, axis. Then we go down 5 units and make a dot.

To plot $(5,0)$, we locate 5 on the first, or horizontal, axis. Since the second coordinate is 0, we do not move up or down. We make a dot at the point we located on the first axis.

To plot $(-5,0)$, we locate $-5$ on the first, or horizontal, axis. Since the second coordinate is 0, we do not move up or down. We make a dot at the point we located on the first axis.

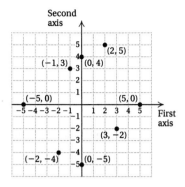

**3.** To plot $(-3, -1)$, we locate $-3$ on the first, or horizontal, axis. Then we go down 1 unit and make a dot.

To plot $(5, 1)$, we locate 5 on the first, or horizontal, axis. Then we go up 1 unit and make a dot.

To plot $(-1, -5)$, we locate $-1$ on the first, or horizontal, axis. Then we go down 5 units and make a dot.

To plot $(0, 0)$, we locate 0 on the first, or horizontal, axis. Since the second coordinate is 0, we do not move up or down. We make a dot at the point we located on the first axis.

To plot $(0, 1)$, we locate 0 on the first, or horizontal, axis. Then we go up 1 unit and make a dot.

To plot $(-4, 0)$, we locate $-4$ on the first, or horizontal, axis. Since the second coordinate is 0, we do not move up or down. We make a dot at the point we located on the first axis.

To plot $\left(2, 3\frac{1}{2}\right)$, we locate 2 on the first, or horizontal, axis. Then we go up $3\frac{1}{2}$ units and make a dot.

To plot $\left(4\frac{1}{2}, -2\right)$, we locate $4\frac{1}{2}$ on the first, or horizontal, axis. Then we go down 2 units and make a dot.

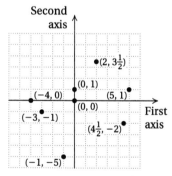

**5.**

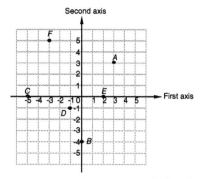

We look below point $A$ to see that its first coordinate is 3. Looking to the left of point $A$, we find that its second coordinate is also 3. Thus, the coordinates of point $A$ are $(3, 3)$.

We look above point $B$ to see that its first coordinate is 0. Looking at the location of point $B$ on the second, or vertical, axis, we find that its second coordinate is $-4$. Thus, the coordinates of point $B$ are $(0, -4)$.

Looking at the location of point $C$ on the first, or horizontal, axis, we see that the first coordinate of point $C$ is $-5$. We look to the right of point $C$ to see that its second coordinate is 0. Thus, the coordinates of point $C$ are $(-5, 0)$.

We look above point $D$ to see that its first coordinate is $-1$. Looking to the right of point $D$, we find that its second coordinate is also $-1$. Thus, the coordinates of point $D$ are $(-1, -1)$.

Looking at the location of point $E$ on the first, or horizontal, axis, we see that the first coordinate of point $E$ is 2. We look to the left of point $E$ to see that its second coordinate is 0. Thus, the coordinates of point $E$ are $(2, 0)$.

We look below point $F$ to see that its first coordinate is $-3$. Looking to the right of point $F$, we find that its second coordinate is 5. Thus, the coordinates of point $F$ are $(-3, 5)$.

**7.**

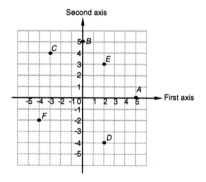

Looking at the location of point $A$ on the first, or horizontal, axis, we see that the first coordinate of point $A$ is 5. We look to the left of point $A$ to see that its second coordinate is 0. Thus, the coordinates of point $A$ are $(5, 0)$.

We look below point $B$ to see that its first coordinate is 0. Looking at the location of point $B$ on the second, or vertical, axis, we find that its second coordinate is 5. Thus, the coordinates of point $B$ are $(0, 5)$.

We look below point $C$ to see that its first coordinate is $-3$. Looking to the right of point $C$, we find that its second coordinate is 4. Thus, the coordinates of point $C$ are $(-3, 4)$.

We look above point $D$ to see that its first coordinate is 2. Looking to the left of point $D$, we find that its second coordinate is $-4$. Thus, the coordinates of point $D$ are $(2, -4)$.

We look below point $E$ to see that its first coordinate is 2. Looking to the left of point $E$, we find that its second coordinate is 3. Thus the coordinates of point $E$ are $(2, 3)$.

We look above point $F$ to see that its first coordinate is $-4$. Looking to the right of point $F$, we find that its second coordinate is $-2$. Thus, the coordinates of point $F$ are $(-4, -2)$.

**9.** Since the first coordinate is negative and the second coordinate positive, the point $(-5, 3)$ is located in quadrant II.

**11.** Since the first coordinate is positive and the second coordinate negative, the point $(100, -1)$ is in quadrant IV.

**13.** Since both coordinates are negative, the point $(-6.5, -1.9)$ is in quadrant III.

**15.** Since both coordinates are positive, the point $\left(3\frac{7}{10}, 9\frac{1}{11}\right)$ is in quadrant I.

**17.** In quadrant IV, first coordinates are always <u>positive</u> and second coordinates are always <u>negative</u>.

**19.** In quadrant <u>III</u>, both coordinates are always negative.

**21.** In quadrant I, the first coordinate is always positive. The other quadrant where this occurs is IV. Thus, the statement should read as follows:

In quadrants I and <u>IV</u>, the first coordinate is always <u>positive</u>.

**23.**  $y = 3x + 1$

$$\begin{array}{c|c} 7 \ ? \ 3 \cdot 2 + 1 & \text{Substituting 2 for } x \text{ and 7 for } y \\ 6 + 1 & \text{(alphabetical order of variables)} \\ 7 \ \big| \ 7 & \text{TRUE} \end{array}$$

The equation becomes true: $(2, 7)$ is a solution.

**25.**    $3x - y = 4$

$$\begin{array}{c|c} 3 \cdot 2 - (-3) \ ? \ 4 & \text{Substituting 2 for } x \text{ and } -3 \text{ for } y \\ 6 + 3 & \\ 9 \ \big| \ 4 & \text{FALSE} \end{array}$$

The equation becomes false; $(2, -3)$ is not a solution.

**27.**    $2c + 3d = -7$

$$\begin{array}{c|c} 2(-2) + 3(-1) \ ? \ -7 & \text{Substituting } -2 \text{ for } c \text{ and } -1 \\ & \text{for } d \\ -4 - 3 & \\ -7 \ \big| \ -7 & \text{TRUE} \end{array}$$

The equation becomes true; $(-2, -1)$ is a solution.

**29.**     $3x + y = 19$

$$\begin{array}{c|c} 3 \cdot 5 + (-4) \ ? \ 19 & \text{Substituting 5 for } x \text{ and } -4 \text{ for } y \\ 15 - 4 & \\ 11 \ \big| \ 19 & \text{FALSE} \end{array}$$

The equation becomes false; $(5, -4)$ is not a solution.

**31.**     $2q - 3p = 3$

$$\begin{array}{c|c} 2 \cdot 5 - 3\left(2\frac{1}{3}\right) \ ? \ 3 & \text{Substituting } 2\frac{1}{3} \text{ for } p \text{ and 5 for } q \\ 10 - 3 \cdot \frac{7}{3} & \\ 10 - 7 & \\ 3 \ \big| \ 3 & \text{TRUE} \end{array}$$

The equation becomes true; $\left(2\frac{1}{3}, 5\right)$ is a solution.

**33.**   $y = 5x - 6.3$

$$\begin{array}{c|c} 0.7 \ ? \ 5(2.4) - 6.3 & \text{Substituting 2.4 for } x \text{ and 0.7} \\ & \text{for } y \\ 12 - 6.3 & \\ 0.7 \ \big| \ 5.7 & \text{FALSE} \end{array}$$

The equation becomes false; $(2.4, 0.7)$ is not a solution.

**35.**     $3x - 4 = 17$

$$3x - 4 + 4 = 17 + 4 \quad \text{Adding 4 on both sides}$$
$$3x = 21$$
$$\frac{3x}{3} = \frac{21}{3} \quad \text{Dividing by 3 on both sides}$$
$$x = 7$$

The solution is 7.

**37.**     $5(x - 2) = 3x - 4$
$$5x - 10 = 3x - 4$$
$$5x - 10 + 10 = 3x - 4 + 10 \quad \text{Adding 10 on both sides}$$
$$5x = 3x + 6$$
$$5x - 3x = 3x + 6 - 3x \quad \text{Subtracting } 3x \text{ on both sides}$$
$$2x = 6$$
$$\frac{2x}{2} = \frac{6}{2} \quad \text{Dividing by 2 on both sides}$$
$$x = 3$$

The solution is 3.

**39.** $7\frac{2}{11}a - 5\frac{1}{3}a = \left(7\frac{2}{11} - 5\frac{1}{3}\right)a = \left(7\frac{6}{33} - 5\frac{11}{33}\right)a = \left(6\frac{39}{33} - 5\frac{11}{33}\right)a = 1\frac{28}{33}a$

**41.** ◈

**43.** ◈

**45.**      $6.5x - 7.2y = -94.36$

$$\begin{array}{c|c} 6.5(4.16) - 7.2(-9.35) \ ? \ -94.36 & \\ 27.04 + 67.32 & \\ 94.36 \ \big| \ -94.36 & \text{FALSE} \end{array}$$

The equation becomes false, so $(4.16, -9.35)$ is not a solution.

**47.**

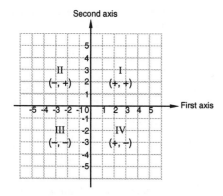

If the second coordinate is negative, then the point must be in either quadrant III or quadrant IV.

**49.** If the first coordinate is the opposite of the second coordinate, then the coordinates have different signs. The point must be in quadrant II or quadrant IV. (See the graph in Exercise 47.)

**51.**

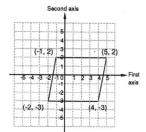

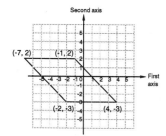

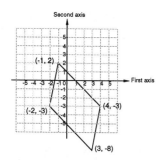

The coordinates of the fourth vertex are $(5, 2)$, $(-7, 2)$, or $(3, -8)$.

**53.** Answers may vary.

We select eight points such that the first coordinate minus the second coordinate of each point is 1.

| | |
|---|---|
| $(-3, -4)$ | $-3 - (-4) = 1$ |
| $(-2, -3)$ | $-2 - (-3) = 1$ |
| $(-1, -2)$ | $-1 - (-2) = 1$ |
| $(0, -1)$ | $0 - (-1) = 1$ |
| $(1, 0)$ | $1 - 0 = 1$ |
| $(2, 1)$ | $2 - 1 = 1$ |
| $(3, 2)$ | $3 - 2 = 1$ |
| $(4, 3)$ | $4 - 3 = 1$ |

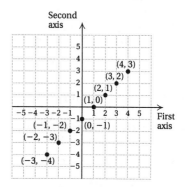

**55.**

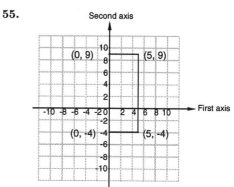

The length is 13 and the width is 5.
$A = l \times w = 13 \times 5 = 65$

## Exercise Set 6.4

**1.**
$$x - y = 3$$
$$8 - y = 3 \quad \text{Substituting 8 for } x$$
$$-y = -5 \quad \text{Subtracting 8 on both sides}$$
$$-1 \cdot y = -5 \quad \text{Recall that } -a = -1 \cdot a.$$
$$y = 5 \quad \text{Dividing by } -1 \text{ on both sides}$$

The pair $(8, 5)$ is a solution of $x - y = 3$.

**3.**
$$2x + y = 7$$
$$2 \cdot 3 + y = 7 \quad \text{Substituting 3 for } x$$
$$6 + y = 7$$
$$y = 1 \quad \text{Subtracting 6 on both sides}$$

The pair $(3, 1)$ is a solution of $2x + y = 7$.

**5.**
$$y = 3x - 1$$
$$y = 3 \cdot 5 - 1 \quad \text{Substituting 5 for } x$$
$$y = 15 - 1$$
$$y = 14$$

The pair $(5, 14)$ is a solution of $y = 3x - 1$.

**7.** $x + 3y = 1$
$7 + 3y = 1$    Substituting 7 for $x$
$3y = -6$    Subtracting 7 on both sides
$y = -2$    Dividing by 3 on both sides

The pair $(7, -2)$ is a solution of $x + 3y = 1$.

**9.**    $2x + 5y = 17$
$2 \cdot 1 + 5y = 17$    Substituting 1 for $x$
$2 + 5y = 17$
$5y = 15$    Subtracting 2 on both sides
$y = 3$    Dividing by 5 on both sides

The pair $(1, 3)$ is a solution of $2x + 5y = 17$.

**11.**      $3x - 2y = 8$
$3x - 2(-1) = 8$    Substituting $-1$ for $y$
$3x + 2 = 8$
$3x = 6$    Subtracting 2 on both sides
$x = 2$    Dividing by 3 on both sides

The pair $(2, -1)$ is a solution of $3x - 2y = 8$.

**13.** To complete the pair $(\ ,8)$, we replace $y$ with 8 and solve for $x$.
$x + y = 7$
$x + 8 = 7$    Substituting 8 for $y$
$x = -1$    Subtracting 8 on both sides

Thus, $(-1, 8)$ is a solution of $x + y = 7$.

To complete the pair $(4,\ )$, we replace $x$ with 4 and solve for $y$.
$x + y = 7$
$4 + y = 7$    Substituting 4 for $x$
$y = 3$    Subtracting 4 on both sides

Thus, $(4, 3)$ is a solution of $x + y = 7$.

**15.** To complete the pair $(\ ,3)$, we replace $y$ with 3 and solve for $x$.
$x - y = 4$
$x - 3 = 4$    Substituting 3 for $y$
$x = 7$    Adding 3 on both sides

Thus, $(7, 3)$ is a solution of $x - y = 4$.

To complete the pair $(10,\ )$, we replace $x$ with 10 and solve for $y$.
$x - y = 4$
$10 - y = 4$    Substituting 10 for $x$
$-y = -6$    Subtracting 10 on both sides
$-1 \cdot y = -6$    Recall that $-a = -1 \cdot a$.
$y = 6$    Dividing by $-1$ on both sides

Thus, $(10, 6)$ is a solution of $x - y = 4$.

**17.** To complete the pair $(3,\ )$, we replace $x$ with 3 and solve for $y$.
$2x + 3y = 15$
$2 \cdot 3 + 3y = 15$    Substituting 3 for $x$
$6 + 3y = 15$
$3y = 9$    Subtracting 6 on both sides
$y = 3$    Dividing by 3 on both sides

Thus, $(3, 3)$ is a solution of $2x + 3y = 15$.

To complete the pair $(\ ,1)$, we replace $y$ with 1 and solve for $x$.
$2x + 3y = 15$
$2x + 3 \cdot 1 = 15$    Substituting 1 for $y$
$2x + 3 = 15$
$2x = 12$    Subtracting 3 on both sides
$x = 6$    Dividing by 2 on both sides

Thus, $(6, 1)$ is a solution of $2x + 3y = 15$.

**19.** To complete the pair $(3,\ )$, we replace $x$ with 3 and solve for $y$.
$5x + 2y = 11$
$5 \cdot 3 + 2y = 11$    Substituting 3 for $x$
$15 + 2y = 11$
$2y = -4$    Subtracting 15 on both sides
$y = -2$    Dividing by 2 on both sides

Thus, $(3, -2)$ is a solution of $5x + 2y = 11$.

To complete the pair $(\ ,3)$, we replace $y$ with 3 and solve for $x$.
$5x + 2y = 11$
$5x + 2 \cdot 3 = 11$    Substituting 3 for $y$
$5x + 6 = 11$
$5x = 5$    Subtracting 6 on both sides
$x = 1$    Dividing by 5 on both sides

Thus, $(1, 3)$ is a solution of $5x + 2y = 11$.

**21.** To complete the pair $(\ ,4)$, we replace $y$ with 4 and solve for $x$.
$y = 4x$
$4 = 4x$    Substituting 4 for y
$1 = x$    Dividing by 4 on both sides

Thus, $(1, 4)$ is a solution of $y = 4x$.

To complete the pair $(-2,\ )$, we replace $x$ with $-2$ and solve for $y$.
$y = 4x$
$y = 4(-2)$    Substituting $-2$ for $x$
$y = -8$

Thus, $(-2, -8)$ is a solution of $y = 4x$.

**23.** To complete the pair $(0,\ )$, we replace $x$ with 0 and solve for $y$.
$2x + 5y = 3$
$2 \cdot 0 + 5y = 3$    Substituting 0 for $x$
$5y = 3$
$y = \dfrac{3}{5}$    Dividing by 5 on both sides

Thus, $\left(0, \dfrac{3}{5}\right)$ is a solution of $2x + 5y = 3$.

To complete the pair $(\ ,0)$, we replace $y$ with 0 and solve for $x$.
$2x + 5y = 3$
$2x + 5 \cdot 0 = 3$    Substituting 0 for $y$
$2x = 3$
$x = \dfrac{3}{2}$    Dividing by 2 on both sides

Thus, $\left(\dfrac{3}{2}, 0\right)$ is a solution of $2x + 5y = 3$.

**25.** We are free to choose any number as a replacement for $x$ or $y$. To find one solution we choose to replace $x$ with 0.

$$x + y = 12$$
$$0 + y = 12 \quad \text{Substituting 0 for } x$$
$$y = 12$$

Thus, $(0, 12)$ is one solution of $x + y = 12$.

To find a second solution we can replace $y$ with 5.

$$x + y = 12$$
$$x + 5 = 12 \quad \text{Substituting 5 for } y$$
$$x = 7 \quad \text{Subtracting 5 on both sides}$$

Thus, $(7, 5)$ is a second solution of $x + y = 12$.

To find a third solution we can replace $x$ with $-3$.

$$x + y = 12$$
$$-3 + y = 12 \quad \text{Substituting } -3 \text{ for } x$$
$$y = 15 \quad \text{Adding 3 on both sides}$$

Thus, $(-3, 15)$ is a third solution of $x + y = 12$.

**27.** The solutions $(1, 4)$ and $(-2, -8)$ were found in Exercise 21. To find a third solution we can replace $x$ with 0.

$$y = 4x$$
$$y = 4 \cdot 0 \quad \text{Substituting 0 for } x$$
$$y = 0$$

Thus, $(0, 0)$ is a third solution of $y = 4x$.

**29.** We are free to choose any number as a replacement for $x$ or $y$. To find one solution we choose to replace $x$ with 0.

$$3x + y = 13$$
$$3 \cdot 0 + y = 13 \quad \text{Substituting 0 for } x$$
$$0 + y = 13$$
$$y = 13$$

Thus, $(0, 13)$ is one solution of $3x + y = 13$.

To find a second solution we can replace $y$ with 10.

$$3x + y = 13$$
$$3x + 10 = 13 \quad \text{Substituting 10 for } y$$
$$3x = 3 \quad \text{Subtracting 10 on both sides}$$
$$x = 1 \quad \text{Dividing by 3 on both sides}$$

Thus, $(1, 10)$ is a second solution of $3x + y = 13$.

To find a third solution we can replace $x$ with 2.

$$3x + y = 13$$
$$3 \cdot 2 + y = 13 \quad \text{Substituting 2 for } x$$
$$6 + y = 13$$
$$y = 7 \quad \text{Subtracting 6 on both sides}$$

Thus, $(2, 7)$ is a third solution of $3x + y = 13$.

**31.** We are free to choose any number as a replacement for $x$ or $y$. Since $y$ is isolated it is generally easiest to substitute for $x$ and then calculate $y$. To find one solution we choose to replace $x$ with $-1$.

$$y = 3x - 1$$
$$y = 3(-1) - 1 \quad \text{Substituting } -1 \text{ for } x$$
$$y = -3 - 1$$
$$y = -4$$

Thus, $(-1, -4)$ is one solution of $y = 3x - 1$.

To find a second solution we can replace $x$ with 0.

$$y = 3x - 1$$
$$y = 3 \cdot 0 - 1 \quad \text{Substituting 0 for } x$$
$$y = -1$$

Thus, $(0, -1)$ is a second solution of $y = 3x - 1$.

To find a third solution we can replace $x$ with 2.

$$y = 3x - 1$$
$$y = 3 \cdot 2 - 1 \quad \text{Substituting 2 for } x$$
$$y = 6 - 1$$
$$y = 5$$

Thus, $(2, 5)$ is a third solution of $y = 3x - 1$.

**33.** We are free to choose any number as a replacement for $x$ or $y$. Since $y$ is isolated it is generally easiest to substitute for $x$ and then calculate $y$. To find one solution we choose to replace $x$ with 0.

$$y = -5x$$
$$y = -5 \cdot 0 \quad \text{Substituting 0 for } x$$
$$y = 0$$

Thus, $(0, 0)$ is one solution of $y = -5x$.

To find a second solution we can replace $x$ with 1.

$$y = -5x$$
$$y = -5 \cdot 1 \quad \text{Substituting 1 for } x$$
$$y = -5$$

Thus, $(1, -5)$ is a second solution of $y = -5x$.

To find a third solution we can replace $x$ with $-1$.

$$y = -5x$$
$$y = -5(-1) \quad \text{Substituting } -1 \text{ for x}$$
$$y = 5$$

Thus, $(-1, 5)$ is a third solution of $y = -5x$.

**35.** We are free to choose any number as a replacement for $x$ or $y$. Since $x$ is isolated it is easiest to substitute for $y$ and then calculate $x$. To find one solution we choose to replace $y$ with $-4$.

$$4 + y = x$$
$$4 + (-4) = x \quad \text{Substituting } -4 \text{ for } y$$
$$0 = x$$

Thus, $(0, -4)$ is one solution of $4 + y = x$.

To find a second solution we can replace $y$ with 0.

$$4 + y = x$$
$$4 + 0 = x \quad \text{Substituting 0 for } y$$
$$4 = x$$

Thus, $(4, 0)$ is a second solution of $4 + y = x$.

To find a third solution we can replace $y$ with $-3$.

$$4 + y = x$$
$$4 + (-3) = x \quad \text{Substituting } -3 \text{ for } y$$
$$1 = x$$

Thus, $(1, -3)$ is a third solution of $4 + y = x$.

**37.** We are free to choose any number as a replacement for $x$ or $y$. To find one solution we choose to replace $x$ with 0.

$$3x + 2y = 12$$
$$3 \cdot 0 + 2y = 12 \quad \text{Substituting 0 for } x$$
$$2y = 12$$
$$y = 6 \quad \text{Dividing by 2 on both sides}$$

Thus, $(0, 6)$ is one solution of $3x + 2y = 12$.

To find a second solution we can replace $y$ with 0.

$$3x + 2y = 12$$
$$3x + 2 \cdot 0 = 12 \quad \text{Substituting 0 for } y$$
$$3x = 12$$
$$x = 4 \quad \text{Dividing by 3 on both sides}$$

Thus, $(4, 0)$ is a second solution of $3x + 2y = 12$.

To find a third solution we can replace $x$ with 1.

$$3x + 2y = 12$$
$$3 \cdot 1 + 2y = 12 \quad \text{Substituting 1 for } x$$
$$3 + 2y = 12$$
$$2y = 9 \quad \text{Subtracting 3 on both sides}$$
$$y = \frac{9}{2} \quad \text{Dividing by 2 on both sides}$$

Thus, $\left(1, \frac{9}{2}\right)$ is a third solution of $3x + 2y = 12$.

**39.** We are free to choose any number as a replacement for $x$ or $y$. Since $y$ is isolated it is generally easiest to substitute for $x$ and then calculate $y$. Note that when $x$ is a multiple of 3, fractional values for $y$ are avoided. To find one solution we choose to replace $x$ with $-3$.

$$y = \frac{1}{3}x + 2$$
$$y = \frac{1}{3}(-3) + 2 \quad \text{Substituting } -3 \text{ for } x$$
$$y = -1 + 2$$
$$y = 1$$

Thus, $(-3, 1)$ is one solution of $y = \frac{1}{3}x + 2$.

To find a second solution we can replace $x$ with 0.

$$y = \frac{1}{3}x + 2$$
$$y = \frac{1}{3} \cdot 0 + 2 \quad \text{Substituting 0 for } x$$
$$y = 2$$

Thus, $(0, 2)$ is a second solution of $y = \frac{1}{3}x + 2$.

To find a third solution we can replace $x$ with 3.

$$y = \frac{1}{3}x + 2$$
$$y = \frac{1}{3} \cdot 3 + 2 \quad \text{Substituting 3 for } x$$
$$y = 1 + 2$$
$$y = 3$$

Thus, $(3, 3)$ is a third solution of $y = \frac{1}{3}x + 2$.

**41.** Graph: $x + y = 4$

We make a table of solutions. Then we plot the points, draw the line, and label it.

When $x = 0$:     $0 + y = 4$
                  $y = 4$

When $y = -2$:   $x + (-2) = 4$
                  $x = 6$

When $x = 4$:     $4 + y = 4$
                  $y = 0$

| $x$ | $y$ $x + y = 4$ | $(x, y)$ |
|---|---|---|
| 0 | 4 | $(0, 4)$ |
| 6 | $-2$ | $(6, -2)$ |
| 4 | 0 | $(4, 0)$ |

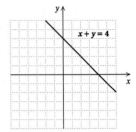

**43.** Graph: $x - 1 = y$

We make a table of solutions. Then we plot the points, draw the line, and label it.

If $x = 5$, then $y = 5 - 1 = 4$.
If $x = 0$, then $y = 0 - 1 = -1$.
If $x = -2$, then $y = -2 - 1 = -3$.

| $x$ | $y$ $x - 1 = y$ | $(x, y)$ |
|---|---|---|
| 5 | 4 | $(5, 4)$ |
| 0 | $-1$ | $(0, -1)$ |
| $-2$ | $-3$ | $(-2, -3)$ |

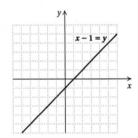

**45.** Graph: $y = x - 3$

We make a table of solutions. Then we plot the points, draw the line, and label it.

If $x = -2$, $y = -2 - 3 = -5$.
If $x = 0$, $y = 0 - 3 = -3$.
If $x = 3$, $y = 3 - 3 = 0$.

| $x$ | $y$ $y = x - 3$ | $(x, y)$ |
|---|---|---|
| $-2$ | $-5$ | $(-2, -5)$ |
| 0 | $-3$ | $(0, -3)$ |
| 3 | 0 | $(3, 0)$ |

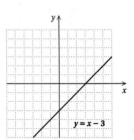

**47.** Graph: $y = \frac{1}{3}x$

We make a table of solutions. Note that when $x$ is a multiple of 3, fractional values for $y$ are avoided. We plot the points, draw the line, and label it.

If $x = -3$, $y = \frac{1}{3}(-3) = -1$.

If $x = 0$, $y = \frac{1}{3} \cdot 0 = 0$.

If $x = 3$, $y = \frac{1}{3} \cdot 3 = 1$.

| $x$ | $y$ $y = \frac{1}{3}x$ | $(x,y)$ |
|---|---|---|
| $-3$ | $-1$ | $(-3,-1)$ |
| $0$ | $0$ | $(0,0)$ |
| $3$ | $1$ | $(3,1)$ |

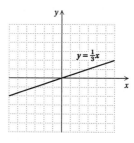

**49.** Graph: $y = x$

We make a table of solutions. Then we plot the points, draw the line, and label it.

If $x = -3$, $y = -3$.
If $x = 0$, $y = 0$.
If $x = 2$, $y = 2$.

| $x$ | $y$ $y = x$ | $(x,y)$ |
|---|---|---|
| $-3$ | $-3$ | $(-3,-3)$ |
| $0$ | $0$ | $(0,0)$ |
| $2$ | $2$ | $(2,2)$ |

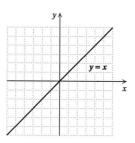

**51.** Graph: $y = 2x - 1$

We make a table of solutions. Then we plot the points, draw the line, and label it.

If $x = -1$, $y = 2(-1) - 1 = -2 - 1 = -3$.
If $x = 0$, $y = 2 \cdot 0 - 1 = -1$.
If $x = 2$, $y = 2 \cdot 2 - 1 = 4 - 1 = 3$.

| $x$ | $y$ $y = 2x - 1$ | $(x,y)$ |
|---|---|---|
| $-1$ | $-3$ | $(-1,-3)$ |
| $0$ | $-1$ | $(0,-1)$ |
| $2$ | $3$ | $(2,3)$ |

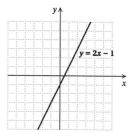

**53.** Graph: $y = 2x - 7$

We make a table of solutions. Then we plot the points, draw the line, and label it.

If $x = 0$, $y = 2 \cdot 0 - 7 = 0 - 7 = -7$.
If $x = 2$, $y = 2 \cdot 2 - 7 = 4 - 7 = -3$.
If $x = 5$, $y = 2 \cdot 5 - 7 = 10 - 7 = 3$.

| $x$ | $y$ $y = 2x - 7$ | $(x,y)$ |
|---|---|---|
| $0$ | $-7$ | $(0,-7)$ |
| $2$ | $-3$ | $(2,-3)$ |
| $5$ | $3$ | $(5,3)$ |

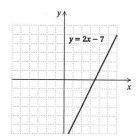

**55.** Graph: $y = \frac{2}{5}x$

We make a table of solutions. Note that when $x$ is a multiple of 5, fractional values for $y$ are avoided. We plot the points, draw the line, and label it.

If $x = -5$, $y = \frac{2}{5}(-5) = -2$.

If $x = 0$, $y = \frac{2}{5} \cdot 0 = 0$.

If $x = 5$, $y = \frac{2}{5} \cdot 5 = 2$.

| $x$ | $y$ $y = \frac{2}{5}x$ | $(x,y)$ |
|---|---|---|
| $-5$ | $-2$ | $(-5,-2)$ |
| $0$ | $0$ | $(0,0)$ |
| $5$ | $2$ | $(5,2)$ |

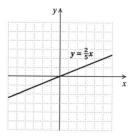

**57.** Graph: $y = -x + 4$

We make a table of solutions. Then we plot the points, draw the line, and label it.

If $x = -1$, $y = -(-1) + 4 = 1 + 4 = 5$.
If $x = 1$, $y = -1 + 4 = 3$.
If $x = 4$, $y = -4 + 4 = 0$.

| $x$ | $y$ $y = -x + 4$ | $(x,y)$ |
|---|---|---|
| $-1$ | $5$ | $(-1,5)$ |
| $1$ | $3$ | $(1,3)$ |
| $4$ | $0$ | $(4,0)$ |

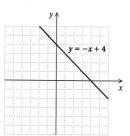

**59.** *Familiarize.* We let $y =$ the amount of vinegar needed to make $2\frac{1}{2}$ batches of chili.

*Translate.* We write a multiplication sentence that fits the situation.

$$y = 2\frac{1}{2} \cdot \left(\frac{3}{4}\right)$$

*Solve.* We do the computation.

$$y = 2\frac{1}{2} \cdot \left(\frac{3}{4}\right)$$

$$y = \frac{5}{2} \cdot \frac{3}{4} \qquad \text{Writing } 2\frac{1}{2} \text{ as } \frac{5}{2}$$

$$y = \frac{15}{8}, \text{ or } 1\frac{7}{8}$$

*Check*. We repeat the computation. The result checks.

*State*. $1\frac{7}{8}$ cups of vinegar are needed to make $2\frac{1}{2}$ batches of chili.

**61.** $\begin{aligned} -8 - 5^2 \cdot 2(3-4) &= -8 - 5^2 \cdot 2(-1) \\ &= -8 - 5 \cdot 5 \cdot 2(-1) \\ &= -8 - 25 \cdot 2(-1) \\ &= -8 - 50(-1) \\ &= -8 + 50 \\ &= 42 \end{aligned}$

**63.**
$$4.8 - 1.5x = 0.9$$
$$4.8 - 1.5x - 4.8 = 0.9 - 4.8 \quad \text{Subtracting 4.8 on both sides}$$
$$-1.5x = -3.9$$
$$\frac{-1.5x}{-1.5} = \frac{-3.9}{-1.5} \quad \text{Dividing by } -5 \text{ on both sides}$$
$$x = 2.6$$

The solution is 2.6.

**65.** ◈

**67.** ◈

**69.** Replace $x$ with $-3$:
$$\begin{aligned} 50x + 75y &= 180 \\ 50(-3) + 75y &= 180 \\ -150 + 75y &= 180 \\ 75y &= 330 \\ y &= 4.4 \end{aligned}$$

$(-3, 4.4)$ is one solution.

Replace $y$ with 5:
$$\begin{aligned} 50x + 75y &= 180 \\ 50x + 75 \cdot 5 &= 180 \\ 50x + 375 &= 180 \\ 50x &= -195 \\ x &= -3.9 \end{aligned}$$

$(-3.9, 5)$ is a second solution.

Replace $x$ with 3:
$$\begin{aligned} 50x + 75y &= 180 \\ 50 \cdot 3 + 75y &= 180 \\ 150 + 75y &= 180 \\ 75y &= 30 \\ y &= 0.4 \end{aligned}$$

$(3, 0.4)$ is a third solution.

We graph the equation.

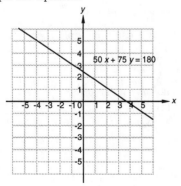

**71.** On the graph we locate three points, other than those already listed, whose coordinates are easily determined. We see that three solutions of $y = x + 2$ are $(2, 4)$, $(-3, -1)$, and $(-5, -3)$.

**73.** Answers may vary.

We substitute three negative values for $x$ to find three solutions in the second quadrant.

If $x = -4$, $y = |-4| = 4$.
If $x = -2$, $y = |-2| = 2$.
If $x = -1$, $y = |-1| = 1$.

Thus, three solutions in the second quadrant are $(-4, 4)$, $(-2, 2)$, and $(-1, 1)$.

We substitute three positive values for $x$ to find three solutions in the first quadrant.

If $x = 1$, $y = |1| = 1$.
If $x = 3$, $y = |3| = 3$.
If $x = 5$, $y = |5| = 5$.

Thus, three solutions in the first quadrant are $(1, 1)$, $(3, 3)$, and $(5, 5)$.

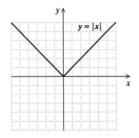

## Exercise Set 6.5

**1.** To find the mean, add the numbers. Then divide by the number of addends.

$$\frac{16 + 18 + 29 + 14 + 29 + 19 + 15}{7} = \frac{140}{7} = 20$$

The mean is 20.

To find the median, first list the numbers in order from smallest to largest. Then locate the middle number.

$$14, 15, 16, 18, 19, 29, 29$$
$$\uparrow$$
$$\text{Middle number}$$

The median is 18.

Find the mode:

The number that occurs most often is 29. The mode is 29.

**3.** To find the mean, add the numbers. Then divide by the number of addends.

$$\frac{5 + 30 + 20 + 20 + 35 + 5 + 25}{7} = \frac{140}{7} = 20$$

The mean is 20.

To find the median, first list the numbers in order from smallest to largest. Then locate the middle number.

$$5, 5, 20, 20, 25, 30, 35$$
$$\uparrow$$
$$\text{Middle number}$$

The median is 20.

Find the mode:

There are two numbers that occur most often, 5 and 20. Thus the modes are 5 and 20.

**5.** Find the mean:

$$\frac{1.2 + 4.3 + 5.7 + 7.4 + 7.4}{5} = \frac{26}{5} = 5.2$$

The mean is 5.2.

Find the median:

$$1.2, 4.3, 5.7, 7.4, 7.4$$
$$\uparrow$$
$$\text{Middle number}$$

The median is 5.7.

Find the mode:

The number that occurs most often is 7.4. The mode is 7.4.

**7.** Find the mean:

$$\frac{134 + 128 + 128 + 129 + 128 + 178}{6} = \frac{825}{6} = 137.5$$

The mean is 137.5.

Find the median:

$$128, 128, 128, 129, 134, 178$$
$$\uparrow$$
$$\text{Middle number}$$

The median is halfway between 128 and 129.

$$\frac{128 + 129}{2} = \frac{257}{2} = 128.5$$

The median is 128.5.

Find the mode:

The number that occurs most often is 128. The mode is 128.

**9.** Find the mean:

$$\frac{23 + 21 + 19 + 23 + 20}{5} = \frac{106}{5} = 21.2$$

The mean is 21.2.

Find the median:

$$19, 20, 21, 23, 23$$
$$\uparrow$$
$$\text{Middle number}$$

The median is 21.

Find the mode:

The number that occurs most often is 23. The mode is 23.

**11.** We divide the total number of miles, 297, by the number of gallons, 9.

$$\frac{297}{9} = 33$$

The mean was 33 miles per gallon.

**13.** To find the GPA we first add the grade point values for each hour taken. This is done by first multiplying the grade point value by the number of hours in the course and then adding as follows:

$$
\begin{array}{llr}
\text{B} & 3.0 \cdot 4 = & 12 \\
\text{A} & 4.0 \cdot 3 = & 12 \\
\text{B} & 3.0 \cdot 3 = & 9 \\
\text{C} & 2.0 \cdot 4 = & 8 \\
\hline
& & 41 \ \text{(Total)}
\end{array}
$$

The total number of hours taken is

$$4 + 3 + 3 + 4, \text{ or } 14.$$

We divide 41 by 14 and round to the nearest tenth.

$$\frac{41}{14} \approx 2.9$$

The student's grade point average is 2.9.

**15.** Find the mean price per pound:

$$\frac{\$7.99 + \$9.49 + \$9.99 + \$7.99 + \$10.49}{5} = \frac{\$45.95}{5} = \$9.19$$

The mean price per pound of Atlantic salmon was \$9.19.

Find the median price per pound:

List the prices in order:

$$\$7.99, \$7.99, \$9.49, \$9.99, \$10.49$$
$$\uparrow$$
$$\text{Middle number}$$

The median is \$9.49.

Find the mode:

The number that occurs most often is \$7.99. The mode is \$7.99.

**17.** We can find the total of the five scores needed as follows:

$$80 + 80 + 80 + 80 + 80 = 400.$$

The total of the scores on the first four tests is

$$80 + 74 + 81 + 75 = 310.$$

Thus Rich needs to get at least

$$400 - 310, \text{ or } 90$$

to get a B. We can check this as follows:

$$\frac{80 + 74 + 81 + 75 + 90}{5} = \frac{400}{5} = 80.$$

**19.** We can find the total number of days needed as follows:

$$266 + 266 + 266 + 266 = 1064.$$

The total number of days for Marta's first three pregnancies is

$$270 + 259 + 272 = 801.$$

Thus, Marta's fourth pregnancy must last

$$1064 - 801 = 263 \text{ days}$$

in order to equal the worldwide average.

We can check this as follows:

$$\frac{270 + 259 + 272 + 263}{4} = \frac{1064}{4} = 266.$$

**21.**
$$\begin{array}{r} 1\,4 \\ \times\,1\,4 \\ \hline 5\,6 \\ 1\,4\,0 \\ \hline 1\,9\,6 \end{array}$$

**23.**
$$\begin{array}{rl} 1.\,4 & \text{(1 decimal place)} \\ \times\,1.\,4 & \text{(1 decimal place)} \\ \hline 5\,6 & \\ 1\,4\,0 & \\ \hline 1.\,9\,6 & \text{(2 decimal places)} \end{array}$$

**25.** *Familiarize.* Let $h$ = the number of hours the disc jockey can work for \$165. The amount charged for setting-up and working $h$ hours is given by $40 + 50h$.

*Translate.*

$$\underbrace{\text{Amount charged}}_{\displaystyle 40 + 50h} \underset{\displaystyle =}{\text{ is }} \underset{\displaystyle 165}{\$165.}$$

*Solve.* We solve the equation.

$$40 + 50h = 165$$
$$40 + 50h - 40 = 165 - 40$$
$$50h = 125$$
$$h = \frac{125}{50} = 2.5$$

*Check.* At \$50 per hour, the disc jockey charges $50(2.5)$, or \$125, for working 2.5 hr. The total cost is \$125 plus the set-up fee of \$40, $\$125 + \$40$, or \$165. The answer checks.

*State.* The disc jockey can work for 2.5 hr.

**27.** ◈

**29.** ◈

**31.** Use a calculator to divide the total by the number of games.

$$\frac{4621}{27} \approx 171.15$$

Drop the amount to the right of the decimal point. The bowler's average was 171.

**33.** We can find the total number of home runs needed over Aaron's 22-yr career as follows:

$$22 \cdot 34\frac{7}{22} = 22 \cdot \frac{755}{22} = \frac{22 \cdot 755}{22} = \frac{22}{22} \cdot \frac{755}{1} = 755.$$

The total number of home runs during the first 21 years of Aaron's career was

$$21 \cdot 35\frac{10}{21} = 21 \cdot \frac{745}{21} = \frac{21 \cdot 745}{21} = \frac{21}{21} \cdot \frac{745}{1} = 745.$$

Then Aaron hit

$$755 - 745 = 10 \text{ home runs}$$

in his final year.

**35.** Amy's second offer: $\dfrac{\$3600 + \$3200}{2} = \$3400$

Jim's second offer: $\dfrac{\$3400 + \$3600}{2} = \$3500$

Amy's third offer: $\dfrac{\$3500 + \$3400}{2} = \$3450$

Jim's third offer: $\dfrac{\$3500 + \$3450}{2} = \$3475$

Amy will pay \$3475 for the car.

## Exercise Set 6.6

**1.** We interpolate by finding the average of the data values for 17 hours of study and 21 hours of study.
$$\frac{80 + 86}{2} = \frac{166}{2} = 83$$
The missing data value is 83.

We could have also used a graph to find this value, as in Example 1.

**3.** Use the line graph in Exercise Set 6.2, Exercise 37, to extrapolate. Drawing a "representative" line through the data and beyond gives an estimate of about 3112 parts per billion for the missing data value. Answers will vary according to the placement of the representative line.

**5.** Graph the data and use the graph to extrapolate. Drawing a "representative" line through the data and beyond gives an estimate of about \$148.8 billion for the missing data value. Answers will vary according to the accuracy of the graph and the placement of the representative line.

**7.** We interpolate by finding the average of the rates for 6 lb and 8 lb.
$$\frac{\$24.90 + \$29.70}{2} = \frac{\$54.60}{2} = \$27.30$$
The missing data values is \$27.30.

We could have also used a graph to find this value, as in Example 1.

**9.** Since 1, 2, 3, 4, 5, or 6 are equally likely to occur, the probability that a 3 is rolled is $\dfrac{1}{6}$, or $0.1\overline{6}$.

**11.** Since 1, 2, 3, 4, 5, or 6 are equally likely to occur and 3 of these possibilities are odd numbers, the probability of rolling an odd number is:

$$\frac{\text{Number of ways to roll an odd number}}{\text{Number of equally likely outcomes}}$$
$$= \frac{3}{6}$$
$$= \frac{1}{2}, \text{ or } 0.5$$

**13.** The probability that the card is the ace of spades is:

$$\frac{\text{Number of ways to select the ace of spades}}{\text{Number of ways to select any card}}$$

$$= \frac{1}{52}$$

**15.** The probability that an eight or six is selected is:

$$\frac{\text{Number of ways to select an 8 or a 6}}{\text{Number of ways to select any card}}$$

$$= \frac{8}{52}$$

$$= \frac{2}{13}$$

**17.** The probability of selecting a black picture card is:

$$\frac{\text{Number of ways to select a black picture card}}{\text{Number of ways to select any card}}$$

$$= \frac{6}{52}$$

$$= \frac{3}{26}$$

**19.** The probability that a cherry gumdrop is selected is:

$$\frac{\text{Number of ways to select a cherry gumdrop}}{\text{Number of ways to select any gumdrop}}$$

$$= \frac{4}{39}$$

**21.** The probability that a gumdrop that is not lime is selected is:

$$\frac{\text{Number of ways to select a non-lime gumdrop}}{\text{Number of ways to select any gumdrop}}$$

$$= \frac{34}{39}$$

**23.**

$$-3x + 8 = 2x - 7$$

$$-3x + 8 - 8 = 2x - 7 - 8 \qquad \text{Subtracting 8 on both sides}$$

$$-3x = 2x - 15$$

$$-3x - 2x = 2x - 15 - 2x \qquad \text{Subtracting } 2x \text{ on both sides}$$

$$-5x = -15$$

$$\frac{-5x}{-5} = \frac{-15}{-5} \qquad \text{Dividing by } -5 \text{ on both sides}$$

$$x = 3$$

The solution is 3.

**25.**

$$-7 + 3x - 5 = 8x - 1$$

$$3x - 12 = 8x - 1$$

$$3x - 12 + 12 = 8x - 1 + 12 \qquad \text{Adding 12 on both sides}$$

$$3x = 8x + 11$$

$$3x - 8x = 8x + 11 - 8x \qquad \text{Subtracting } 8x \text{ on both sides}$$

$$-5x = 11$$

$$\frac{-5x}{-5} = \frac{11}{-5} \qquad \text{Dividing by } -5 \text{ on both sides}$$

$$x = -\frac{11}{5}$$

The solution is $-\frac{11}{5}$.

**27.** $\dfrac{17}{15} = 17 \div 15$

$$
\begin{array}{r}
1.133 \\
15\overline{\smash)17.000} \\
\underline{15\phantom{.000}} \\
20\phantom{00} \\
\underline{15\phantom{00}} \\
50\phantom{0} \\
\underline{45\phantom{0}} \\
50 \\
\underline{45} \\
5
\end{array}
$$

$$\frac{17}{15} = 1.1\overline{3}$$

**29.** ◈

**31.** ◈

**33.** The probability that the first flip produces a head and the second a tail is $\frac{1}{2} \cdot \frac{1}{2}$, or $\frac{1}{4}$.

The probability that the first flip produces a tail and the second a head is $\frac{1}{2} \cdot \frac{1}{2}$, or $\frac{1}{4}$. Then the probability that one head and one tail will occur is $\frac{1}{4} + \frac{1}{4}$, or $\frac{1}{2}$, or 0.5.

**35.** There are 31 days in July and 366 days in a leap year. Thus, the probability that a day chosen randomly during a leap year is in July is $\frac{31}{366}$.

# Chapter 7

# Ratio and Proportion

## Exercise Set 7.1

**1.** The ratio of 4 to 5 is $\dfrac{4}{5}$.

**3.** The ratio of 178 to 572 is $\dfrac{178}{572}$.

**5.** The ratio of 0.4 to 12 is $\dfrac{0.4}{12}$.

**7.** The ratio of 3.8 to 7.4 is $\dfrac{3.8}{7.4}$.

**9.** The ratio of 56.78 to 98.35 is $\dfrac{56.78}{98.35}$.

**11.** The ratio of $8\dfrac{3}{4}$ to $9\dfrac{5}{6}$ is $\dfrac{8\frac{3}{4}}{9\frac{5}{6}}$.

**13.** The ratio of those who play an instrument to the total number of people is $\dfrac{1}{4}$.

If one person in four plays an instrument, then $4 - 1$, or 3, do not play an instrument. Thus the ratio of those who do not play an instrument to those who do is $\dfrac{3}{1}$.

**15.** If four of every five fatal accidents involving a Corvette do not involve another vehicle, then $5 - 4$, or 1, involves a Corvette and at least one other vehicle. Thus, the ratio of fatal accidents involving just a Corvette to those involving a Corvette and at least one other vehicle is $\dfrac{4}{1}$.

**17.** The ratio of length to width is $\dfrac{478}{213}$.

The ratio of width to length is $\dfrac{213}{478}$.

**19.** The ratio of 4 to 6 is $\dfrac{4}{6} = \dfrac{2 \cdot 2}{2 \cdot 3} = \dfrac{2}{2} \cdot \dfrac{2}{3} = \dfrac{2}{3}$.

**21.** The ratio of 18 to 24 is $\dfrac{18}{24} = \dfrac{3 \cdot 6}{4 \cdot 6} = \dfrac{3}{4} \cdot \dfrac{6}{6} = \dfrac{3}{4}$.

**23.** The ratio of 4.8 to 10 is $\dfrac{4.8}{10} = \dfrac{4.8}{10} \cdot \dfrac{10}{10} = \dfrac{48}{100} = \dfrac{4 \cdot 12}{4 \cdot 25} = \dfrac{4}{4} \cdot \dfrac{12}{25} = \dfrac{12}{25}$.

**25.** The ratio of 2.8 to 3.6 is $\dfrac{2.8}{3.6} = \dfrac{2.8}{3.6} \cdot \dfrac{10}{10} = \dfrac{28}{36} = \dfrac{4 \cdot 7}{4 \cdot 9} = \dfrac{4}{4} \cdot \dfrac{7}{9} = \dfrac{7}{9}$.

**27.** The ratio is $\dfrac{20}{30} = \dfrac{2 \cdot 10}{3 \cdot 10} = \dfrac{2}{3} \cdot \dfrac{10}{10} = \dfrac{2}{3}$.

**29.** The ratio is $\dfrac{56}{100} = \dfrac{4 \cdot 14}{4 \cdot 25} = \dfrac{4}{4} \cdot \dfrac{14}{25} = \dfrac{14}{25}$.

**31.** The ratio is $\dfrac{128}{256} = \dfrac{1 \cdot 128}{2 \cdot 128} = \dfrac{1}{2} \cdot \dfrac{128}{128} = \dfrac{1}{2}$.

**33.** The ratio is $\dfrac{0.48}{0.64} = \dfrac{0.48}{0.64} \cdot \dfrac{100}{100} = \dfrac{48}{64} = \dfrac{3 \cdot 16}{4 \cdot 16} = \dfrac{3}{4} \cdot \dfrac{16}{16} = \dfrac{3}{4}$.

**35.** The ratio is $\dfrac{51}{49}$. It cannot be simplified further.

**37.** The ratio is $\dfrac{6.4}{20.2} = \dfrac{6.4}{20.2} \cdot \dfrac{10}{10} = \dfrac{64}{202} = \dfrac{2 \cdot 32}{2 \cdot 101} = \dfrac{2}{2} \cdot \dfrac{32}{101} = \dfrac{32}{101}$.

**39.** $-\dfrac{5}{6} \ \square \ -\dfrac{3}{4}$, or $\dfrac{-5}{6} \ \square \ \dfrac{-3}{4}$

The LCD is 12.

$\dfrac{-5}{6} \cdot \dfrac{2}{2} = \dfrac{-10}{12}, \ \dfrac{-3}{4} \cdot \dfrac{3}{3} = \dfrac{-9}{12}$

Since $-10 < -9$, it follows that $\dfrac{-10}{12} < \dfrac{-9}{12}$, or $-\dfrac{5}{6} < -\dfrac{3}{4}$.

**41.** $\dfrac{5}{9} \ \square \ \dfrac{6}{11}$

The LCD is 99.

$\dfrac{5}{9} \cdot \dfrac{11}{11} = \dfrac{55}{99}, \ \dfrac{6}{11} \cdot \dfrac{9}{9} = \dfrac{54}{99}$

Since $55 > 54$, it follows that $\dfrac{55}{99} > \dfrac{54}{99}$, or $\dfrac{5}{9} > \dfrac{6}{11}$.

**43.** *Familiarize*. We let $h$ = Rocky's excess height.

*Translate*. We have a "how much more" situation.

| Height of daughter | plus | How much more height | is | Rocky's height |
|:---:|:---:|:---:|:---:|:---:|
| ↓ | ↓ | ↓ | ↓ | ↓ |
| $180\frac{3}{4}$ | $+$ | $h$ | $=$ | $187\frac{1}{10}$ |

*Solve*. We solve the equation as follows:

$$h = 187\frac{1}{10} - 180\frac{3}{4}$$

$$187\ \boxed{\dfrac{1}{10} \cdot \dfrac{2}{2}} = 187\frac{2}{20}$$

$$180\ \boxed{\dfrac{3}{4} \cdot \dfrac{5}{5}} = 180\frac{15}{20}$$

$$
\begin{array}{rcrcr}
187\frac{1}{10} & = & 187\frac{2}{20} & = & 186\frac{22}{20} \\
-\,180\frac{3}{4} & = & -\,180\frac{15}{20} & = & -\,180\frac{15}{20} \\
\hline
& & & & 6\frac{7}{20}
\end{array}
$$

Thus, $h = 6\dfrac{7}{20}$.

*Check*. We add Rocky's excess height to his daughter's height:

$$180\frac{3}{4} + 6\frac{7}{20} = 180\frac{15}{20} + 6\frac{7}{20} = 186\frac{22}{20} = 187\frac{2}{20} = 187\frac{1}{10}$$

The answer checks.

*State*. Rocky is $6\frac{7}{20}$ cm taller.

**45.** ◈

**47.** ◈

**49.** $\dfrac{\$937,905,284}{\$927,334,416} \approx 1.011399197$

The ratio is 1.011399197 to 1.

**51.** Potassium to nitrogen: $\dfrac{5}{15} = \dfrac{5 \cdot 1}{5 \cdot 3} = \dfrac{1}{3}$

Nitrogen to phosphorus: $\dfrac{15}{10} = \dfrac{3 \cdot 5}{2 \cdot 5} = \dfrac{3}{2}$

## Exercise Set 7.2

**1.** $\dfrac{120 \text{ km}}{3 \text{ hr}}$, or $40 \dfrac{\text{km}}{\text{hr}}$

**3.** $\dfrac{440 \text{ m}}{40 \text{ sec}}$, or $11 \dfrac{\text{m}}{\text{sec}}$

**5.** $\dfrac{342 \text{ yd}}{2.25 \text{ days}}$, or $152 \dfrac{\text{yd}}{\text{day}}$

$$
\begin{array}{r}
1\;5\;2.\phantom{0} \\
2.2\,5_\wedge\overline{)\;3\;4\;2.0\;0_\wedge} \\
2\;2\;5\;0\;0 \\
\overline{\phantom{00}1\;1\;7\;0\;0} \\
1\;1\;2\;5\;0 \\
\overline{\phantom{000}4\;5\;0} \\
4\;5\;0 \\
\overline{\phantom{0000}0}
\end{array}
$$

**7.** $\dfrac{500 \text{ mi}}{20 \text{ hr}} = 25 \dfrac{\text{mi}}{\text{hr}}$

$\dfrac{20 \text{ hr}}{500 \text{ mi}} = 0.04 \dfrac{\text{hr}}{\text{mi}}$

**9.** $\dfrac{\$5.75}{10 \text{ min}} = \dfrac{575\cancel{c}}{10 \text{ min}} = 57.5 \dfrac{\cancel{c}}{\text{min}}$

**11.** $\dfrac{623 \text{ gal}}{1000 \text{ sq ft}} = 0.623 \text{ gal/ft}^2$

**13.** $\dfrac{66,000 \text{ ft}}{1 \text{ min}} = \dfrac{66,000 \text{ ft}}{60 \text{ sec}} = 1100 \text{ ft/sec}$

**15.** $\dfrac{310 \text{ km}}{2.5 \text{ hr}} = 124 \dfrac{\text{km}}{\text{hr}}$

**17.** $\dfrac{2660 \text{ mi}}{4.75 \text{ hr}} = 560 \dfrac{\text{mi}}{\text{hr}}$

**19.** Unit price $= \dfrac{\text{Price}}{\text{Number of units}} = \dfrac{\$165.75}{8.5 \text{ yd}} = 19.5 \dfrac{\text{dollars}}{\text{yd}}$, or \$19.50/yd

**21.** We need to find the number of ounces in 2 pounds:

$$2 \text{ lb} = 2 \times 1 \text{ lb} = 2 \times 16 \text{ oz} = 32 \text{ oz}$$

Unit price $= \dfrac{\text{Price}}{\text{Number of units}} = \dfrac{\$6.59}{32 \text{ oz}} = \dfrac{659\cancel{c}}{32 \text{ oz}} \approx 20.59\dfrac{\cancel{c}}{\text{oz}}$

**23.** $\dfrac{\$2.89}{\frac{2}{3} \text{ lb}} = 2.89 \times \dfrac{3}{2} \times \dfrac{\text{dollars}}{\text{lb}} \approx 4.34\dfrac{\text{dollars}}{\text{lb}}$, or \$4.34/lb

**25.** Compare the unit prices.

For Tico's: $\dfrac{\$3.79}{18 \text{ oz}} \approx \$0.211/\text{oz}$

For Sure Fire: $\dfrac{\$3.49}{16 \text{ oz}} \approx \$0.218/\text{oz}$

Thus, Tico's has the lower unit price.

**27.** Compare the unit prices. Recall that 1 qt = 32 oz, so 2 qt = 2 × 1 qt = 2 × 32 oz = 64 oz.

For Sunbeam: $\dfrac{\$2.79}{64 \text{ oz}} \approx \$0.04359/\text{oz}$

For Dell's: $\dfrac{\$2.09}{48 \text{ oz}} \approx \$0.04354/\text{oz}$

Thus, Dell's has the lower unit price.

**29.** Compare the unit prices.

For Shine: $\dfrac{\$2.19}{3 \text{ bars}} = \$0.73/\text{bar}$

For Pristine: $\dfrac{\$1.58}{2 \text{ bars}} = \$0.79/\text{bar}$

Thus, Shine has the lower unit price.

**31.** Compare the unit prices.

For Tina's: $\dfrac{\$1.19}{6\frac{1}{8} \text{ oz}} \approx \$0.194/\text{oz}$

For Big Net: $\dfrac{\$1.11}{6 \text{ oz}} \approx \$0.185/\text{oz}$

Thus, Big Net has the lower unit price.

**33.** Six 10-oz bottles contain 6 × 10 oz = 60 oz of sparkling water; four 12-oz bottles contain 4 × 12 oz = 48 oz of sparkling water.

Compare the unit prices.

Six 10-oz bottles: $\dfrac{\$3.09}{60 \text{ oz}} = \$0.0515/\text{oz}$

Four 12-oz bottles: $\dfrac{\$2.39}{48 \text{ oz}} \approx \$0.0498/\text{oz}$

Thus, four 12-oz bottles have the lower unit price.

**35.** Compare the unit prices.

For Package A: $\dfrac{99\cancel{c}}{5.9 \text{ oz}} \approx 16.8\cancel{c}/\text{oz}$

For Package B: $\dfrac{63\cancel{c}}{3.9 \text{ oz}} \approx 16.2\cancel{c}/\text{oz}$

Thus, Package B has the lower unit price.

**37.** Compare the unit prices. Note that two 10.5-oz cans contain 2(10.5 oz), or 21 oz, of soup.

Big Chunk: $\dfrac{\$1.59}{21 \text{ oz}} \approx \$0.076/\text{oz}$

Bert's: $\dfrac{\$0.82}{11 \text{ oz}} \approx \$0.075/\text{oz}$

Thus, Bert's has the lower unit price.

**39.** *Familiarize.* We visualize the situation. We let $p =$ the number by which the number of piano players exceeds the number of guitar players, in millions.

| 18.9 million | $p$ |
|---|---|
| 20.6 million | |

*Translate.* This is a "how-much-more" situation.

$$\underbrace{\text{Number of guitar players}}_{18.9} + \underbrace{\text{Additional number of piano players}}_{p} = \underbrace{\text{Number of piano players}}_{20.6}$$

*Solve.* To solve the equation we subtract 18.9 on both sides.

$$p = 20.6 - 18.9$$
$$p = 1.7$$

$$\begin{array}{r} {\scriptstyle 1\ 9\ 16} \\ \cancel{2\,0.6} \\ -\ 1\,8.9 \\ \hline 1.7 \end{array}$$

*Check.* We repeat the calculation.

*State.* There are 1.7 million more piano players than guitar players.

**41.**
$$\begin{array}{r} 4\,5.6\,7 \\ \times\qquad 2.4 \\ \hline 1\,8\,2\,6\,8 \\ 9\,1\,3\,4\,0 \\ \hline 1\,0\,9.6\,0\,8 \end{array}$$

**43.** The first coordinate is positive and the second coordinate is negative, so the point is in quadrant IV.

**45.** ◈

**47.** ◈

**49.** The unit price of the 16-oz bottle was $\dfrac{64¢}{16 \text{ oz}} = 4\dfrac{¢}{\text{oz}}$. The unit price of the 20-oz bottle when it was introduced was $\dfrac{64¢}{20 \text{ oz}} = 3.2\dfrac{¢}{\text{oz}}$. After about a month the unit price of the 20-oz bottle became $\dfrac{80¢}{20 \text{ oz}} = 4\dfrac{¢}{\text{oz}}$. Thus, although there was no net change in the unit price, for about a month the unit price decreased by $4\dfrac{¢}{\text{oz}} - 3.2\dfrac{¢}{\text{oz}}$, or $0.8\dfrac{¢}{\text{oz}}$.

**51.** We find the area of a 14-in. pizza:

$r = 14 \text{ in.} \div 2 = 7 \text{ in.}$

$A = \pi r^2 = 3.14(7 \text{ in.})^2 = 3.14(49 \text{ in}^2) = 153.86 \text{ in}^2$

Now we find the area of a 16-in. pizza:

$r = 16 \text{ in.} \div 2 = 8 \text{ in.}$

$A = \pi r^2 = 3.14(8 \text{ in.})^2 = 3.14(64 \text{ in}^2) = 200.96 \text{ in}^2$

We compare the unit prices.

14-in. pizza: $\dfrac{\$10.50}{153.86 \text{ in}^2} \approx \$0.068/\text{in}^2$

16-in. pizza: $\dfrac{\$11.95}{200.96 \text{ in}^2} \approx \$0.059/\text{in}^2$

The 16-in. pizza is a better buy.

**53.** $\dfrac{509 \text{ at-bats}}{70 \text{ home runs}} \approx 7.27$ at-bats per home run

## Exercise Set 7.3

**1.** We can use cross-products:

$$5 \cdot 9 = 45 \quad \begin{matrix} 5 & 7 \\ 6 & 9 \end{matrix} \quad 6 \cdot 7 = 42$$

Since the cross-products are not the same, $45 \neq 42$, we know that the numbers are not proportional.

**3.** We can use cross-products:

$$1 \cdot 20 = 20 \quad \begin{matrix} 1 & 10 \\ 2 & 20 \end{matrix} \quad 2 \cdot 10 = 20$$

Since the cross-products are the same, $20 = 20$, we know that $\dfrac{1}{2} = \dfrac{10}{20}$, so the numbers are proportional.

**5.** We can use cross-products:

$$2.4 \cdot 2.7 = 6.48 \quad \begin{matrix} 2.4 & 1.8 \\ 3.6 & 2.7 \end{matrix} \quad 3.6 \cdot 1.8 = 6.48$$

Since the cross-products are the same, $6.48 = 6.48$, we know that $\dfrac{2.4}{3.6} = \dfrac{1.8}{2.7}$, so the numbers are proportional.

**7.** We can use cross-products:

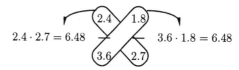

$$5\tfrac{1}{3} \cdot 9\tfrac{1}{2} = 50\tfrac{2}{3} \qquad 8\tfrac{1}{4} \cdot 2\tfrac{1}{5} = 18\tfrac{3}{20}$$

Since the cross-products are not the same, $50\tfrac{2}{3} \neq 18\tfrac{3}{20}$, we know that the numbers are not proportional.

**9.**
$$\frac{18}{4} = \frac{x}{10}$$

$18 \cdot 10 = 4 \cdot x$     Equating cross-products

$\dfrac{18 \cdot 10}{4} = x$    Dividing by 4

$\dfrac{180}{4} = x$    Multiplying

$45 = x$    Dividing

The solution is 45.

**11.** $\dfrac{x}{8} = \dfrac{9}{6}$

$6 \cdot x = 8 \cdot 9$   Equating cross-products

$x = \dfrac{8 \cdot 9}{6}$   Dividing by 6

$x = \dfrac{72}{6}$   Multiplying

$x = 12$   Dividing

The solution is 12.

**13.** $\dfrac{t}{12} = \dfrac{5}{6}$

$6 \cdot t = 12 \cdot 5$

$t = \dfrac{12 \cdot 5}{6}$

$t = \dfrac{60}{6}$

$t = 10$

The solution is 10.

**15.** $\dfrac{2}{5} = \dfrac{8}{n}$

$2 \cdot n = 5 \cdot 8$

$n = \dfrac{5 \cdot 8}{2}$

$n = \dfrac{40}{2}$

$n = 20$

The solution is 20.

**17.** $\dfrac{n}{15} = \dfrac{10}{30}$

$30 \cdot n = 15 \cdot 10$

$n = \dfrac{15 \cdot 10}{30}$

$n = \dfrac{150}{30}$

$n = 5$

The solution is 5.

**19.** $\dfrac{16}{12} = \dfrac{24}{x}$

$16 \cdot x = 12 \cdot 24$

$x = \dfrac{12 \cdot 24}{16}$

$x = \dfrac{288}{16}$

$x = 18$

The solution is 18.

**21.** $\dfrac{6}{11} = \dfrac{12}{x}$

$6 \cdot x = 11 \cdot 12$

$x = \dfrac{11 \cdot 12}{6}$

$x = \dfrac{132}{6}$

$x = 22$

The solution is 22.

**23.** $\dfrac{20}{7} = \dfrac{80}{x}$

$20 \cdot x = 7 \cdot 80$

$x = \dfrac{7 \cdot 80}{20}$

$x = \dfrac{560}{20}$

$x = 28$

The solution is 28.

**25.** $\dfrac{12}{9} = \dfrac{x}{7}$

$12 \cdot 7 = 9 \cdot x$

$\dfrac{12 \cdot 7}{9} = x$

$\dfrac{84}{9} = x$

$\dfrac{28}{3} = x$   Simplifying

$9\dfrac{1}{3} = x$   Writing a mixed numeral

The solution is $9\dfrac{1}{3}$.

**27.** $\dfrac{x}{13} = \dfrac{2}{9}$

$9 \cdot x = 13 \cdot 2$

$x = \dfrac{13 \cdot 2}{9}$

$x = \dfrac{26}{9}$, or $2\dfrac{8}{9}$

The solution is $\dfrac{26}{9}$, or $2\dfrac{8}{9}$.

**29.** $\dfrac{t}{0.16} = \dfrac{0.15}{0.40}$

$0.40 \times t = 0.16 \times 0.15$

$t = \dfrac{0.16 \times 0.15}{0.40}$

$t = \dfrac{0.024}{0.40}$

$t = 0.06$

The solution is 0.06.

**31.** $\dfrac{100}{25} = \dfrac{20}{n}$

$100 \cdot n = 25 \cdot 20$

$n = \dfrac{25 \cdot 20}{100}$

$n = \dfrac{500}{100}$

$n = 5$

The solution is 5.

**33.** $\dfrac{7}{\frac{1}{4}} = \dfrac{28}{x}$

$7 \cdot x = \dfrac{1}{4} \cdot 28$

$x = \dfrac{\frac{1}{4} \cdot 28}{7}$

$x = \dfrac{7}{7}$

$x = 1$

The solution is 1.

**35.** $\dfrac{\frac{1}{4}}{\frac{1}{2}} = \dfrac{\frac{1}{2}}{x}$

$\dfrac{1}{4} \cdot x = \dfrac{1}{2} \cdot \dfrac{1}{2}$

$x = \dfrac{\frac{1}{2} \cdot \frac{1}{2}}{\frac{1}{4}}$

$x = \dfrac{\frac{1}{4}}{\frac{1}{4}}$

$x = 1$

The solution is 1.

**37.** $\dfrac{x}{\frac{4}{5}} = \dfrac{0}{\frac{9}{11}}$

$\dfrac{9}{11} \cdot x = \dfrac{4}{5} \cdot 0$

$x = \dfrac{\frac{4}{5} \cdot 0}{\frac{9}{11}}$

$x = \dfrac{0}{\frac{9}{11}}$

$x = 0$

The solution is 0.

**39.** $\dfrac{2\frac{1}{2}}{3\frac{1}{3}} = \dfrac{x}{4\frac{1}{4}}$

$2\dfrac{1}{2} \cdot 4\dfrac{1}{4} = 3\dfrac{1}{3} \cdot x$

$\dfrac{5}{2} \cdot \dfrac{17}{4} = \dfrac{10}{3} \cdot x$

$\dfrac{3}{10} \cdot \dfrac{5}{2} \cdot \dfrac{17}{4} = x \qquad \text{Dividing by } \dfrac{10}{3}$

$\dfrac{3}{\cancel{5} \cdot 2} \cdot \dfrac{\cancel{5}}{2} \cdot \dfrac{17}{4} = x$

$\dfrac{3 \cdot 17}{2 \cdot 2 \cdot 4} = x$

$\dfrac{51}{16} = x, \text{ or}$

$3\dfrac{3}{16} = x$

The solution is $\dfrac{51}{16}$, or $3\dfrac{3}{16}$.

**41.** $\dfrac{1.28}{3.76} = \dfrac{4.28}{y}$

$1.28 \times y = 3.76 \times 4.28$

$y = \dfrac{3.76 \times 4.28}{1.28}$

$y = \dfrac{16.0928}{1.28}$

$y = 12.5725$

The solution is 12.5725.

**43.** $\dfrac{10\frac{3}{8}}{12\frac{2}{3}} = \dfrac{5\frac{3}{4}}{y}$

$10\dfrac{3}{8} \cdot y = 12\dfrac{2}{3} \cdot 5\dfrac{3}{4}$

$\dfrac{83}{8} \cdot y = \dfrac{38}{3} \cdot \dfrac{23}{4}$

$y = \dfrac{38}{3} \cdot \dfrac{23}{4} \cdot \dfrac{8}{83} \qquad \text{Dividing by } \dfrac{83}{3}$

$y = \dfrac{38}{3} \cdot \dfrac{23}{\cancel{4}} \cdot \dfrac{2 \cdot \cancel{4}}{83}$

$y = \dfrac{38 \cdot 23 \cdot 2}{3 \cdot 83}$

$y = \dfrac{1748}{249}, \text{ or } 7\dfrac{5}{249}$

The solution is $\dfrac{1748}{249}$, or $7\dfrac{5}{249}$.

**45.** Both coordinates are negative, so the point is in quadrant III.

**47.** The first coordinate is positive and the second is negative, so the point is in quadrant IV.

**49.** $260 \div (-5)$

First consider $260 \div 5$.

```
     5 2
5 | 2 6 0
    2 5 0
    ─────
      1 0
      1 0
    ─────
        0
```

Since a positive number divided by a negative number is negative, the answer is $-52$.

**51.**
```
        2 9 0. 5
16 | 4 6 4 8. 0
     3 2 0 0
     ───────
     1 4 4 8
     1 4 4 0
     ───────
         8 0
         8 0
     ───────
          0
```

The answer is 290.5.

**53.** ◈

**55.** ◈

**57.**
$$\frac{328.56}{627.48} = \frac{y}{127.66}$$
$$328.56 \times 127.66 = 627.48 \times y$$
$$\frac{328.56 \times 127.66}{627.48} = y$$
$$66.85 \approx y$$

The solution is about 66.85.

**59.**
$$\frac{x+3}{5} = \frac{x}{7}$$
$$7(x+3) = 5 \cdot x$$
$$7x + 21 = 5x$$
$$7x + 21 - 7x = 5x - 7x$$
$$21 = -2x$$
$$\frac{21}{-2} = \frac{-2x}{-2}$$
$$-10.5 = x$$

The solution is $-10.5$.

---

## Exercise Set 7.4

---

**1.** Let $g =$ the number of gallons of gasoline needed to travel 126 mi.

Miles $\;\rightarrow\; \dfrac{84}{6.5} = \dfrac{126}{g} \;\leftarrow\;$ Miles
Gallons $\rightarrow$ $\phantom{\dfrac{84}{6.5}}$ $\leftarrow$ Gallons

Solve:  $84 \cdot g = 6.5 \cdot 126$   Equating cross products

$\qquad g = \dfrac{6.5 \cdot 126}{84}$   Dividing by 84

$\qquad g = \dfrac{819}{84}$

$\qquad g = 9.75$

9.75 gallons of gasoline are needed to travel 126 mi.

**3.** Let $d =$ the number of defective bulbs in a lot of 2500.

Defective bulbs $\rightarrow \dfrac{7}{100} = \dfrac{d}{2500} \leftarrow$ Defective bulbs
Bulbs in lot $\;\;\rightarrow$ $\phantom{\dfrac{7}{100}}$ $\leftarrow$  Bulbs in lot

Solve:  $7 \cdot 2500 = 100 \cdot d$

$\qquad \dfrac{7 \cdot 2500}{100} = d$

$\qquad \dfrac{7 \cdot 25 \cdot 100}{100} = d$

$\qquad 7 \cdot 25 = d$

$\qquad 175 = d$

There will be 175 defective bulbs in a lot of 2500.

**5.** Let $s =$ the number of square feet of siding that Fred can paint with 7 gal of paint.

Gallons $\rightarrow \dfrac{3}{1275} = \dfrac{7}{s} \leftarrow$ Gallons
Siding $\;\rightarrow$ $\phantom{\dfrac{3}{1275}}$ $\leftarrow$ Siding

Solve:  $3 \cdot s = 1275 \cdot 7$

$\qquad s = \dfrac{1275 \cdot 7}{3}$

$\qquad s = \dfrac{3 \cdot 425 \cdot 7}{3}$

$\qquad s = 425 \cdot 7$

$\qquad s = 2975$

Fred can paint 2975 ft$^2$ of siding with 7 gal of paint.

**7.** Let $p =$ the number of published pages in a 540-page manuscript.

Published pages $\rightarrow \dfrac{5}{6} = \dfrac{p}{540} \leftarrow$ Published pages
Manuscript $\;\;\rightarrow$ $\phantom{\dfrac{5}{6}}$ $\leftarrow$  Manuscript

Solve:  $5 \cdot 540 = 6 \cdot p$

$\qquad \dfrac{5 \cdot 540}{6} = p$

$\qquad \dfrac{5 \cdot 6 \cdot 90}{6} = p$

$\qquad 5 \cdot 90 = p$

$\qquad 450 = p$

A 540-page manuscript will become 450 published pages.

**9.** Let $s =$ the number of students estimated to be in the class.

Class size $\rightarrow \dfrac{40}{6} = \dfrac{s}{9} \leftarrow$ Class size
Lefties $\;\;\rightarrow$ $\phantom{\dfrac{40}{6}}$ $\leftarrow$  Lefties

Solve:  $40 \cdot 9 = 6 \cdot s$

$\qquad \dfrac{40 \cdot 9}{6} = s$

$\qquad \dfrac{2 \cdot 20 \cdot 3 \cdot 3}{2 \cdot 3} = s$

$\qquad 20 \cdot 3 = s$

$\qquad 60 = s$

If a class includes 9 lefties, we estimate that there are 60 students in the class.

**11.** Let $m =$ the number of miles the car will be driven in 1 year. Note that 1 year $=$ 12 months.

Months $\rightarrow \dfrac{8}{9000} = \dfrac{12}{m} \leftarrow$ Months
Miles $\;\;\rightarrow$ $\phantom{\dfrac{8}{9000}}$ $\leftarrow$  Miles

Solve:  $8 \cdot m = 9000 \cdot 12$

$\qquad m = \dfrac{9000 \cdot 12}{8}$

$\qquad m = \dfrac{2 \cdot 4500 \cdot 3 \cdot 4}{2 \cdot 4}$

$\qquad m = 4500 \cdot 3$

$\qquad m = 13{,}500$

At the given rate, the car will be driven 13,500 mi in one year.

**13.** Let $z =$ the number of pounds of zinc.

Zinc $\;\rightarrow \dfrac{3}{13} = \dfrac{z}{520} \leftarrow$ Zinc
Copper $\rightarrow$ $\phantom{\dfrac{3}{13}}$ $\leftarrow$  Copper

Solve:  $3 \cdot 520 = 13 \cdot z$

$\qquad \dfrac{3 \cdot 520}{13} = z$

$\qquad \dfrac{3 \cdot 13 \cdot 40}{13} = z$

$\qquad 3 \cdot 40 = z$

$\qquad 120 = z$

There are 120 lb of zinc in the alloy.

**15.** Let $p =$ the number of gallons of paint Helen should buy.

$$\text{Area} \rightarrow \frac{950}{2} = \frac{30{,}000}{p} \leftarrow \text{Area}$$
$$\text{Paint} \rightarrow \qquad\quad\; \leftarrow \text{Paint}$$

Solve: $950 \cdot p = 2 \cdot 30{,}000$

$$p = \frac{2 \cdot 30{,}000}{950}$$

$$p = \frac{2 \cdot 50 \cdot 600}{19 \cdot 50}$$

$$p = \frac{2 \cdot 600}{19}$$

$$p = \frac{1200}{19}, \text{ or } 63\frac{3}{19}$$

Assuming that Helen is buying paint in one gallon cans, she will have to buy 64 cans of paint.

**17.** Let $d =$ the distance between the cities in reality, in miles.

$$\text{Map distance} \rightarrow \frac{\frac{1}{4}}{50} = \frac{3\frac{1}{4}}{d} \leftarrow \text{Map distance}$$
$$\text{Actual distance} \rightarrow \qquad\quad \leftarrow \text{Actual distance}$$

Solve:
$$\frac{1}{4} \cdot d = 50 \cdot 3\frac{1}{4}$$
$$\frac{1}{4} \cdot d = 50 \cdot \frac{13}{4}$$
$$4 \cdot \frac{1}{4} \cdot d = 4 \cdot 50 \cdot \frac{13}{4}$$
$$d = 50 \cdot 13$$
$$d = 650$$

The cities are 650 mi apart in reality.

**19.** Let $p =$ the number of pews needed to seat 44 people.

$$\text{Pews} \rightarrow \frac{2}{14} = \frac{p}{44} \leftarrow \text{Pews}$$
$$\text{People} \rightarrow \qquad\quad \leftarrow \text{People}$$

Solve:
$$2 \cdot 44 = 14 \cdot p$$
$$\frac{2 \cdot 44}{14} = p$$
$$\frac{2 \cdot 44}{2 \cdot 7} = p$$
$$\frac{44}{7} = p$$
$$6\frac{2}{7} = p$$

We round up to the next entire pew. For a wedding party of 44 people, 7 pews will be needed.

**21.** To plot $(-3, 2)$, we locate $-3$ on the first, or horizontal, axis. Then we go up 2 units and make a dot.

To plot $(4, 5)$, we locate 4 on the first, or horizontal, axis. Then we go up 5 units and make a dot.

To plot $(-4, -1)$, we locate $-4$ on the first, or horizontal, axis. Then we go down 1 unit and make a dot.

To plot $(0, 3)$, we locate 0 on the first, or horizontal, axis. Then we go up 3 units and make a dot.

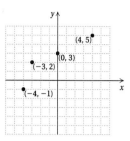

**23.** First consider $13.11 \div 5.7$:

$$
\begin{array}{r}
2\,.\,3 \\
5.7_\wedge\overline{)\,1\,3\;.\,1_\wedge 1\,} \\
\underline{1\,1\;4\;0} \\
1\;7\;1 \\
\underline{1\;7\;1} \\
0
\end{array}
$$

Since a negative number divided by a positive number is negative, the answer is $-2.3$.

**25.** $-19.7 + 12.5$

The difference in absolute values is $19.7 - 12.5$, or $7.2$. Since the negative number has the larger absolute value, the answer is negative.

$-19.7 + 12.5 = -7.2$

**27.** ◈

**29.** ◈

**31.** The amount of Sheri's raise is $\$29{,}380 - \$26{,}000$, or $\$3380$. Let $r =$ the amount of the raise Tim should request.

$$\text{Raise} \rightarrow \frac{3380}{26{,}000} = \frac{r}{23{,}000} \leftarrow \text{Raise}$$
$$\text{Old salary} \rightarrow \qquad\qquad \leftarrow \text{Old salary}$$

Solve:
$$3380 \cdot 23{,}000 = 26{,}000 \cdot r$$
$$\frac{3380 \cdot 23{,}000}{26{,}000} = r$$
$$2990 = r \quad \text{Using a calculator}$$

Tim should ask for a \$2990 raise.

**33.** Let $p =$ the selling price in Austin of a home that sells for \$450,000 in San Francisco.

$$\text{Austin price} \rightarrow \frac{89{,}000}{286{,}000} = \frac{p}{450{,}000} \leftarrow \text{Austin price}$$
$$\text{S.F. price} \rightarrow \qquad\qquad \leftarrow \text{S.F. price}$$

Solve:
$$89{,}000 \cdot 450{,}000 = 286{,}000 \cdot p$$
$$\frac{89{,}000 \cdot 450{,}000}{286{,}000} = p$$
$$140{,}000 \approx p$$

A house that sells for \$450,000 in San Francisco would sell for about \$140,000 in Austin.

## Exercise Set 7.5

**1.** The ratio of $h$ to 5 is the same as the ratio of 45 to 9. We have the proportion

$$\frac{h}{5} = \frac{45}{9}.$$

Solve: $9 \cdot h = 5 \cdot 45$     Equating cross-products

$\qquad h = \dfrac{5 \cdot 45}{9}$     Dividing by 9 on both sides

$\qquad h = 25$     Simplifying

The missing length $h$ is 25.

**3.** The ratio of $x$ to 2 is the same as the ratio of 2 to 3. We have the proportion

$$\frac{x}{2} = \frac{2}{3}.$$

Solve: $3 \cdot x = 2 \cdot 2$     Equating cross-products

$\qquad x = \dfrac{2 \cdot 2}{3}$     Dividing by 3 on both sides

$\qquad x = \dfrac{4}{3}$, or $1\dfrac{1}{3}$

The missing length $x$ is $\dfrac{4}{3}$, or $1\dfrac{1}{3}$. We could also have used $\dfrac{x}{2} = \dfrac{1}{1\frac{1}{2}}$ to find $x$.

**5.** First we find $x$. The ratio of $x$ to 9 is the same as the ratio of 6 to 8. We have the proportion

$$\frac{x}{9} = \frac{6}{8}.$$

Solve: $8 \cdot x = 9 \cdot 6$

$\qquad x = \dfrac{9 \cdot 6}{8}$

$\qquad x = \dfrac{27}{4}$, or $6\dfrac{3}{4}$

The missing length $x$ is $\dfrac{27}{4}$, or $6\dfrac{3}{4}$.

Next we find $y$. The ratio of $y$ to 12 is the same as the ratio of 6 to 8. We have the proportion

$$\frac{y}{12} = \frac{6}{8}.$$

Solve: $8 \cdot y = 12 \cdot 6$

$\qquad y = \dfrac{12 \cdot 6}{8}$

$\qquad y = 9$

The missing length $y$ is 9.

**7.** First we find $x$. The ratio of $x$ to 2.5 is the same as the ratio of 2.1 to 0.7. We have the proportion

$$\frac{x}{2.5} = \frac{2.1}{0.7}.$$

Solve: $0.7 \cdot x = 2.5 \cdot 2.1$

$\qquad x = \dfrac{2.5 \cdot 2.1}{0.7}$

$\qquad x = 7.5$

The missing length $x$ is 7.5.

Next we find $y$. The ratio of $y$ to 2.4 is the same as the ratio of 2.1 to 0.7. We have the proportion

$$\frac{y}{2.4} = \frac{2.1}{0.7}.$$

Solve: $0.7 \cdot y = 2.4 \cdot 2.1$

$\qquad y = \dfrac{2.4 \cdot 2.1}{0.7}$

$\qquad y = 7.2$

The missing length $y$ is 7.2.

**9.** If we use the sun's rays to represent the third side of a triangle in a drawing of the situation, we see that we have similar triangles. We let $s =$ the length of a shadow cast by a person 2 m tall.

The ratio of $s$ to 5 is the same as the ratio of 2 to 8. We have the proportion

$$\frac{s}{5} = \frac{2}{8}.$$

Solve: $8 \cdot s = 5 \cdot 2$

$\qquad s = \dfrac{5 \cdot 2}{8}$

$\qquad s = \dfrac{5}{4}$, or 1.25

The length of a shadow cast by a person 2 m tall is 1.25 m.

**11.** If we use the sun's rays to represent the third side of a triangle in a drawing of the situation, we see that we have similar triangles. We let $h =$ the height of the tree.

The ratio of $h$ to 4 is the same as the ratio of 27 to 3. We have the proportion

$$\frac{h}{4} = \frac{27}{3}.$$

Solve: $3 \cdot h = 4 \cdot 27$

$\qquad h = \dfrac{4 \cdot 27}{3}$

$\qquad h = 36$

The tree is 36 ft tall.

**13.** The ratio of $h$ to 7 ft is the same as the ratio of 6 ft to 6 ft. We have the proportion

$$\frac{h}{7} = \frac{6}{6}.$$

Solve: $6 \cdot h = 7 \cdot 6$

$\qquad h = \dfrac{7 \cdot 6}{6}$

$\qquad h = 7$

The wall is 7 ft high.

**15.** Since the ratio of $d$ to 25 ft is the same as the ratio of 40 ft to 10 ft, we have the proportion

$$\frac{d}{25} = \frac{40}{10}.$$

Solve: $10 \cdot d = 25 \cdot 40$

$$d = \frac{25 \cdot 40}{10}$$

$$d = 100$$

The distance across the river is 100 ft.

**17.**  Width $\rightarrow$ $\dfrac{6}{9} = \dfrac{x}{6}$ $\leftarrow$ Width
Length $\rightarrow$ $\phantom{\dfrac{6}{9}}$ $\leftarrow$ Length

Solve: $\dfrac{2}{3} = \dfrac{x}{6}$    Rewriting $\dfrac{6}{9}$ as $\dfrac{2}{3}$

$2 \cdot 6 = 3 \cdot x$    Equating cross-products

$$\frac{2 \cdot 6}{3} = x$$

$$\frac{2 \cdot 2 \cdot 3}{3} = x$$

$$2 \cdot 2 = x$$

$$4 = x$$

The missing length $x$ is 4.

**19.**  Width $\rightarrow$ $\dfrac{4}{7} = \dfrac{6}{x}$ $\leftarrow$ Width
Length $\rightarrow$ $\phantom{\dfrac{4}{7}}$ $\leftarrow$ Length

Solve: $4 \cdot x = 7 \cdot 6$    Equating cross-products

$$x = \frac{7 \cdot 6}{4}$$

$$x = \frac{7 \cdot 2 \cdot 3}{2 \cdot 2}$$

$$x = \frac{7 \cdot 3}{2}$$

$$x = \frac{21}{2}, \text{ or } 10\frac{1}{2}$$

The missing length $x$ is $10\frac{1}{2}$.

**21.** First we find $x$. The ratio of $x$ to 8 is the same as the ratio of 3 to 4. We have the proportion

$$\frac{x}{8} = \frac{3}{4}.$$

Solve:  $4 \cdot x = 8 \cdot 3$

$$x = \frac{8 \cdot 3}{4}$$

$$x = 6$$

The missing length $x$ is 6.

Next we find $y$. The ratio of $y$ to 7 is the same as the ratio of 3 to 4. We have the proportion

$$\frac{y}{7} = \frac{3}{4}.$$

Solve:  $4 \cdot y = 7 \cdot 3$

$$y = \frac{7 \cdot 3}{4}$$

$$y = \frac{21}{4}, \text{ or } 5\frac{1}{4}, \text{ or } 5.25$$

The missing length $y$ is $\dfrac{21}{4}$, or $5\dfrac{1}{4}$, or 5.25.

Finally we find $z$. The ratio of $z$ to 4 is the same as the ratio of 3 to 4. This statement tells us that $z$ must be 3. We could also calculate this using the proportion

$$\frac{z}{4} = \frac{3}{4}.$$

The missing length $z$ is 3.

**23.** First we find $x$. The ratio of $x$ to 8 is the same as the ratio of 2 to 3. We have the proportion

$$\frac{x}{8} = \frac{2}{3}.$$

Solve: $3 \cdot x = 8 \cdot 2$

$$x = \frac{8 \cdot 2}{3}$$

$$x = \frac{16}{3}, \text{ or } 5\frac{1}{3}$$

The missing length $x$ is $\dfrac{16}{3}$, or $5\dfrac{1}{3}$, or $5.\overline{3}$.

Next we find $y$. The ratio of $y$ to 7 is the same as the ratio of 2 to 3. We have the proportion

$$\frac{y}{7} = \frac{2}{3}.$$

Solve: $3 \cdot y = 7 \cdot 2$

$$y = \frac{7 \cdot 2}{3}$$

$$y = \frac{14}{3} = 4\frac{2}{3}$$

The missing length $y$ is $\dfrac{14}{3}$, or $4\dfrac{2}{3}$, or $4.\overline{6}$.

Finally we find $z$. The ratio of $z$ to 8 is the same as the ratio of 2 to 3. We have the proportion

$$\frac{z}{9} = \frac{2}{3}.$$

This is the same proportion we solved above when we found $x$. Then the missing length $z$ is $5\dfrac{1}{3}$, or $5.\overline{3}$.

**25.**  Height $\rightarrow$ $\dfrac{h}{32} = \dfrac{5}{8}$ $\leftarrow$ Height
Width $\rightarrow$ $\phantom{\dfrac{h}{32}}$ $\leftarrow$ Width

Solve: $8 \cdot h = 32 \cdot 5$

$$h = \frac{32 \cdot 5}{8}$$

$$h = \frac{4 \cdot 8 \cdot 5}{8}$$

$$h = 4 \cdot 5$$

$$h = 20$$

The missing length is 20 ft.

**27.** **Familiarize**. This is a multistep problem.

First we find the total cost of the purchases. We let $c =$ this amount.

**Translate and Solve.**

| Price of book | plus | Price of CD | plus | Price of sweatshirt | is | Total cost |
|:---:|:---:|:---:|:---:|:---:|:---:|:---:|
| $\downarrow$ | $\downarrow$ | $\downarrow$ | $\downarrow$ | $\downarrow$ | $\downarrow$ | $\downarrow$ |
| $49.95 | + | $14.88 | + | $29.95 | = | $c$ |

To solve the equation we carry out the addition.

$$
\begin{array}{r}
{\scriptstyle 2\ \ 2\ \ 1}\\
4\,9.\,9\,5\\
1\,4.\,8\,8\\
+\,2\,9.\,9\,5\\
\hline
9\,4.\,7\,8
\end{array}
$$

Thus, $c = \$94.78$.

Now we find how much more money the student needs to make these purchases. We let $m =$ this amount.

| Money student has | plus | How much more money | is | Total cost of purchases |
|:---:|:---:|:---:|:---:|:---:|
| ↓ | ↓ | ↓ | ↓ | ↓ |
| \$34.97 | + | $m$ | = | \$94.78 |

To solve the equation we subtract 34.97 on both sides.

$$m = 94.78 - 34.97$$
$$m = 59.81$$

$$
\begin{array}{r}
{\scriptstyle 13}\\
{\scriptstyle 8\ \ \not3\ 17}\\
\not9\,\not4.\,\not7\,8\\
-\,3\,4.\,9\,7\\
\hline
5\,9.\,8\,1
\end{array}
$$

**Check.** We repeat the calculations.

**State.** The student needs \$59.81 more to make the purchases.

**29.** Multiplying the absolute values, we have

$$
\begin{array}{r}
{\scriptstyle 7\ \ 7\ \ 1}\\
{\scriptstyle 3\ \ 3}\\
8\,0.\,8\,9\,2\\
\times\quad\ \ 8.\,4\\
\hline
3\,2\,3\,5\,6\,8\\
6\,4\,7\,1\,3\,6\,0\\
\hline
6\,7\,9.\,4\,9\,2\,8
\end{array}
$$

Since the product of a negative number and a positive number is negative, the answer is $-679.4928$.

**31.** $\underline{1\,00} \times 274.568 \qquad 274.56.8$

2 zeros    Move 2 places to the right.

$100 \times 274.568 = 27{,}456.8$

**33.**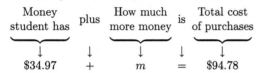

**35.**

**37.**

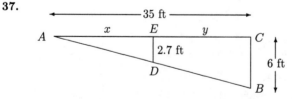

We note that triangle $ADE$ is similar to triangle $ABC$ and use this information to find the length $x$.

$$\frac{x}{35} = \frac{2.7}{6}$$
$$6 \cdot x = 35 \cdot 2.7$$
$$x = \frac{35 \cdot 2.7}{6}$$
$$x = 15.75$$

Thus the goalie should be 15.75 ft from point $A$. We subtract to find how far from the goal the goalie should be located.

$$35 - 15.75 = 19.25$$

The goalie should stand 19.25 ft from the goal.

**39.** First we find $x$. The ratio of $x$ to $19.35 + x$ is the same as the ratio of 0.3 to 16.8. We have the proportion

$$\frac{x}{19.35 + x} = \frac{0.3}{16.8}$$

Solve:
$$16.8 \cdot x = 0.3(19.35 + x)$$
$$16.8x = 5.805 + 0.3x$$
$$16.8x - 0.3x = 5.805 + 0.3x - 0.3x$$
$$16.5x = 5.805$$
$$\frac{16.5x}{16.5} = \frac{5.805}{16.5}$$
$$x \approx 0.35$$

The missing length $x$ is approximately 0.35.

Now we find $y$. The ratio of $y$ to 22.4 is the same as the ratio of 0.3 to 16.8. We have the proportion

$$\frac{y}{22.4} = \frac{0.3}{16.8}.$$

Solve: $16.8 \cdot y = 22.4(0.3)$
$$y = \frac{22.4(0.3)}{16.8}$$
$$y = 0.4$$

The missing length $y$ is 0.4.

**41.** Let $s =$ the length of a side of the model diamond.

Model width → $\dfrac{12}{116} = \dfrac{s}{90}$ ← Model side

Actual width →        ← Actual side

Solve:
$$12 \cdot 90 = 116 \cdot s$$
$$\frac{12 \cdot 90}{116} = s$$
$$\frac{3 \cdot 4 \cdot 90}{4 \cdot 29} = s$$
$$\frac{3 \cdot 90}{29} = s$$
$$\frac{270}{29} = s$$

The length of a side of the model diamond is $\dfrac{270}{29}$ cm. Now we use a calculator to find the area of the model diamond.

$$A = s^2 = \left(\frac{270}{29}\right)^2 \approx 86.68 \text{ cm}^2$$

The area of the model diamond is about 86.68 cm².

# Chapter 8

# Percent Notation

## Exercise Set 8.1

**1.** Since the circle is divided into 100 sections, we can think of it as a pie cut into 100 equally sized pieces. We shade a wedge equal in size to 32 of these pieces to represent 32%. Then we shade wedges equal in size to 33, 2, 4, 10, and 19 of these pieces to represent 33%, 2%, 4%, 10%, and 19%, respectively.

**Reasons for Drinking Coffee**

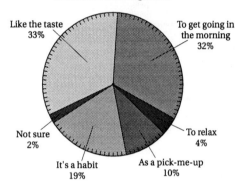

**3.** $90\% = \dfrac{90}{100}$     A ratio of 90 to 100

$90\% = 90 \times \dfrac{1}{100}$     Replacing % with $\times \dfrac{1}{100}$

$90\% = 90 \times 0.01$     Replacing % with $\times 0.01$

**5.** $12.5\% = \dfrac{12.5}{100}$     A ratio of 12.5 to 100

$12.5\% = 12.5 \times \dfrac{1}{100}$     Replacing % with $\times \dfrac{1}{100}$

$12.5\% = 12.5 \times 0.01$     Replacing % with $\times 0.01$

**7.** 12%

a) Replace the percent symbol with ×0.01.

    12 × 0.01

b) Multiply to move the decimal point two places to the left.

    0.12.

Thus, 12% = 0.12.

**9.** 34.7%

a) Replace the percent symbol with ×0.01.

    34.7 × 0.01

b) Multiply to move the decimal point two places to the left.

    0.34.7

Thus, 34.7% = 0.347.

**11.** 59.01%

a) Replace the percent symbol with ×0.01.

    59.01 × 0.01

b) Multiply to move the decimal point two places to the left.

    0.59.01

Thus, 59.01% = 0.5901.

**13.** 10%

a) Replace the percent symbol with ×0.01.

    10 × 0.01

b) Multiply to move the decimal point two places to the left.

    0.10.

Thus, 10% = 0.1.

**15.** 1%

a) Replace the percent symbol with ×0.01.

    1 × 0.01

b) Multiply to move the decimal point two places to the left.

    0.01.

Thus, 1% = 0.01.

**17.** 300%

a) Replace the percent symbol with ×0.01.

    300 × 0.01

b) Multiply to move the decimal point two places to the left.

    3.00.

Thus, 300%=3.

**19.** 0.6%

a) Replace the percent symbol with ×0.01.

    0.6 × 0.01

b) Multiply to move the decimal point two places to the left.

    0.00.6

Thus, 0.6% = 0.006.

**21.** 0.23%

a) Replace the percent symbol with ×0.01.

   0.23 × 0.01

b) Multiply to move the decimal point two places to the left.

   0.00.23

Thus, 0.23% = 0.0023.

**23.** 105.24%

a) Replace the percent symbol with ×0.01.

   105.24 × 0.01

b) Multiply to move the decimal point two places to the left.

   1.05.24

Thus, 105.24% = 1.0524.

**25.** 40%

a) Replace the percent symbol with ×0.01.

   40 × 0.01

b) Move the decimal point two places to the left.

   0.40.

Thus, 40% = 0.4.

**27.** 2.5%

a) Replace the percent symbol with ×0.01.

   2.5 × 0.01

b) Move the decimal point two places to the left.

   0.02.5

Thus, 2.5% = 0.025.

**29.** 62.2%

a) Replace the percent symbol with ×0.01.

   62.2 × 0.01

b) Move the decimal point two places to the left.

   0.62.2

Thus, 62.2% = 0.622.

**31.** 0.47

a) Move the decimal point two places to the right.

   0.47.

b) Write a percent symbol: 47%

Thus, 0.47 = 47%.

**33.** 0.03

a) Move the decimal point two places to the right.

   0.03.

b) Write a percent symbol: 3%

Thus, 0.03 = 3%.

**35.** 8.7

a) Move the decimal point two places to the right.

   8.70.

b) Write a percent symbol: 870%

Thus, 8.7 = 870%.

**37.** 0.334

a) Move the decimal point two places to the right.

   0.33.4

b) Write a percent symbol: 33.4%

Thus, 0.334 = 33.4%.

**39.** 0.75

a) Move the decimal point two places to the right.

   0.75.

b) Write a percent symbol: 75%

Thus, 0.75 = 75%.

**41.** 0.4

a) Move the decimal point two places to the right.

   0.40.

b) Write a percent symbol: 40%

Thus, 0.4 = 40%.

**43.** 0.8925

a) Move the decimal point two places to the right.

   0.89.25

b) Write a percent symbol: 89.25%

Thus, 0.8925 = 89.25%.

**45.** 0.000104

a) Move the decimal point two places to the right.

   0.00.0104

b) Write a percent symbol: 0.0104%

Thus, 0.000104 = 0.0104%.

**47.** 0.24

    a) Move the decimal point two places to the right.

    0.24.

    └─↑

    b) Write a percent symbol: 24%

    Thus, 0.24 = 24%.

**49.** 0.326

    a) Move the decimal point two places to the right.

    0.32.6

    └─↑

    b) Write a percent symbol: 32.6%

    Thus, 0.326 = 32.6%.

**51.** We use the definition of percent as a ratio.

$$\frac{41}{100} = 41\%$$

**53.** We use the definition of percent as a ratio.

$$\frac{5}{100} = 5\%$$

**55.** We multiply by 1 to get 100 in the denominator.

$$\frac{2}{10} = \frac{2}{10} \cdot \frac{10}{10} = \frac{20}{100} = 20\%$$

**57.** We multiply by 1 to get 100 in the denominator.

$$\frac{7}{25} = \frac{7}{25} \cdot \frac{4}{4} = \frac{28}{100} = 28\%$$

**59.** We multiply by 1 to get 100 in the denominator.

$$\frac{3}{20} = \frac{3}{20} \cdot \frac{5}{5} = \frac{15}{100} = 15\%$$

**61.** Find decimal notation by division.

$$\begin{array}{r} 0.6\,2\,5 \\ 8\overline{)5.0\,0\,0} \\ \underline{4\,8}\phantom{00} \\ 2\,0\phantom{0} \\ \underline{1\,6}\phantom{0} \\ 4\,0 \\ \underline{4\,0} \\ 0 \end{array}$$

$$\frac{5}{8} = 0.625$$

Convert to percent notation.

    0.62.5

    └─↑

$$\frac{5}{8} = 62.5\%, \text{ or } 62\frac{1}{2}\%$$

**63.** $\dfrac{4}{5} = \dfrac{4}{5} \cdot \dfrac{20}{20} = \dfrac{80}{100} = 80\%$

**65.** Find decimal notation by division.

$$\begin{array}{r} 0.6\,6\,6 \\ 3\overline{)2.0\,0\,0} \\ \underline{1\,8}\phantom{00} \\ 2\,0\phantom{0} \\ \underline{1\,8}\phantom{0} \\ 2\,0 \\ \underline{1\,8} \\ 2 \end{array}$$

We get a repeating decimal: $\dfrac{2}{3} = 0.66\overline{6}$

Convert to percent notation.

    0.66.$\overline{6}$

    └─↑

$$\frac{2}{3} = 66.\overline{6}\%, \text{ or } 66\frac{2}{3}\%$$

**67.** $\dfrac{29}{50} = \dfrac{29}{50} \cdot \dfrac{2}{2} = \dfrac{58}{100} = 58\%$

**69.** $\dfrac{9}{25} = \dfrac{9}{25} \cdot \dfrac{4}{4} = \dfrac{36}{100} = 36\%$

**71.** $85\% = \dfrac{85}{100}$      Definition of percent

$$\left.\begin{aligned} &= \frac{5 \cdot 17}{5 \cdot 20} \\ &= \frac{5}{5} \cdot \frac{17}{20} \\ &= \frac{17}{20} \end{aligned}\right\} \text{Simplifying}$$

**73.** $62.5\% = \dfrac{62.5}{100}$      Definition of percent

$= \dfrac{62.5}{100} \cdot \dfrac{10}{10}$    Multiplying by 1 to eliminate the decimal point in the numerator

$= \dfrac{625}{1000}$

$$\left.\begin{aligned} &= \frac{5 \cdot 125}{8 \cdot 125} \\ &= \frac{5}{8} \cdot \frac{125}{125} \\ &= \frac{5}{8} \end{aligned}\right\} \text{Simplifying}$$

**75.** $33\dfrac{1}{3}\% = \dfrac{100}{3}\%$    Converting from mixed numeral to fractional notation

$= \dfrac{100}{3} \times \dfrac{1}{100}$    Definition of percent

$= \dfrac{100 \cdot 1}{3 \cdot 100}$    Multiplying

$$\left.\begin{aligned} &= \frac{1}{3} \cdot \frac{100}{100} \\ &= \frac{1}{3} \end{aligned}\right\} \text{Simplifying}$$

**77.**  $16.\overline{6}\% = 16\frac{2}{3}\%$     $(16.\overline{6} = 16\frac{2}{3})$

$= \frac{50}{3}\%$     Converting from mixed numeral to fractional notation

$= \frac{50}{3} \times \frac{1}{100}$     Definition of percent

$= \frac{50 \cdot 1}{3 \cdot 50 \cdot 2}$     Multiplying

$\left. \begin{array}{l} = \dfrac{1}{2 \cdot 3} \cdot \dfrac{50}{50} \\[2mm] = \dfrac{1}{6} \end{array} \right\}$     Simplifying

**79.**  $7.25\% = \frac{7.25}{100} = \frac{7.25}{100} \cdot \frac{100}{100}$

$= \frac{725}{10,000} = \frac{29 \cdot 25}{400 \cdot 25} = \frac{29}{400} \cdot \frac{25}{25}$

$= \frac{29}{400}$

**81.**  $0.8\% = \frac{0.8}{100} = \frac{0.8}{100} \cdot \frac{10}{10}$

$= \frac{8}{1000} = \frac{1 \cdot 8}{125 \cdot 8} = \frac{1}{125} \cdot \frac{8}{8}$

$= \frac{1}{125}$

**83.**  $\frac{21}{100} = 21\%$     Using the definition of percent as a ratio

**85.**  $\frac{6}{25} = \frac{6}{25} \cdot \frac{4}{4} = \frac{24}{100} = 24\%$

**87.**  $55\% = \frac{55}{100} = \frac{5 \cdot 11}{5 \cdot 20} = \frac{5}{5} \cdot \frac{11}{20} = \frac{11}{20}$

**89.**  $38\% = \frac{38}{100} = \frac{2 \cdot 19}{2 \cdot 50} = \frac{2}{2} \cdot \frac{19}{50} = \frac{19}{50}$

**91.**  $11\% = \frac{11}{100}$

**93.**  $26.4\% = \frac{26.4}{100} = \frac{26.4}{100} \cdot \frac{10}{10} = \frac{264}{1000} = \frac{8 \cdot 33}{8 \cdot 125} =$

$\frac{8}{8} \cdot \frac{33}{125} = \frac{33}{125}$

**95.**  $45.7\% = \frac{45.7}{100} = \frac{45.7}{100} \cdot \frac{10}{10} = \frac{457}{1000}$

**97.**  $\frac{1}{8} = 1 \div 8$

$$\begin{array}{r} 0.1\,2\,5 \\ 8\,\overline{)\,1.0\,0\,0} \\ \underline{8}\phantom{.000} \\ 2\,0\phantom{0} \\ \underline{1\,6}\phantom{0} \\ 4\,0 \\ \underline{4\,0} \\ 0 \end{array}$$

$\frac{1}{8} = 0.125 = 12\frac{1}{2}\%,$ or $12.5\%$

$\frac{1}{6} = 1 \div 6$

$$\begin{array}{r} 0.1\,6\,6 \\ 6\,\overline{)\,1.0\,0\,0} \\ \underline{6}\phantom{.000} \\ 4\,0\phantom{0} \\ \underline{3\,6}\phantom{0} \\ 4\,0 \\ \underline{3\,6} \\ 4 \end{array}$$

We get a repeating decimal: $0.1\overline{6}$

$0.16.\overline{6}$          $0.1\overline{6} = 16.\overline{6}\%$

$\frac{1}{6} = 0.1\overline{6} = 16.\overline{6}\%,$ or $16\frac{2}{3}\%$

$20\% = \frac{20}{100} = \frac{1}{5} \cdot \frac{20}{20} = \frac{1}{5}$

$0.20.$          $20\% = 0.2$

$\frac{1}{5} = 0.2 = 20\%$

$0.25.$          $0.25 = 25\%$

$25\% = \frac{25}{100} = \frac{1}{4} \cdot \frac{25}{25} = \frac{1}{4}$

$\frac{1}{4} = 0.25 = 25\%$

$33\frac{1}{3}\% = \frac{100}{3}\% = \frac{100}{3} \times \frac{1}{100} = \frac{100}{300} = \frac{1}{3} \cdot \frac{100}{100} = \frac{1}{3}$

$0.33.\overline{3}$          $33.\overline{3}\% = 0.33\overline{3},$ or $0.\overline{3}$

$\frac{1}{3} = 0.\overline{3} = 33\frac{1}{3}\%,$ or $33.\overline{3}\%$

$37.5\% = \frac{37.5}{100} = \frac{37.5}{100} \cdot \frac{10}{10} = \frac{375}{1000} = \frac{3}{8} \cdot \frac{125}{125} = \frac{3}{8}$

$0.37.5$          $37.5\% = 0.375$

$\frac{3}{8} = 0.375 = 37\frac{1}{2}\%,$ or $37.5\%$

$40\% = \frac{40}{100} = \frac{2}{5} \cdot \frac{20}{20} = \frac{2}{5}$

$0.40.$          $40\% = 0.4$

$\frac{2}{5} = 0.4 = 40\%$

**99.** $\quad$ 0.50. $\qquad$ 0.5 = 50%

$$50\% = \frac{50}{100} = \frac{1}{2} \cdot \frac{50}{50} = \frac{1}{2}$$

$$\mathbf{\frac{1}{2} = 0.5 = 50\%}$$

$$\frac{1}{3} = 1 \div 3$$

$$\begin{array}{r} 0.3 \\ 3\overline{)1.0} \\ \underline{9} \\ 1 \end{array}$$

We get a repeating decimal: $0.\overline{3}$

$\quad$ 0.33.$\overline{3}$ $\qquad$ $0.\overline{3} = 33.\overline{3}\%$

$$\mathbf{\frac{1}{3} = 0.\overline{3} = 33.\overline{3}\%, \ or \ 33\frac{1}{3}\%}$$

$$25\% = \frac{25}{100} = \frac{25}{25} \cdot \frac{1}{4} = \frac{1}{4}$$

0.25. $\qquad$ 25% = 0.25

$$\mathbf{\frac{1}{4} = 0.25 = 25\%}$$

$$16\frac{2}{3}\% = \frac{50}{3}\% = \frac{50}{3} \times \frac{1}{100} = \frac{50 \cdot 1}{3 \cdot 2 \cdot 50} = \frac{50}{50} \cdot \frac{1}{6} = \frac{1}{6}$$

$$\frac{1}{6} = 1 \div 6$$

$$\begin{array}{r} 0.1\,6 \\ 6\overline{)1.0\,0} \\ \underline{6} \\ 4\,0 \\ \underline{3\,6} \\ 4 \end{array}$$

We get a repeating decimal: $0.1\overline{6}$

$$\mathbf{\frac{1}{6} = 0.1\overline{6} = 16\frac{2}{3}\%, \ or \ 16.\overline{6}\%}$$

0.12.5 $\qquad$ 0.125 = 12.5%

$$12.5\% = \frac{12.5}{100} = \frac{12.5}{100} \cdot \frac{10}{10} = \frac{125}{1000} = \frac{125}{125} \cdot \frac{1}{8} = \frac{1}{8}$$

$$\mathbf{\frac{1}{8} = 0.125 = 12.5\%, \ or \ 12\frac{1}{2}\%}$$

$$\frac{3}{4} = \frac{3}{4} \cdot \frac{25}{25} = \frac{75}{100} = 75\%$$

0.75. $\qquad$ 75% = 0.75

$$\mathbf{\frac{3}{4} = 0.75 = 75\%}$$

$0.8\overline{3} = 0.83.\overline{3}$ $\qquad$ $0.8\overline{3} = 83.\overline{3}\%$

$$83.\overline{3}\% = 83\frac{1}{3}\% = \frac{250}{3}\% = \frac{250}{3} \times \frac{1}{100} = \frac{5 \cdot 50}{3 \cdot 2 \cdot 50} =$$
$$\frac{5}{6} \cdot \frac{50}{50} = \frac{5}{6}$$

$$\mathbf{\frac{5}{6} = 0.8\overline{3} = 83.\overline{3}\%, \ or \ 83\frac{1}{3}\%}$$

$$\frac{3}{8} = 3 \div 8$$

$$\begin{array}{r} 0.3\,7\,5 \\ 8\overline{)3.0\,0\,0} \\ \underline{2\,4} \\ 6\,0 \\ \underline{5\,6} \\ 4\,0 \\ \underline{4\,0} \\ 0 \end{array}$$

$$\frac{3}{8} = 0.375$$

0.37.5 $\qquad$ 0.375 = 37.5%

$$\mathbf{\frac{3}{8} = 0.375 = 37.5\%, \ or \ 37\frac{1}{2}\%}$$

**101.** To convert $\dfrac{100}{3}$ to a mixed numeral, we divide.

$$\begin{array}{r} 3\,3 \\ 3\overline{)1\,0\,0} \\ \underline{9\,0} \\ 1\,0 \\ \underline{9} \\ 1 \end{array} \qquad \frac{100}{3} = 33\frac{1}{3}$$

**103.** $\quad 0.05 \times b = 20$

$$\frac{0.05 \times b}{0.05} = \frac{20}{0.05}$$

$$b = 400$$

**105.** $\quad \dfrac{24}{37} = \dfrac{15}{x}$

$24 \cdot x = 37 \cdot 15 \qquad$ Equating cross-products

$$x = \frac{37 \cdot 15}{24}$$

$$x = 23.125$$

**107.** ◈

**109.** ◈

**111.** Use a calculator.

$$\frac{54}{999} = 0.05.\overline{405} = 5.\overline{405}\%$$

**113.** $3.2\overline{93847} = 3.29384793847\ldots = 3.29\overline{3847.9}$

$$3.29\overline{38479} = 3.29.\overline{38479} = 329.\overline{38479}\%$$

**115.** $\frac{19}{12}\% = \frac{19}{12} \times \frac{1}{100} = \frac{19}{1200}$

To find decimal notation for $\frac{19}{2000}$ we divide.

```
            0.0 1 5 8 3 3
  1 2 0 0 | 1 9.0 0 0 0 0 0
            1 2 0 0
            ─────────
              7 0 0 0
              6 0 0 0
            ─────────
              1 0 0 0 0
                9 6 0 0
              ─────────
                4 0 0 0
                3 6 0 0
                ─────────
                  4 0 0 0
                  3 6 0 0
                  ─────────
                    4 0 0
```

We get a repeating decimal: $\frac{19}{12}\% = 0.0158\overline{3}$.

**117.** $\frac{637}{6}\% = \frac{637}{6} \times \frac{1}{100} = \frac{637}{600}$

To find decimal notation for $\frac{637}{600}$ we divide.

```
            1.0 6 1 6 6
  6 0 0 | 6 3 7.0 0 0 0 0
          6 0 0
          ─────────
          3 7 0 0
          3 6 0 0
          ─────────
          1 0 0 0
            6 0 0
          ─────────
            4 0 0 0
            3 6 0 0
            ─────────
              4 0 0 0
              3 6 0 0
              ─────────
                4 0 0
```

We get a repeating decimal: $\frac{637}{6}\% = 1.061\overline{6}$.

---

## Exercise Set 8.2

**1.** What is    37%    of 74?

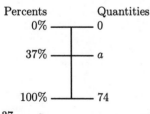

$$\frac{37}{100} = \frac{a}{74}$$

**3.** 4.3 is what percent of 5.9?

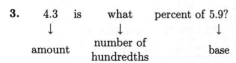

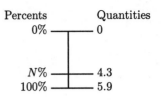

$$\frac{N}{100} = \frac{4.3}{5.9}$$

**5.** 14 is 25% of what?

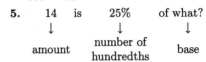

$$\frac{25}{100} = \frac{14}{b}$$

**7.** 9% of what is 37?

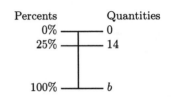

$$\frac{9}{100} = \frac{37}{b}$$

**9.** 70% of 660 is what?

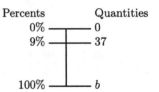

$$\frac{70}{100} = \frac{a}{660}$$

**11.** What is 4% of 1000?

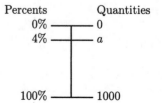

*Translate:* $\dfrac{4}{100} = \dfrac{a}{1000}$

*Solve:* $4 \cdot 1000 = 100 \cdot a$

$$\dfrac{4000}{100} = \dfrac{100a}{100}$$
$$40 = a$$

40 is 4% of 1000. The answer is 40.

**13.**

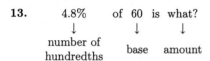

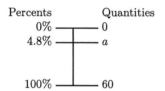

*Translate:* $\dfrac{4.8}{100} = \dfrac{a}{60}$

*Solve:* $4.8 \cdot 60 = 100 \cdot a$

$$\dfrac{288}{100} = \dfrac{100a}{100}$$
$$2.88 = a$$

4.8% of 60 is 2.88. The answer is 2.88.

**15.**

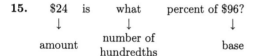

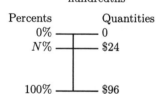

*Translate:* $\dfrac{N}{100} = \dfrac{24}{96}$

*Solve:* $96 \cdot N = 100 \cdot 24$

$$\dfrac{96N}{96} = \dfrac{2400}{96}$$
$$N = 25$$

$24 is 25% of $96. The answer is 25%.

**17.**

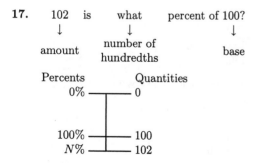

*Translate:* $\dfrac{N}{100} = \dfrac{102}{100}$

*Solve:* $100 \cdot N = 100 \cdot 102$

$$\dfrac{100N}{100} = \dfrac{10,200}{100}$$
$$N = 102$$

102 is 102% of 100. The answer is 102%.

**19.**

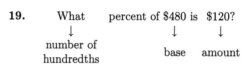

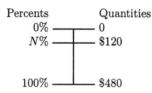

*Translate:* $\dfrac{N}{100} = \dfrac{120}{480}$

*Solve:* $480 \cdot N = 100 \cdot 120$

$$\dfrac{480N}{480} = \dfrac{12,000}{480}$$
$$N = 25$$

25% of $480 is $120. The answer is 25%.

**21.**

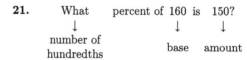

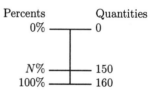

*Translate:* $\dfrac{N}{100} = \dfrac{150}{160}$

*Solve:* $160 \cdot N = 100 \cdot 150$

$$\dfrac{160N}{160} = \dfrac{15,000}{160}$$
$$N = 93.75$$

93.75% of 160 is 150. The answer is 93.75%.

**23.**

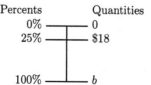

*Translate:* $\dfrac{25}{100} = \dfrac{18}{b}$

*Solve:* $25 \cdot b = 100 \cdot 18$

$$\dfrac{25b}{25} = \dfrac{1800}{25}$$
$$b = 72$$

$18 is 25% of $72.  The answer is $72.

**25.**     60%        of what is    $54?
             ↓              ↓          ↓
         number of       base      amount
         hundredths

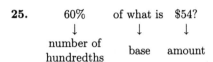

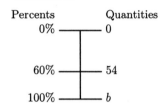

*Translate:* $\dfrac{60}{100} = \dfrac{54}{b}$

*Solve:* $60 \cdot b = 100 \cdot 54$

$$\dfrac{60b}{60} = \dfrac{5400}{60}$$
$$b = 90$$

60% of 90 is 54.  The answer is 90.

**27.**   65.12   is      74%      of what?
           ↓                ↓          ↓
         amount        number of    base
                       hundredths

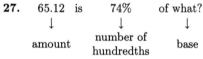

*Translate:* $\dfrac{74}{100} = \dfrac{65.12}{b}$

*Solve:* $74 \cdot b = 100 \cdot 65.12$

$$\dfrac{74b}{74} = \dfrac{6512}{74}$$
$$b = 88$$

65.12 is 74% of 88.  The answer is 88.

**29.**     80%        of what is    16?
             ↓              ↓          ↓
         number of       base      amount
         hundredths

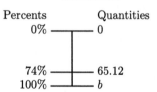

*Translate:* $\dfrac{80}{100} = \dfrac{16}{b}$

*Solve:* $80 \cdot b = 100 \cdot 16$

$$\dfrac{80b}{80} = \dfrac{1600}{80}$$
$$b = 20$$

80% of 20 is 16.  The answer is 20.

**31.**   What  is    $62\frac{1}{2}$%    of 40?
           ↓             ↓               ↓
         amount      number of         base
                     hundredths

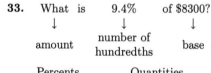

*Translate:* $\dfrac{62\frac{1}{2}}{100} = \dfrac{a}{40}$

*Solve:*  $62\dfrac{1}{2} \cdot 40 = 100 \cdot a$

$$\dfrac{125}{2} \cdot \dfrac{40}{1} = 100a$$
$$2500 = 100a$$
$$\dfrac{2500}{100} = \dfrac{100a}{100}$$
$$25 = a$$

25 is $62\frac{1}{2}$% of 40.  The answer is 25.

**33.**   What  is     9.4%     of $8300?
           ↓             ↓          ↓
         amount      number of     base
                     hundredths

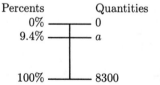

*Translate:* $\dfrac{9.4}{100} = \dfrac{a}{8300}$

*Solve:* $9.4 \cdot 8300 = 100 \cdot a$

$$\dfrac{78,020}{100} = \dfrac{100a}{100}$$
$$780.2 = a$$

$780.20 is 9.4% of $8300.  The answer is $780.20.

**35.**   9.48   is      120%     of what?
           ↓              ↓          ↓
         amount       number of    base
                      hundredths

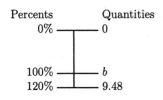

Percents     Quantities

Translate: $\dfrac{120}{100} = \dfrac{9.48}{b}$

Solve: $120 \cdot b = 100 \cdot 9.48$

$$\frac{120b}{120} = \frac{948}{120}$$

$$b = 7.9$$

9.48 is 120% of 7.9. The answer is 7.9.

**37.** Graph: $y = -\dfrac{1}{2}x$

We make a table of solutions. Note that when $x$ is a multiple of 2, fractional values for $y$ are avoided. Next we plot the points, draw the line, and label it.

When $x = -4$, $y = -\dfrac{1}{2}(-4) = \dfrac{4}{2} = 2$.

When $x = 0$, $y = -\dfrac{1}{2} \cdot 0 = 0$.

When $x = 2$, $y = -\dfrac{1}{2} \cdot 2 = -\dfrac{2}{2} = -1$.

| $x$ | $y = -\frac{1}{2}x$ | $(x,y)$ |
|---|---|---|
| $-4$ | $2$ | $(-4,2)$ |
| $0$ | $0$ | $(0,0)$ |
| $2$ | $-1$ | $(2,-1)$ |

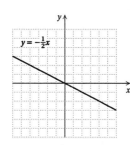

**39.** Graph: $y = 2x - 4$

We make a table of solutions. Then we plot the points, draw the line, and label it.

When $x = -1$, $y = 2(-1) - 4 = -2 - 4 = -6$.
When $x = 1$, $y = 2 \cdot 1 - 4 = 2 - 4 = -2$.
When $x = 4$, $y = 2 \cdot 4 - 4 = 8 - 4 = 4$.

| $x$ | $y = 2x - 4$ | $(x,y)$ |
|---|---|---|
| $-1$ | $-6$ | $(-1,-6)$ |
| $1$ | $-2$ | $(1,-2)$ |
| $4$ | $4$ | $(4,4)$ |

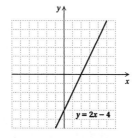

**41.** *Familiarize.* Let $q$ = the number of quarts of liquid ingredients the recipe calls for.

*Translate.*

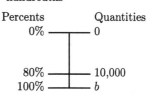

| Butter-milk | plus | Skim milk | plus | Oil | is | Total liquid ingredients |
|---|---|---|---|---|---|---|
| $\frac{1}{2}$ | $+$ | $\frac{1}{3}$ | $+$ | $\frac{1}{16}$ | $=$ | $q$ |

*Solve.* We carry out the addition. The LCM of the denominators is 48, so the LCD is 48.

$$\frac{1}{2} \cdot \frac{24}{24} + \frac{1}{3} \cdot \frac{16}{16} + \frac{1}{16} \cdot \frac{3}{3} = q$$

$$\frac{24}{48} + \frac{16}{48} + \frac{3}{48} = q$$

$$\frac{43}{48} = q$$

*Check.* We repeat the calculation. The answer checks.

*State.* The recipe calls for $\dfrac{43}{48}$ qt of liquid ingredients.

**43.** ◈

**45.** ◈

**47.** Estimate: Round 78.8% to 80% and 9809.024 to 10,000.

80%    of what is 10,000?

number of hundredths    base    amount

Percents     Quantities

Translate: $\dfrac{80}{100} = \dfrac{10,000}{b}$

Solve: $80 \cdot b = 100 \cdot 10,000$

$$\frac{80b}{80} = \frac{1,000,000}{80}$$

$$b = 12,500$$

We estimate that 78.8% of 12,500 is about 9809.024. (Answers may vary.)

Calculate:

78.8%    of what is 9809.024?

number of hundredths    base    amount

Percents     Quantities

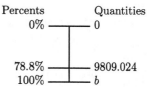

Translate: $\dfrac{78.8}{100} = \dfrac{9809.0924}{b}$

*Solve:* $78.8 \cdot b = 100 \cdot 9809.024$

$$\frac{78.8b}{78.8} = \frac{980,902.4}{78.8}$$

$$b = 12,448$$

We calculate that 78.8% of 12,448 is 9809.024.

## Exercise Set 8.3

**1.** What is 32% of 78?

   $\downarrow \quad \downarrow \quad \downarrow \quad \downarrow \quad \downarrow$

   $a \quad = 32\% \quad \cdot \quad 78$

**3.** 89 is what percent of 99?

   $\downarrow \downarrow \qquad \downarrow \qquad \downarrow \downarrow$

   $89 = \qquad n \qquad \cdot \ 99$

**5.** 13 is 25% of what?

   $\downarrow \downarrow \quad \downarrow \quad \downarrow \qquad \downarrow$

   $13 = 25\% \quad \cdot \qquad b$

**7.** What is 85% of 276?

*Translate:* $a = 85\% \cdot 276$

*Solve:* The letter is by itself. To solve the equation we convert 85% to decimal notation and multiply.

$a = 0.85(276)$

$$\begin{array}{r} 2\ 7\ 6 \\ \times\ 0.8\ 5 \\ \hline 1\ 3\ 8\ 0 \\ 2\ 2\ 0\ 8\ 0 \\ \hline 2\ 3\ 4.6\ 0 \end{array} \quad (85\% = 0.85)$$

$a = 234.6$

234.6 is 85% of 276. The answer is 234.6.

**9.** 150% of 30 is what?

*Translate:* $150\% \cdot 30 = a$

*Solve:* Convert 150% to decimal notation and multiply.

$a = 1.5(30)$

$$\begin{array}{r} 3\ 0 \\ \times\ 1.5 \\ \hline 1\ 5\ 0 \\ 3\ 0\ 0 \\ \hline 4\ 5.0 \end{array} \quad (150\% = 1.5)$$

$a = 45$

150% of 30 is 45. The answer is 45.

**11.** What is 6% of $300?

*Translate:* $a = 6\% \cdot \$300$

*Solve:* Convert 6% to decimal notation and multiply.

$a = 0.06(300)$

$$\begin{array}{r} \$\ 3\ 0\ 0 \\ \times\ 0.0\ 6 \\ \hline \$\ 1\ 8.0\ 0 \end{array} \quad (6\% = 0.06)$$

$a = 18$

$18 is 6% of $300. The answer is $18.

**13.** 3.8% of 50 is what?

*Translate:* $3.8\% \cdot 50 = a$

*Solve:* Convert 3.8% to decimal notation and multiply.

$a = 0.038(50)$

$$\begin{array}{r} 5\ 0 \\ \times\ 0.0\ 3\ 8 \\ \hline 4\ 0\ 0 \\ 1\ 5\ 0\ 0 \\ \hline 1.9\ 0\ 0 \end{array} \quad (3.8\% = 0.038)$$

$a = 1.9$

3.8% of 50 is 1.9. The answer is 1.9.

**15.** $39 is what percent of $50?

*Translate:* $39 = n \cdot 50$

*Solve:* To solve the equation we divide on both sides by 50 and convert the answer to percent notation.

$$n \cdot 50 = 39$$

$$\frac{n \cdot 50}{50} = \frac{39}{50}$$

$$n = 0.78, \text{ or } n = 78\%$$

$39 is 78% of $50. The answer is 78%.

**17.** 20 is what percent of 10?

*Translate:* $20 = n \cdot 10$

*Solve:* To solve the equation we divide on both sides by 10 and convert the answer to percent notation.

$$n \cdot 10 = 20$$

$$\frac{n \cdot 10}{10} = \frac{20}{10}$$

$$n = 2, \text{ or } n = 200\%$$

20 is 200% of 10. The answer is 200%.

**19.** What percent of $300 is $150?

*Translate:* $n \cdot 300 = 150$

*Solve:* $n \cdot 300 = 150$

$$\frac{n \cdot 300}{300} = \frac{150}{300}$$

$$n = 0.5, \text{ or } n = 50\%$$

50% of $300 is $150. The answer is 50%.

**21.** What percent of 80 is 100?

*Translate:* $n \cdot 80 = 100$

*Solve:* $n \cdot 80 = 100$

$$\frac{n \cdot 80}{80} = \frac{100}{80}$$

$$n = 1.25, \text{ or } n = 125\%$$

125% of 80 is 100. The answer is 125%.

**23.** 20 is 50% of what?

*Translate:* $20 = 50\% \cdot b$

*Solve:* To solve the equation we convert 50% to 0.5 and divide by 0.5 on both sides:

$$\frac{20}{0.5} = \frac{0.5 \cdot b}{0.5}$$

$$\frac{20}{0.5} = b$$

$$40 = b$$

```
              4 0 .
0. 5∧)2 0. 0 ∧
       2 0 0
           0
           0
           0
```

20 is 50% of 40. The answer is 40.

**25.** 40% of what is $16?

*Translate:* $40\% \times b = 16$

*Solve:* To solve the equation we convert 40% to 0.4 and divide by 0.4 on both sides:

$$\frac{0.40 \cdot b}{0.4} = \frac{16}{0.4}$$

$$b = \frac{16}{0.4}$$

$$b = 40$$

```
              4 0 .
0. 4∧)1 6. 0 ∧
       1 6 0
           0
           0
           0
```

40% of $40 is $16. The answer is $40.

**27.** 56.32 is 64% of what?

*Translate:* $56.32 = 64\% \cdot b$

*Solve:* $\dfrac{56.32}{0.64} = \dfrac{0.64 \cdot b}{0.64}$

$$\frac{56.32}{0.64} = b$$

$$88 = b$$

```
                8 8 .
0. 6 4∧)5 6. 3 2 ∧
        5 1 2 0
          5 1 2
          5 1 2
              0
```

56.32 is 64% of 88. The answer is 88.

**29.** 70% of what is 14?

*Translate:* $70\% \cdot b = 14$

*Solve:* $\dfrac{0.7 \cdot b}{0.7} = \dfrac{14}{0.7}$

$$b = \frac{14}{0.7}$$

$$b = 20$$

```
                2 0 .
0. 7 ∧)1 4. 0 ∧
        1 4 0
            0
            0
            0
```

70% of 20 is 14. The answer is 20.

**31.** What is $62\frac{1}{2}\%$ of 10?

*Translate:* $a = 62\frac{1}{2}\% \cdot 10$

*Solve:* $a = 0.625 \cdot 10 \quad (62\frac{1}{2}\% = 0.625)$

$\qquad a = 6.25 \qquad$ Multiplying

6.25 is $62\frac{1}{2}\%$ of 10. The answer is 6.25.

**33.** What is 8.3% of $10,200?

*Translate:* $a = 8.3\% \cdot 10,200$

*Solve:* $a = 8.3\% \cdot 10,200$

$\qquad a = 0.083 \cdot 10,200 \quad (8.3\% = 0.083)$

$\qquad a = 846.6 \qquad$ Multiplying

$846.60 is 8.3% of $10,200. The answer is $846.60.

**35.** $66\frac{2}{3}\%$ of what is 27.4?

*Translate:* $66\frac{2}{3}\% \cdot b = 27.4$

*Solve:* We convert $66\frac{2}{3}\%$ to $\frac{2}{3}$. Then we divide by $\frac{2}{3}$, or multiply by $\frac{3}{2}$ on both sides.

$$\frac{2}{3} \cdot b = 27.4$$

$$\frac{3}{2} \cdot \frac{2}{3} \cdot b = \frac{3}{2} \cdot 27.4$$

$$b = 41.1$$

$66\frac{2}{3}\%$ of 41.1 is 27.4. The answer is 41.1.

**37.** $\underset{\substack{\text{3 decimal}\\\text{places}}}{0.\underline{623}} = \underset{\substack{\text{3 zeros}}}{\dfrac{623}{1000}}$

**39.** $\underset{\substack{\text{2 decimal}\\\text{places}}}{2.\underline{37}} = \underset{\substack{\text{2 zeros}}}{\dfrac{237}{100}}$

**41.** $\underset{\substack{\text{2 zeros}}}{\dfrac{39}{100}} \qquad \underset{\substack{\text{Move 2 places}}}{0.\underline{39}.}$

$$\frac{39}{100} = 0.39$$

**43.** ◈

**45.** ◈

**47.** Estimate: Round 50,951.775 to 50,000 and 78,995 to 80,000. Then translate:

50,000 is what percent of 80,000?

$\downarrow \quad \downarrow \qquad \downarrow \qquad \downarrow \quad \downarrow$
$50,000 = \qquad n \qquad \cdot \; 80,000$

To solve we use a calculator to divide by 80,000 on both sides and convert the result to percent notation.

$$50,000 = n \cdot 80,000$$

$$\frac{50,000}{80,000} = \frac{n \cdot 80,000}{80,000}$$

$$0.625 = n, \text{ or } n = 62.5\%$$

We estimate that 50,951.775 is about 62.5% of 78,995. (Answers may vary.)

Calculate: First we translate:

50,951.775 is what percent of 78,995?

$$\downarrow \quad \downarrow \quad \downarrow \quad \downarrow \quad \downarrow$$
$$50,951.775 = \quad n \quad \cdot \quad 78,995$$

To solve we use a calculator to divide by 78,995 on both sides and convert the result to percent notation.

$$50,951.775 = n \cdot 78,995$$

$$\frac{50,951.775}{78,995} = \frac{n \cdot 78,995}{78,995}$$

$$0.645 = n, \text{ or } n = 64.5\%$$

50,951.775 is 64.5% of 78,995.

**49.** 40% of $18\frac{3}{4}$% of $25,000 is what?

*Translate:* $40\% \cdot 18\frac{3}{4}\% \cdot 25,000 = a$

*Solve:* We convert 40% to 0.4 and $18\frac{3}{4}$% to 0.1875 and then carry out the calculation.

$$0.4(0.1875)(25,000) = a$$
$$0.075(25,000) = a$$
$$1875 = a$$

40% of $18\frac{3}{4}$% of $25,000 is $1875. The answer is $1875.

---

## Exercise Set 8.4

**1. Familiarize.** First we find the number of bowlers who would be expected to be left-handed. Let $a$ represent this number.

**Translate.** We rephrase the question and translate.

What is 17% of 120?

$$\downarrow \quad \downarrow \quad \downarrow \quad \downarrow \quad \downarrow$$
$$a \quad = 17\% \quad \cdot \quad 120$$

**Solve:** We convert 17% to decimal notation and multiply.

$$a = 17\% \cdot 120 = 0.17 \cdot 120 = 20.4 \approx 20$$

We can subtract to find the number of bowlers who would not be expected to be left-handed:

$$120 - 20 = 100$$

**Check.** We can repeat the calculations. We also observe that, since 17% of bowlers are expected to be left-handed, 83% would not be expected to be left-handed. Because 83% of 120 = $0.83 \cdot 120 = 99.6 \approx 100$, our answer checks.

**State.** You would expect 20 bowlers to be left-handed and 100 not to be left-handed.

**3. Familiarize.** Let $a$ = the number of moviegoers in the 12-29 age group.

**Translate.** We rephrase the question and translate.

What is 67% of 600?

$$\downarrow \quad \downarrow \quad \downarrow \quad \downarrow \quad \downarrow$$
$$a \quad = 67\% \quad \cdot \quad 600$$

**Solve.** We convert 67% to decimal notation and multiply.

$$a = 67\% \cdot 600 = 0.67 \cdot 600 = 402$$

**Check.** We can repeat the calculation. The answer checks.

**State.** 402 moviegoers were in the 12-29 age group.

**5. Familiarize.** The question asks for percents. We know that 10% of 40 is 4. Since $13 \approx 3 \cdot 4$, we would expect the percent of at bats that are hits to be close to 30%. Then we would also expect the percent of at bats that are not hits to be 70%. First we find $n$, the percent that are hits.

**Translate.** We rephrase the question and translate.

13 is what percent of 40?

$$\downarrow \downarrow \quad \downarrow \quad \downarrow \downarrow$$
$$13 = \quad n \quad \cdot \quad 40$$

**Solve.** We divide on both sides by 40 and convert the result to percent notation.

$$13 = n \cdot 40$$
$$\frac{13}{40} = \frac{n \cdot 40}{40}$$
$$0.325 = n$$
$$32.5\% = n \qquad \text{Finding percent notation}$$

Now we find the percent of at bats that are not hits. We let $p$ = this percent.

| 100% of hits | minus | percent of hits | is | percent that are not hits |
|---|---|---|---|---|
| $\downarrow$ | $\downarrow$ | $\downarrow$ | $\downarrow$ | $\downarrow$ |
| 100% | − | 32.5% | = | $p$ |

To solve the equation we carry out the subtraction.

$$p = 100\% - 32.5\% = 67.5\%$$

**Check.** Note that the answer, 32.5%, is close to 30% as estimated in the Familiarize step. We also note that the percent of at bats that are not hits, 67.5%, is close to 70% as estimated

**State.** 32.5% of at bats are hits, and 67.5% are not hits.

**7. Familiarize.** First we find the amount of the solution that is acid. We let $a$ = this amount.

**Translate.** We rephrase the question and translate.

What is 3% of 680?

$$\downarrow \quad \downarrow \quad \downarrow \quad \downarrow \quad \downarrow$$
$$a \quad = 3\% \quad \cdot \quad 680$$

**Solve.** We convert 3% to decimal notation and multiply.

$$a = 3\% \cdot 680 = 0.03 \cdot 680 = 20.4$$

Now we find the amount that is water. We let $w =$ this amount.

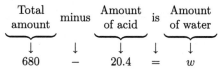

To solve the equation we carry out the subtraction.

$$w = 680 - 20.4 = 659.6$$

**Check**. We can repeat the calculations. Also, observe that, since 3% of the solution is acid, 97% is water. Because 97% of $680 = 0.97 \cdot 680 = 659.6$, our answer checks.

**State**. The solution contains 20.4 mL of acid and 659.6 mL of water.

9. **Familiarize**. We let $n =$ the percent of time that television sets are on.

**Translate**. We rephrase the question and translate.

2190 is what percent of 8760?

$$2190 = n \cdot 8760$$

**Solve**. We divide on both sides by 8760 and convert the result to percent notation.

$$2190 = n \cdot 8760$$
$$\frac{2190}{8760} = \frac{n \cdot 8760}{8760}$$
$$0.25 = n$$
$$25\% = n$$

**Check**. To check we find 25% of 8760:

$25\% \cdot 8760 = 0.25 \cdot 8760 = 2190$. The answer checks.

**State**. Television sets are on for 25% of the year.

11. First we find the maximum heart rate for a 25 year old person.

**Familiarize**. Note that $220 - 25 = 195$. We let $x =$ the maximum heart rate for a 25 year old person.

**Translate**. We rephrase the question and translate.

What is 85% of 195?

$$x = 85\% \cdot 195$$

**Solve**. We convert 85% to a decimal and simplify.

$$x = 0.85 \cdot 195 = 165.75 \approx 166$$

**Check**. We can repeat the calculations. Also, 85% of $195 \approx 0.85 \cdot 200 = 170 \approx 166$. The answer checks.

**State**. The maximum heart rate for a 25 year old person is 166 beats per minute.

Next we find the maximum heart rate for a 36 year old person.

**Familiarize**. Note that $220 - 36 = 184$. We let $x =$ the maximum heart rate for a 36 year old person.

**Translate**. We rephrase the question and translate.

What is 85% of 184?

$$x = 85\% \cdot 184$$

**Solve**. We convert 85% to a decimal and simplify.

$$x = 0.85 \cdot 184 = 156.4 \approx 156$$

**Check**. We can repeat the calculations. Also, 85% of $184 \approx 0.9 \cdot 180 = 162 \approx 156$. The answer checks.

**State**. The maximum heart rate for a 36 year old person is 156 beats per minute.

Next we find the maximum heart rate for a 48 year old person.

**Familiarize**. Note that $220 - 48 = 172$. We let $x =$ the maximum heart rate for a 48 year old person.

**Translate**. We rephrase the question and translate.

What is 85% of 172?

$$x = 85\% \cdot 172$$

**Solve**. We convert 85% to a decimal and simplify.

$$x = 0.85 \cdot 172 = 146.2 \approx 146$$

**Check**. We can repeat the calculations. Also, 85% of $172 \approx 0.9 \cdot 170 = 153 \approx 146$. The answer checks.

**State**. The maximum heart rate for a 48 year old person is 146 beats per minute.

We find the maximum heart rate for a 55 year old person.

**Familiarize**. Note that $220 - 55 = 165$. We let $x =$ the maximum heart rate for a 55 year old person.

**Translate**. We rephrase the question and translate.

What is 85% of 165?

$$x = 85\% \cdot 165$$

**Solve**. We convert 85% to a decimal and simplify.

$$x = 0.85 \cdot 165 = 140.25 \approx 140$$

**Check**. We can repeat the calculations. Also, 85% of $165 \approx 0.9 \cdot 160 = 144 \approx 140$. The answer checks.

**State**. The maximum heart rate for a 55 year old person is 140 beats per minute.

Finally we find the maximum heart rate for a 76 year old person.

**Familiarize**. Note that $220 - 76 = 144$. We let $x =$ the maximum heart rate for a 76 year old person.

**Translate**. We rephrase the question and translate.

What is 85% of 144?

$$x = 85\% \cdot 144$$

**Solve**. We convert 85% to a decimal and simplify.

$$x = 0.85 \cdot 144 = 122.4 \approx 122$$

**Check**. We can repeat the calculations. Also, 85% of $144 \approx 0.9 \cdot 140 = 126 \approx 122$. The answer checks.

**State**. The maximum heart rate for a 76 year old person is 122 beats per minute.

13. **Familiarize**. Use the drawing in the text to visualize the situation. Note that the increase in the amount was $12.

Let $n =$ the percent of increase.

*Translate*. We rephrase the question and translate.

$12 is what percent of $150?

$$12 = n \cdot 150$$

*Solve*. We divide by 150 on both sides and convert the result to percent notation.

$$12 = n \cdot 150$$
$$\frac{12}{150} = \frac{n \cdot 150}{150}$$
$$0.08 = n$$
$$8\% = n$$

*Check*. Find 8% of 150: $8\% \cdot 150 = 0.08 \cdot 150 = 12$. Since this is the amount of the increase, the answer checks.

*State*. The percent of increase was 8%.

**15. Familiarize.** We use the drawing in the text to visualize the situation. Note that the reduction is $18.

We let $n$ = the percent of decrease.

*Translate*. We rephrase the question and translate.

$18 is what percent of $90?

$$18 = n \cdot 90$$

*Solve*. To solve the equation, we divide on both sides by 90 and convert the result to percent notation.

$$\frac{18}{90} = \frac{n \cdot 90}{90}$$
$$0.2 = n$$
$$20\% = n$$

*Check*. We find 20% of 90: $20\% \cdot 90 = 0.2 \cdot 90 = 18$. Since this is the price decrease, the answer checks.

*State*. The percent of decrease was 20%.

**17. Familiarize.** We note that the amount of the raise can be found and then added to the old salary. A drawing helps us visualize the situation.

| $28,600 | $ ? |
|---|---|
| 100% | 5% |

We let $x$ = the new salary.

*Translate*. We rephrase the question and translate.

What is the old salary plus 5% of the old salary?

$$x = 28,600 + 5\% \cdot 28,600$$

*Solve*. We convert 5% to a decimal and simplify.

$$x = 28,600 + 0.05 \cdot 28,600$$
$$= 28,600 + 1430 \qquad \text{The raise is } \$1430.$$
$$= 30,030$$

*Check*. To check, we note that the new salary is 100% of the old salary plus 5% of the old salary, or 105% of the old salary. Since $1.05 \cdot 28,600 = 30,030$, our answer checks.

*State*. Rachel's new salary is $30,030.

**19. Familiarize.** We visualize the situation.

| $18,000 | |
|---|---|
| | $ ? |
| 100% | |
| 70% | 30% |

This is a two-step problem. First we find the amount of the decrease. Let $a$ represent this amount.

*Translate*. We rephrase the question and translate.

What is 30% of $18,000?

$$a = 30\% \cdot 18,000$$

*Solve*. We convert to decimal notation and multiply:

$$a = 0.3 \cdot 18,000 = 5400$$

To find the value after one year we subtract:

$$\$18,000 - \$5400 = \$12,600$$

*Check*. Note that with a 30% decrease, the reduced value should be 70% of the original value. Since 70% of $18,000 = 0.7 \cdot 18,000 = 12,600$, the answer checks.

*State*. The value of the car is $12,600 after one year.

**21. Familiarize.** This is a multi-step problem. To find the population in 2000, we first find the increase over the population in 1999. Let $a$ represent this increase in billions.

*Translate*. We rephrase the question and translate.

What is 1.6% of 6.0 billion?

$$a = 1.6\% \cdot 6.0$$

We convert to decimal notation and multiply:

$$a = 0.016 \cdot 6.0 = 0.096$$

Now we add to find the population in 2000:

$$6.0 + 0.096 = 6.096$$

To find the population in 2001, we first find the increase $a$ over the population in 2000.

What is 1.6% of 6.096 billion?

$$a = 1.6\% \cdot 6.096$$

We convert to decimal notation and multiply:

$$a = 0.016 \cdot 6.096 = 0.097536$$

Then we add to find the population in 2001:

$$6.096 + 0.097536 = 6.193536 \approx 6.194$$

To find the population in 2002, we first find the increase $a$ over the population in 2001.

What is 1.6% of 6.194 billion?

$$a = 1.6\% \cdot 6.194$$

Again, we convert to decimal notation and multiply:

$$a = 0.016 \cdot 6.194 = 0.099104$$

Then we add to find the population in 2002:

$$6.194 + 0.099104 = 6.293104 \approx 6.293$$

**Check**. Note that the population each year is 101.6% of the population the previous year. Since

$$101.6\% \text{ of } 6.0 = 1.016 \cdot 6.0 = 6.096,$$

$$101.6\% \text{ of } 6.096 = 1.016 \cdot 6.096 \approx 6.194, \text{ and}$$

$$101.6\% \text{ of } 6.096 = 1.016 \cdot 6.194 \approx 6.293,$$

the results check.

**State**. In 2000 the world population will be 6.096 billion, in 2001 it will be about 6.194 billion, and in 2002 it will be about 6.293 billion.

**23.** **Familiarize**. Since the car depreciates 30% in the first year, its value after the first year is 100% − 30%, or 70%, of the original value. To find the decrease in value, we ask:

$$\$11,480 \text{ is } 70\% \text{ of what?}$$

Let $b$ = the original cost.

**Translate**. We rephrase the question and translate.

$$\$25,480 \text{ is } 70\% \text{ of what?}$$
$$\downarrow \quad \downarrow \quad \downarrow \quad \downarrow \quad \downarrow$$
$$\$25,480 = 70\% \cdot \quad b$$

**Solve**.

$$25,480 = 70\% \cdot b$$

$$\frac{25,480}{70\%} = \frac{70\% \cdot b}{70\%}$$

$$\frac{25,480}{0.7} = b$$

$$36,400 = b$$

**Check**. We find 30% of 36,400 and then subtract this amount from 36,400:

$$0.3 \cdot 36,400 = 10,920 \text{ and}$$

$$36,400 - 10,920 = 25,480$$

The answer checks.

**State**. The original cost was $36,400.

**25.** **Familiarize**. This is a multistep problem. First we find the amount of each tip. Then we add that amount to the corresponding cost of the meal. Let $x$, $y$, and $z$ represent the amounts of the tips on the $15, $34, and $49 meals, respectively.

**Translate**. We rephrase the questions and translate.

$$\text{What is } 15\% \text{ of } \$15?$$
$$\downarrow \quad \downarrow \quad \downarrow \quad \downarrow \quad \downarrow$$
$$x \quad = 15\% \cdot \quad 15$$

$$\text{What is } 15\% \text{ of } \$34?$$
$$\downarrow \quad \downarrow \quad \downarrow \quad \downarrow \quad \downarrow$$
$$y \quad = 15\% \cdot \quad 34$$

$$\text{What is } 15\% \text{ of } \$49?$$
$$\downarrow \quad \downarrow \quad \downarrow \quad \downarrow \quad \downarrow$$
$$z \quad = 15\% \cdot \quad 49$$

**Solve**. We convert to decimal notation and multiply.

$$x = 0.15 \cdot 15 = 2.25$$

$$y = 0.15 \cdot 34 = 5.10$$

$$z = 0.15 \cdot 49 = 7.35$$

Now we add to find the amounts charged.

For the $15 meal: $15 + $2.25 = $17.25

For the $34 meal: $34 + $5.10 = $39.10

For the $49 meal: $49 + $7.35 = $56.35

**Check**. Note that the amount charged in each case is 115% of the cost of the meal. Since

$$115\% \text{ of } \$15 = 1.15 \cdot \$15 = \$17.25,$$

$$115\% \text{ of } \$34 = 1.15 \cdot \$34 = \$39.10, \text{ and}$$

$$115\% \text{ of } \$49 = 1.15 \cdot \$49 = \$56.35,$$

the answer checks.

**State**. The total amounts charged are $17.25 for the $15 meal, $39.10 for the $34 meal, and $56.35 for the $49 meal.

**27.** a) **Familiarize**. Note that the increase in deaths from 1994 to 1995 is $17,274 - 16,589$, or 685. Let $n$ = the percent of increase.

**Translate**. We rephrase the question and translate.

$$685 \text{ is } \underbrace{\text{what percent}} \text{ of } 16,589?$$
$$\downarrow \quad \downarrow \qquad \downarrow \qquad \downarrow \quad \downarrow$$
$$685 = \qquad n \qquad \cdot \quad 16,589$$

**Solve**. We divide on both sides by 16,589 and convert to percent notation.

$$\frac{685}{16,589} = \frac{n \cdot 16,589}{16,589}$$

$$0.041 \approx n$$

$$4.1\% \approx n$$

**Check**. Note that the number of deaths in 1995 will be 104.1% of the number in 1994. Since $1.041 \cdot 16,589 = 17,269.149 \approx 17,274$, the answer checks.

**State**. The percent of increase in alcohol-related deaths from 1994 to 1995 was about 4.1%.

b) **Familiarize**. Note that the decrease in deaths from 1986 to 1994 is $24,045 - 16,589$, or 7456. Let $n$ = the percent of decrease.

**Translate**. We rephrase the question and translate.

$$7456 \text{ is } \underbrace{\text{what percent}} \text{ of } 24,045?$$
$$\downarrow \quad \downarrow \qquad \downarrow \qquad \downarrow \quad \downarrow$$
$$7456 = \qquad n \qquad \cdot \quad 24,045$$

**Solve**. We divide on both sides by 7456 and convert to percent notation.

$$\frac{7456}{24,045} = \frac{n \cdot 24,045}{24,045}$$

$$0.310 \approx n$$

$$31.0\% \approx n$$

**Check**. We find 31.0% of 24,045 and subtract this number from 24,045.

$$0.310 \cdot 24,045 \approx 7454$$

$$24,045 - 7454 = 16,591 \approx 16,589$$

The answer checks.

**State**. The percent of decrease in alcohol-related deaths from 1986 to 1994 was about 31.0%.

**29.** **Familiarize**. First we use the formula $A = l \cdot w$ to find the area of the strike zone:

$$A = 40 \cdot 17 = 680 \text{ in}^2$$

When a 2-in. border is added to the outside of the strike zone, the dimensions of the larger zone are 19 in. by 44 in. The area of this zone is

$$A = 44 \cdot 21 = 924 \text{ in}^2$$

We subtract to find the increase in area:

$$924 \text{ in}^2 - 680 \text{ in}^2 = 244 \text{ in}^2$$

We let $N =$ the percent of increase in the area.

**Translate**. We rephrase the question and translate.

244 is what percent of 680?

$$244 = N \cdot 680$$

**Solve**. We divide by 680 on both sides and convert to percent notation.

$$\frac{244}{680} = \frac{N \cdot 680}{680}$$
$$0.359 = N$$
$$35.9\% = N$$

**Check**. We repeat the calculations.

**State**. The area of the strike zone is increased by 35.9%.

**31.** $\frac{25}{11} = 25 \div 11$

```
      2. 2 7
 1 1 ) 2 5. 0 0
       2 2
       ─────
         3 0
         2 2
         ─────
           8 0
           7 7
           ─────
             3
```

Since the remainders begin to repeat, we have a repeating decimal.

$$\frac{25}{11} = 2.\overline{27}$$

**33.** $\frac{27}{8} = 27 \div 8$

```
      3. 3 7 5
 8 ) 2 7. 0 0 0
     2 4
     ─────
       3 0
       2 4
       ─────
         6 0
         5 6
         ─────
           4 0
           4 0
           ─────
             0
```

$$\frac{27}{8} = 3.375$$

We could also do this conversion as follows:

$$\frac{27}{8} = \frac{27}{8} \cdot \frac{125}{125} = \frac{3375}{1000} = 3.375$$

**35.** $\frac{23}{25} = \frac{23}{25} \cdot \frac{4}{4} = \frac{92}{100} = 0.92$

**37.** $\frac{14}{32} = 14 \div 32$

```
        0. 4 3 7 5
 3 2 ) 1 4. 0 0 0 0
       1 2 8
       ─────
         1 2 0
           9 6
         ─────
           2 4 0
           2 2 4
           ─────
             1 6 0
             1 6 0
             ─────
                 0
```

$$\frac{14}{32} = 0.4375$$

(Note that we could have simplified the fraction first, getting $\frac{7}{16}$ and then found the quotient $7 \div 16$.)

**39.** Since 10,000 has 4 zeros, we move the decimal point in the number in the numerator 4 places to the left.

$$\frac{34,809}{10,000} = 3.4809$$

**41.** ◈

**43.** ◈

**45. Familiarize**. We will express 4 ft, 8 in. as 56 in. (4 ft + 8 in. $= 4 \cdot 12$ in. $+ 8$ in. $= 48$ in. $+ 8$ in. $= 56$ in.) We let $h =$ Cynthia's final adult height.

**Translate**. We rephrase the question and translate.

56 in. is 84.4% of what?

$$56 = 84.4\% \times h$$

**Solve**. First we convert 84.4% to a decimal.

$$56 = 0.844 \times h$$
$$\frac{56}{0.844} = \frac{0.844 \times h}{0.844}$$
$$66 \approx h$$

**Check**. We find 84.4% of 66: $0.844 \times 66 \approx 56$. The answer checks.

**State**. Cynthia's final adult height will be about 66 in., or 5 ft, 6 in.

**47. Familiarize**. If $p$ is 120% of $q$, then $p = 1.2q$. Let $n =$ the percent of $p$ that $q$ represents.

**Translate**. We rephrase the question and translate. We use $1.2q$ for $p$.

$q$ is what percent of $p$?

$$q = n \times 1.2q$$

*Solve.*

$$q = n \times 1.2q$$

$$\frac{q}{1.2q} = \frac{n \times 1.2q}{1.2q}$$

$$\frac{1}{1.2} = n$$

$$0.8\overline{3} = n$$

$$83.\overline{3}\%, \text{ or } 83\frac{1}{3}\% = n$$

**Check.** We find $83\frac{1}{3}\%$ of $1.2q$:

$$0.8\overline{3} \times 1.2q = q$$

The answer checks.

**State.** $q$ is $83.\overline{3}\%$, or $83\frac{1}{3}\%$, of $p$.

**49.** Let $S$ = the original salary. After a 3% raise, the salary becomes $103\% \cdot S$, or $1.03S$. After a 6% raise, the new salary is $1.06\% \cdot 1.03S$, or $1.06(1.03S)$. Finally, after a 9% raise, the salary is $109\% \cdot 1.06(1.03S)$, or $1.09(1.06)(1.03S)$. Multiplying, we get $1.09(1.06)(1.03S) = 1.190062S$. This is equivalent to $119.0062\% \cdot S$, so the original salary has increased by 19.0062%, or about 19%.

# Exercise Set 8.5

**1.** a) The sales tax on an item costing \$586 is

$$\underbrace{\text{Sales tax rate}}_{\downarrow} \cdot \underbrace{\text{Purchase price}}_{\downarrow}$$

$$\quad 5\% \qquad \cdot \qquad \$586,$$

or $0.05 \cdot 586$, or $29.3$. Thus the tax is \$29.30.

b) The total price is given by the purchase price plus the sales tax:

$$\$586 + \$29.30, \text{ or } \$615.30.$$

To check, note that the total price is the purchase price plus 5% of the purchase price. Thus the total price is 105% of the purchase price. Since $1.05 \cdot 586 = 615.3$, we have a check. The total price is \$615.30.

**3.** a) We first find the cost of the telephones. It is

$$5 \cdot \$53 = \$265.$$

b) The sales tax on items costing \$265 is

$$\underbrace{\text{Sales tax rate}}_{\downarrow} \cdot \underbrace{\text{Purchase price}}_{\downarrow}$$

$$\quad 6.25\% \qquad \cdot \qquad \$265,$$

or $0.0625 \cdot 265$, or $16.5625$. Thus the tax is \$16.56.

c) The total price is given by the purchase price plus the sales tax:

$$\$265 + \$16.56, \text{ or } \$281.56.$$

To check, note that the total price is the purchase price plus 6.25% of the purchase price. Thus the total price is 106.25% of the purchase price. Since $1.0625 \cdot \$265 = \$281.56$ (rounded to the nearest cent), we have a check. The total price is \$281.56.

**5.** *Rephrase:* $\underbrace{\text{Sales tax}}$ is $\underbrace{\text{what percent}}$ of $\underbrace{\text{purchase price?}}$

$$\qquad\qquad\quad \downarrow \qquad \downarrow \quad\; \downarrow \quad\; \downarrow \qquad \downarrow$$

*Translate:* $\quad 15.96 \; = \quad r \quad \cdot \qquad 399$

To solve the equation, we divide on both sides by 399.

$$\frac{15.96}{399} = \frac{r \cdot 399}{399}$$

$$0.04 = r$$

$$4\% = r$$

The sales tax rate is 4%.

**7.** *Rephrase:* $\underbrace{\text{Sales tax}}$ is $\underbrace{\text{what percent}}$ of $\underbrace{\text{purchase price?}}$

$$\qquad\qquad\quad \downarrow \qquad \downarrow \quad\; \downarrow \quad\; \downarrow \qquad \downarrow$$

*Translate:* $\quad 44.75 \; = \quad r \quad \cdot \qquad 895$

To solve the equation, we divide on both sides by 895.

$$\frac{44.75}{895} = \frac{r \cdot 895}{895}$$

$$0.05 = r$$

$$5\% = r$$

The sales tax rate is 5%.

**9.** *Rephrase:* $\underbrace{\text{Sales tax}}$ is 5% of what?

$$\qquad\qquad\qquad \downarrow \quad\;\; \downarrow \; \downarrow \; \downarrow \quad \downarrow$$

*Translate:* $\quad\; 250 \;\; = 5\% \; \cdot \quad b, \text{ or}$

$$\qquad\qquad\quad 250 \;\; = 0.05 \; \cdot \quad b$$

To solve the equation, we divide on both sides by 0.05.

$$\frac{250}{0.05} = \frac{0.05 \cdot b}{0.05}$$

$$5000 = b$$

$$\begin{array}{r} 5\,0\,0\,0.\phantom{0} \\ 0.0\,5\,\overline{)\,2\,5\,0.0\,0\,_\wedge} \\ \underline{2\,5\,0\,0\,0\phantom{.}} \\ 0 \end{array}$$

The purchase price is \$5000.

**11.** *Rephrase:* $\underbrace{\text{Sales tax}}$ is 3.5% of what?

$$\qquad\qquad\qquad \downarrow \quad\;\; \downarrow \; \downarrow \; \downarrow \quad \downarrow$$

*Translate:* $\quad\; 28 \;\; = 3.5\% \; \times \quad b, \text{ or}$

$$\qquad\qquad\quad 28 \;\; = 0.035 \; \times \quad b$$

To solve the equation, we divide on both sides by 0.035.

$$\frac{28}{0.035} = \frac{0.035 \times b}{0.035}$$

$$800 = b$$

$$\begin{array}{r} 8\,0\,0.\phantom{0} \\ 0.0\,3\,5\,_\wedge\overline{)\,2\,8.0\,0\,0\,_\wedge} \\ \underline{2\,8\,0\,0\,0\phantom{.}} \\ 0 \end{array}$$

The purchase price is \$800.

**13.** a) We first find the cost of the shower units. It is

$$2 \times \$332.50 = \$665.$$

b) The total tax rate is the city tax rate plus the state tax rate, or $2\% + 6.25\% = 8.25\%$. The sales tax paid on items costing $665 is

$$\underbrace{\text{Sales tax rate}}_{\downarrow} \cdot \underbrace{\text{Purchase price}}_{\downarrow}$$
$$\underset{8.25\%}{\downarrow} \quad \cdot \quad \underset{\$665,}{\downarrow}$$

or $0.0825 \cdot 665$, or $54.8625$. Thus the tax is $54.86.

c) The total price is given by the purchase price plus the sales tax:

$$\$665 + \$54.86 = \$719.86.$$

To check, note that the total price is the purchase price plus 8.25% of the purchase price. Thus the total price is 108.25% of the purchase price. Since $1.0825 \cdot 665 = 719.8625$, we have a check. The total amount paid for the 2 shower units is $719.86.

**15.** *Rephrase:*
$$\underbrace{\text{Sales}}_{\text{tax}} \text{ is } \underbrace{\text{what}}_{\text{percent}} \text{ of } \underbrace{\text{purchase}}_{\text{price?}}$$

*Translate:*
$$1030.40 = \quad r \quad \cdot \quad 18,400$$

To solve the equation, we divide on both sides by 18,400.

$$\frac{1030.40}{18,400} = \frac{r \cdot 18,400}{18,400}$$
$$0.056 = r$$
$$5.6\% = r$$

The sales tax rate is 5.6%.

**17.** Commission = Commission rate · Sales
$$C \quad = \quad 35\% \quad \cdot 2580$$

This tells us what to do. We multiply.

$$\begin{array}{r} 2\,5\,8\,0 \\ \times \quad 0.\,3\,5 \\ \hline 1\,2\,9\,0\,0 \\ 7\,7\,4\,0\,0 \\ \hline 9\,0\,3.\,0\,0 \end{array} \quad (35\% = 0.35)$$

The commission is $903.

**19.** Commission = Commission rate · Sales
$$87 \quad = \quad r \quad \cdot 174$$

To solve this equation we divide on both sides by 174:

$$\frac{87}{174} = \frac{r \cdot 174}{174}$$

We can divide, but this time we simplify by removing a factor of 1:

$$r = \frac{87}{174} = \frac{87 \cdot 1}{87 \cdot 2} = \frac{87}{87} \cdot \frac{1}{2} = 0.5 = 50\%$$

The commission rate is 50%.

**21.** Commission = Commission rate · Sales
$$392 \quad = \quad 40\% \quad \cdot \quad S$$

To solve this equation we divide on both sides by 0.4:

$$\frac{392}{0.4} = \frac{0.4 \cdot S}{0.4}$$
$$980 = S$$

$$\begin{array}{r} 9\,8\,0.\phantom{0} \\ 0.\,4\,\sqrt{\,3\,9\,2.\,0\,}_{\wedge} \\ 3\,6\,0\,0 \\ \hline 3\,2\,0 \\ 3\,2\,0 \\ \hline 0 \\ 0 \\ \hline 0 \end{array}$$

$980 worth of artwork was sold.

**23.** Commission = Commission rate · Sales
$$C \quad = \quad 6\% \quad \cdot 98,000$$

This tells us what to do. We multiply.

$$\begin{array}{r} 9\,8,0\,0\,0 \\ \times \quad 0.\,0\,6 \\ \hline 5\,8\,8\,0.\,0\,0 \end{array} \quad (6\% = 0.06)$$

The commission is $5880.

**25.** Commission = Commission rate · Sales
$$280.80 \quad = \quad r \quad \cdot 2340$$

To solve this equation we divide on both sides by 2340.

$$\frac{280.80}{2340} = \frac{r \cdot 2340}{2340}$$
$$0.12 = r$$
$$12\% = r$$

$$\begin{array}{r} 0.1\,2 \\ 2\,3\,4\,0\,\sqrt{\,2\,8\,0\,8\,0} \\ 2\,3\,4\,0 \\ \hline 4\,6\,8\,0 \\ 4\,6\,8\,0 \\ \hline 0 \end{array}$$

The commission is 12%.

**27.** First we find the commission on the first $2000 of sales.

Commission = Commission rate · Sales
$$C \quad = \quad 5\% \quad \cdot 2000$$

This tells us what to do. We multiply.

$$\begin{array}{r} 2\,0\,0\,0 \\ \times \quad 0.\,0\,5 \\ \hline 1\,0\,0.\,0\,0 \end{array}$$

The commission on the first $2000 of sales is $100.

Next we subtract to find the amount of sales over $2000.

$$\$6000 - \$2000 = \$4000$$

Miguel had $4000 in sales over $2000.

Then we find the commission on the sales over $2000.

Commission = Commission rate · Sales
$$C \quad = \quad 8\% \quad \cdot 4000$$

This tells us what to do. We multiply.

$$\begin{array}{r} 4\,0\,0\,0 \\ \times \quad 0.\,0\,8 \\ \hline 3\,2\,0.\,0\,0 \end{array}$$

The commission on the sales over $2000 is $320.

Finally we add to find the total commission.

$$\$100 + \$320 = \$420$$

The total commission is $420.

**29.** Discount = Rate of discount · Marked price
       $D$     =        10%        ·       \$300

Convert 10% to decimal notation and multiply.

$$\begin{array}{r} 3\,0\,0 \\ \times\ \ 0.\,1 \\ \hline 3\,0.\,0 \end{array} \qquad (10\% = 0.10 = 0.1)$$

The discount is \$30.

Sale price = Marked price − Discount
     $S$     =      300      −     30

We subtract: $\begin{array}{r} 3\,0\,0 \\ -\ \ 3\,0 \\ \hline 2\,7\,0 \end{array}$

To check, note that the sale price is 90% of the marked price: $0.9 \cdot 300 = 270$.

The sale price is \$270.

**31.** Discount = Rate of discount · Marked price
       $D$     =        15%        ·       \$17

Convert 15% to decimal notation and multiply.

$$\begin{array}{r} 1\,7 \\ \times\ 0.\,1\,5 \\ \hline 8\,5 \\ 1\,7\,0 \\ \hline 2.\,5\,5 \end{array} \qquad (15\% = 0.15)$$

The discount is \$2.55.

Sale price = Marked price − Discount
     $S$     =      17      −     2.55

We subtract: $\begin{array}{r} 1\,7.\,0\,0 \\ -\ \ 2.\,5\,5 \\ \hline 1\,4.\,4\,5 \end{array}$

To check, note that the sale price is 85% of the marked price: $0.85 \cdot 17 = 14.45$.

The sale price is \$14.45.

**33.** Discount = Rate of discount · Marked price
       12.50   =        10%        ·       $M$

To solve the equation we divide on both sides by 0.1.

$$\frac{12.50}{0.1} = \frac{0.1 \cdot M}{0.1}$$
$$125 = M$$

The marked price is \$125.

Sale price = Marked price − Discount
     $S$     =     125.00    −    12.50

We subtract: $\begin{array}{r} 1\,2\,5.\,0\,0 \\ -\ \ 1\,2.\,5\,0 \\ \hline 1\,1\,2.\,5\,0 \end{array}$

To check, note that the sale price is 90% of the marked price: $0.9 \cdot 125 = 112.50$.

The sale price is \$112.50.

**35.** Discount = Rate of discount · Marked price
       240    =        $r$        ·       600

To solve the equation we divide on both sides by 600.

$$\frac{240}{600} = \frac{r \cdot 600}{600}$$

We can simplify by removing a factor of 1:

$$r = \frac{240}{600} = \frac{2}{5} \cdot \frac{120}{120} = \frac{2}{5} = 0.4 = 40\%$$

The rate of discount is 40%.

Sale price = Marked price − Discount
     $S$     =      600      −     240

We subtract: $\begin{array}{r} 6\,0\,0 \\ -\ 2\,4\,0 \\ \hline 3\,6\,0 \end{array}$

To check, note that a 40% discount rate means that 60% of the marked price is paid. Since $\frac{360}{600} = 0.6$, or 60%, we have a check.

The sale price is \$360.

**37.** Discount = Marked price − Sale price
       $D$     =      1275      −     888

We subtract: $\begin{array}{r} 1\,2\,7\,5 \\ -\ \ 8\,8\,8 \\ \hline 3\,8\,7 \end{array}$

The discount is \$387.

Discount = Rate of discount · Marked price
   387    =        $R$        ·      1275

To solve the equation we divide on both sides by 1275.

$$\frac{387}{1275} = \frac{R \cdot 1275}{1275}$$
$$0.304 \approx R$$
$$30.4\% \approx R$$

To check, note that a discount rate of 30.4% means that 69.6% of the marked price is paid:

$0.696 \cdot 1275 = 887.4 \approx 888$. Since that is the sale price, the answer checks.

The rate of discount is about 30.4%.

**39.** Discount = Marked price − Sale price
       83    =        $M$        −     377

We add 377 on both sides of the equation:

$$83 + 377 = M$$
$$460 = M$$

The marked price is \$460.

Discount = Rate of discount · Marked price
    83    =        $R$        ·      460

To solve the equation we divide on both sides by 460.

$$\frac{83}{460} = \frac{R \cdot 460}{460}$$
$$0.18 \approx R$$
$$18\% \approx R$$

To check note that a discount rate of 18% means that 82% of the marked price is paid: $0.82 \cdot 460 = 377.2 \approx 377$. Since this is the sale price, the answer checks.

The rate of discount is about 18%.

**41.**   $\dfrac{x}{12} = \dfrac{24}{16}$

$16 \cdot x = 12 \cdot 24$      Equating cross-products

$x = \dfrac{12 \cdot 24}{16}$      Dividing by 16 on both sides

$x = \dfrac{288}{16}$

$x = 18$

The solution is 18.

**43.** Graph: $y = \dfrac{4}{3}x$

We make a table of solutions. Note that when $x$ is a multiple of 3, fractional values for $y$ are avoided. Next we plot the points, draw the line and label it.

When $x = -3$, $y = \dfrac{4}{3}(-3) = -\dfrac{12}{3} = -4$.

When $x = 0$, $y = \dfrac{4}{3} \cdot 0 = 0$.

When $x = 3$, $y = \dfrac{4}{3} \cdot 3 = \dfrac{12}{3} = 4$.

| $x$ | $y$ $y = \frac{4}{3}x$ | $(x, y)$ |
|-----|------------------------|----------|
| $-3$ | $-4$ | $(-3, -4)$ |
| $0$ | $0$ | $(0, 0)$ |
| $3$ | $4$ | $(3, 4)$ |

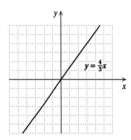

**45.** $\dfrac{5}{9} = 5 \div 9$

```
    0. 5 5
9 ⟌ 5. 0 0
    4 5
    ─────
      5 0
      4 5
      ─────
        5
```

We get a repeating decimal.

$\dfrac{5}{9} = 0.\overline{5}$

**47.** First consider $\dfrac{11}{12}$, or $11 \div 12$.

```
      0. 9 1 6 6
1 2 ⟌ 1 1. 0 0 0 0
      1 0 8
      ───────
          2 0
          1 2
          ─────
            8 0
            7 2
            ─────
              8 0
              7 2
              ─────
                8
```

We get a repeating decimal: $\dfrac{11}{12} = 0.91\overline{6}$.

Since $\dfrac{11}{12} = 0.91\overline{6}$, we have $-\dfrac{11}{12} = -0.91\overline{6}$.

**49.** ◈

**51.** ◈

**53.** Commission = Commission rate $\times$ Sales

$\quad\;\; C \qquad\quad = \qquad 7.5\% \qquad \times\; 98,500$

We multiply.

```
      9 8, 5 0 0
   ×    0. 0 7 5        (7.5% = 0.075)
   ─────────────
      4 9 2 5 0 0
    6 8 9 5 0 0 0
   ─────────────
    7 3 8 7. 5 0 0
```

The commission is $7387.50.

We subtract to find how much the seller gets for the house after paying the commission.

$\$98,500 - \$7387.50 = \$91,112.50$

**55.** First we find the commission on the first $5000 in sales.

Commission = Commission rate $\times$ Sales

$\quad\;\; C \qquad\quad = \qquad 10\% \qquad \times\; 5000$

Using a calculator we find that $0.1 \times 5000 = 500$, so the commission on the first $5000 in sales was $500. We subtract to find the additional commission:

$\$2405 - \$500 = \$1905$

Now we find the amount of sales required to earn $1905 at a commission rate of 15%.

Commission = Commission rate $\times$ Sales

$\quad 1905 \qquad = \qquad 15\% \qquad \times\; S$

Using a calculator to divide 1905 by 15%, or 0.15, we get 12,700.

Finally we add to find the total sales:

$\$5000 + \$12,700 = \$17,700$

**57.** *Familiarize.* Let $x =$ the original price of the plaque that was sold for a profit. Then the plaque was sold for 100% of $x$ plus 20% of $x$, or 120% of $x$. Let $y =$ the original price of the plaque that was sold for a loss. This plaque was sold for 100% of $y$ less 20% of $y$, or 80% of $y$.

*Translate.* First we consider the plaque that was sold for a profit.

Selling price   is 120% of   Original price

$\qquad 200 \quad = \;120\% \; \cdot \qquad x$

Next we consider the plaque that was sold for a loss.

Selling price   is 80% of   Original price

$\qquad 200 \quad = \;80\% \; \cdot \qquad y$

*Solve.* We solve each equation.

$200 = 120\% \cdot x$

$200 = 1.2 \cdot x$

$\dfrac{200}{1.2} = \dfrac{1.2 \cdot x}{1.2}$

$166.\overline{6} = x$, or

$166\dfrac{2}{3} = x$

$$200 = 80\% \cdot y$$
$$200 = 0.8 \cdot y$$
$$\frac{200}{0.8} = \frac{0.8 \cdot y}{0.8}$$
$$250 = y$$

Then the plaques were bought for $\$166\frac{2}{3} + \$250$, or $\$416\frac{2}{3}$ and were sold for $\$200 + \$200$, or $\$400$, so Herb lost money on the sale.

**Check**. 20% of $166\frac{2}{3} = 0.2 \times 166\frac{2}{3} = 33\frac{1}{3}$ and $166\frac{2}{3} + 33\frac{1}{3} = 200$. Also, 20% of $250 = 0.2 \times 250 = 50$ and $250 - 50 = 200$. Since we get the selling price in each case, the answer checks.

**State**. Herb lost money on the sale.

## Exercise Set 8.6

**1.** $I = P \cdot r \cdot t$
$\quad = \$200 \cdot 13\% \cdot 1$
$\quad = \$200 \cdot 0.13$
$\quad = \$26$

$$\begin{array}{r} 2\ 0\ 0 \\ \times\ 0.1\ 3 \\ \hline 6\ 0\ 0 \\ 2\ 0\ 0\ 0 \\ \hline 2\ 6.0\ 0 \end{array}$$

The interest is $26.

**3.** $I = P \cdot r \cdot t$
$\quad = \$2000 \cdot 12.4\% \cdot \frac{1}{2}$
$\quad = \dfrac{\$2000 \cdot 0.124}{2}$
$\quad = \$124$

The interest is $124.

(We could have instead found $\frac{1}{2}$ of 12.4% and then multiplied by 2000.)

**5.** $I = P \cdot r \cdot t$
$\quad = \$4300 \cdot 14\% \cdot \frac{1}{4}$
$\quad = \dfrac{\$4300 \cdot 0.14}{4}$
$\quad = \$150.50$

The interest is $150.50.

(We could have instead found $\frac{1}{4}$ of 14% and then multiplied by 4300.)

**7.** a) We express 60 days as a fractional part of a year and find the interest.
$\quad I = P \cdot r \cdot t$
$\quad\quad = \$10,000 \cdot 9\% \cdot \dfrac{60}{365}$
$\quad\quad = \$10,000 \cdot 0.09 \cdot \dfrac{60}{365}$
$\quad\quad \approx \$147.95 \quad$ Using a calculator

The interest due for 60 days is $147.95.

b) The total amount that must be paid after 60 days is the principal plus the interest.
$$10,000 + 147.95 = 10,147.95$$
The total amount due is $10,147.95.

**9.** a) We express 90 days as a fractional part of a year and find the interest.
$\quad I = P \cdot r \cdot t$
$\quad\quad = \$6500 \cdot 8\% \cdot \dfrac{90}{365}$
$\quad\quad = \$6500 \cdot 0.08 \cdot \dfrac{90}{365}$
$\quad\quad \approx \$128.22 \quad$ Using a calculator

The interest due for 90 days is $128.22.

b) The total amount that must be paid after 90 days is the principal plus the interest.
$$6500 + 128.22 = 6628.22$$
The total amount due is $6628.22.

**11.** a) We express 30 days as a fractional part of a year and find the interest.
$\quad I = P \cdot r \cdot t$
$\quad\quad = \$5600 \cdot 10\% \cdot \dfrac{30}{365}$
$\quad\quad = \$5600 \cdot 0.1 \cdot \dfrac{30}{365}$
$\quad\quad \approx \$46.03 \quad$ Using a calculator

The interest due for 30 days is $46.03.

b) The total amount that must be paid after 30 days is the principal plus the interest.
$$5600 + 46.03 = 5646.03$$
The total amount due is $5646.03.

**13.** a) After 1 year, the account will contain 110% of $400.
$\quad 1.1 \cdot \$400 = \$440$

$$\begin{array}{r} 4\ 0\ 0 \\ \times\ 1.1 \\ \hline 4\ 0\ 0 \\ 4\ 0\ 0\ 0 \\ \hline 4\ 4\ 0.0 \end{array}$$

b) At the end of the second year, the account will contain 110% of $440.
$\quad 1.1 \cdot \$440 = \$484$

$$\begin{array}{r} 4\ 4\ 0 \\ \times\ 1.1 \\ \hline 4\ 4\ 0 \\ 4\ 4\ 0\ 0 \\ \hline 4\ 8\ 4.0 \end{array}$$

The amount in the account after 2 years is $484.

(Note that we could have used the formula
$$A = P \cdot \left(1 + \frac{r}{n}\right)^{n \cdot t},$$ substituting $400 for $P$, 10% for $r$, 1 for $n$, and 2 for $t$.)

**15.** a) After 1 year, the account will contain 108.8% of $200.

$$1.088 \cdot \$200 = \$217.60$$

$$\begin{array}{r} 1.0\,8\,8 \\ \times \quad 2\,0\,0 \\ \hline 2\,1\,7.6\,0\,0 \end{array}$$

b) At the end of the second year, the account will contain 108.8% of $217.60.

$$1.088 \cdot \$217.60 = \$236.7488$$

$$\begin{array}{r} 2\,1\,7.\,6 \\ \times\,1.0\,8\,8 \\ \hline 1\,7\,4\,0\,8 \\ 1\,7\,4\,0\,8\,0 \\ 2\,1\,7\,6\,0\,0\,0 \\ \hline 2\,3\,6.\,7\,4\,8\,8 \end{array}$$

$$\approx \$236.75 \quad \text{Rounding to the nearest cent}$$

The amount in the account after 2 years is $236.75.

(Note that we could have used the formula

$A = P \cdot \left(1 + \dfrac{r}{n}\right)^{n \cdot t}$, substituting $200 for $P$, 15% for $r$,

1 for $n$, and 2 for $t$.)

**17.** We use the compound interest formula, substituting $4000 for $P$, 7% for $r$, 2 for $n$, and 1 for $t$.

$$A = P \cdot \left(1 + \frac{r}{n}\right)^{n \cdot t}$$

$$A = \$4000 \cdot \left(1 + \frac{0.07}{2}\right)^{2 \cdot 1}$$

$$A = \$4000 \cdot (1 + 0.035)^2$$

$$A = \$4000 \cdot (1.035)^2$$

$$A = \$4284.90$$

The amount in the account after 1 year is $4284.90.

**19.** We use the compound interest formula, substituting $2000 for $P$, 9% for $r$, 2 for $n$, and 3 for $t$.

$$A = P \cdot \left(1 + \frac{r}{n}\right)^{n \cdot t}$$

$$A = \$2000 \cdot \left(1 + \frac{0.09}{2}\right)^{2 \cdot 3}$$

$$A = \$2000 \cdot (1 + 0.045)^6$$

$$A = \$2000 \cdot (1.045)^6$$

$$A = \$2604.52$$

The amount in the account after 3 years is $2604.52.

**21.** We use the compound interest formula, substituting $4000 for $P$, 6% for $r$, 12 for $n$, and $\dfrac{5}{12}$ for $t$.

$$A = P \cdot \left(1 + \frac{r}{n}\right)^{n \cdot t}$$

$$A = \$4000 \cdot \left(1 + \frac{0.06}{12}\right)^{12 \cdot \frac{5}{12}}$$

$$A = \$4000 \cdot (1 + 0.005)^5$$

$$A = \$4000 \cdot (1.005)^5$$

$$A \approx \$4101.01$$

The amount in the account after 5 months is $4101.01.

**23.** We use the compound interest formula, substituting $1200 for $P$, 10% for $r$, 4 for $n$, and 1 for $t$.

$$A = P \cdot \left(1 + \frac{r}{n}\right)^{n \cdot t}$$

$$A = \$1200 \cdot \left(1 + \frac{0.1}{4}\right)^{4 \cdot 1}$$

$$A = \$1200 \cdot (1 + 0.025)^4$$

$$A = \$1200 \cdot (1.025)^4$$

$$A \approx \$1324.58$$

The amount in the account after 1 year is $1324.58.

**25.**

$$\frac{9}{10} = \frac{x}{5}$$

$$9 \cdot 5 = 10 \cdot x$$

$$\frac{9 \cdot 5}{10} = x$$

$$\frac{45}{10} = x$$

$$\frac{9}{2} = x, \text{ or}$$

$$4.5 = x$$

**27.**

$$\frac{3}{4} = \frac{6}{x}$$

$$3 \cdot x = 4 \cdot 6$$

$$x = \frac{4 \cdot 6}{3}$$

$$x = \frac{24}{3}$$

$$x = 8$$

**29.** First we consider $\dfrac{64}{17}$.

$$\begin{array}{r} 3 \\ 1\,7\,\overline{)\,6\,4} \\ 5\,1 \\ \hline 1\,3 \end{array} \qquad \frac{64}{17} = 3\frac{13}{17}$$

Since $\dfrac{64}{17} = 3\dfrac{13}{17}$, we have $-\dfrac{64}{17} = -3\dfrac{13}{17}$.

**31.** $1\dfrac{1}{17} = \dfrac{18}{17}$      $(1 \cdot 17 = 17,\ 17 + 1 = 18)$

**33.** ◈

**35.** ◈

**37.** $A = P \cdot \left(1 + \dfrac{r}{n}\right)^{n \cdot t}$

$$A = \$24{,}800 \cdot \left(1 + \frac{0.064}{2}\right)^{2 \cdot 5}$$

$$A = \$24{,}800 \cdot (1 + 0.032)^{10}$$

$$A = \$24{,}800 \cdot (1.032)^{10}$$

$$A \approx \$33{,}981.98$$

The value of the investment is $33,981.98 after 5 years.

**39.** For a principle $P$ invested at 9% compounded monthly, to find the amount in the account at the end of 1 year we would multiply $P$ by $(1 + 0.09/12)^{12}$. Since $(1 + 0.09/12)^{12} = 1.0075^{12} \approx 1.0938$, the effective yield is approximately 9.38%.

**41.** At the end of 1 year, the $20,000 spent on the car has been reduced in value by 30% of $20,000, or $0.3 \times \$20,000$, or $6000.

If the $20,000 is invested at 9%, compounded daily, the amount in the account at the end of 1 year is

$$A = \$20,000\left(1 + \frac{0.09}{365}\right)^{365}$$
$$\approx \$20,000(1.000246575)^{365}$$
$$\approx \$20,000(1.094162144)$$
$$\approx \$21,883.24.$$

Then the $20,000 has increased in value by $21,883.24 - \$20,000$, or $1883.24. All together, the Coniglios have saved the $6000 they would have lost on the value of the car plus the $1883.24 increase in the value of the $20,000 invested at 9%, compounded daily. That is, they have saved $6000 + \$1883.24$, or $7883.24.

# Chapter 9

# Geometry and Measures

**1.** 1 foot = 12 in.

This is the relation stated on page 513 of the text.

**3.** $1 \text{ in.} = 1 \text{ in.} \times \dfrac{1 \text{ ft}}{12 \text{ in.}}$    Multiplying by 1 using $\dfrac{1 \text{ ft}}{12 \text{ in.}}$ to eliminate in.

$= \dfrac{1 \text{ in.}}{12 \text{ in.}} \times 1 \text{ ft}$

$= \dfrac{1}{12} \times \dfrac{\text{in.}}{\text{in.}} \times 1 \text{ ft}$

$= \dfrac{1}{12} \times 1 \text{ ft}$    The $\dfrac{\text{in.}}{\text{in.}}$ acts like 1, so we can omit it.

$= \dfrac{1}{12} \text{ ft}$

**5.** 1 mi = 5280 ft

This is the relation stated on page 513 of the text.

**7.** $4 \text{ yd} = 4 \times 1 \text{ yd}$
$= 4 \times 36 \text{ in.}$    Substituting 36 in. for 1 yd
$= 144 \text{ in.}$    Multiplying

**9.** $84 \text{ in.} = \dfrac{84 \text{ in.}}{1} \times \dfrac{1 \text{ ft}}{12 \text{ in.}}$    Multiplying by 1 using $\dfrac{1 \text{ ft}}{12 \text{ in.}}$

$= \dfrac{84}{12} \times 1 \text{ ft}$

$= 7 \times 1 \text{ ft}$
$= 7 \text{ ft}$

**11.** $18 \text{ in.} = \dfrac{18 \text{ in.}}{1} \times \dfrac{1 \text{ ft}}{12 \text{ in.}}$    Multiplying by 1 using $\dfrac{1 \text{ ft}}{12 \text{ in.}}$

$= \dfrac{18}{12} \times 1 \text{ ft}$

$= \dfrac{3}{2} \times 1 \text{ ft}$

$= \dfrac{3}{2} \text{ ft, or } 1\dfrac{1}{2} \text{ ft}$

**13.** $5 \text{ mi} = 5 \times 1 \text{ mi}$
$= 5 \times 5280 \text{ ft}$    Substituting 5280 ft for 1 mi
$= 26{,}400 \text{ ft}$    Multiplying

**15.** $48 \text{ in.} = \dfrac{48 \text{ in.}}{1} \times \dfrac{1 \text{ ft}}{12 \text{ in.}}$    Multiplying by 1 using $\dfrac{1 \text{ ft}}{12 \text{ in.}}$

$= \dfrac{48}{12} \times 1 \text{ ft}$

$= 4 \times 1 \text{ ft}$
$= 4 \text{ ft}$

**17.** $19 \text{ ft} = 19 \text{ ft} \times \dfrac{1 \text{ yd}}{3 \text{ ft}}$    Multiplying by 1 using $\dfrac{1 \text{ yd}}{3 \text{ ft}}$

$= \dfrac{19}{3} \times 1 \text{ yd}$

$= \dfrac{19}{3} \text{ yd, or } 6\dfrac{1}{3} \text{ yd}$

**19.** $10 \text{ mi} = 10 \times 1 \text{ mi}$
$= 10 \times 5280 \text{ ft}$    Substituting 5280 ft for 1 mi
$= 52{,}800 \text{ ft}$    Multiplying

**21.** $7\dfrac{1}{2} \text{ ft} = 7\dfrac{1}{2} \text{ ft} \times \dfrac{1 \text{ yd}}{3 \text{ ft}}$

$= \dfrac{15}{2} \text{ ft} \times \dfrac{1 \text{ yd}}{3 \text{ ft}}$

$= \dfrac{15}{6} \times 1 \text{ yd}$

$= \dfrac{5}{2} \times 1 \text{ yd}$

$= \dfrac{5}{2} \text{ yd, or } 2\dfrac{1}{2} \text{ yd}$

**23.** $360 \text{ in.} = 360 \text{ in.} \times \dfrac{1 \text{ ft}}{12 \text{ in.}} \times \dfrac{1 \text{ yd}}{3 \text{ ft}}$

$= \dfrac{360}{36} \times 1 \text{ yd}$

$= 10 \times 1 \text{ yd}$
$= 10 \text{ yd}$

**25.** $330 \text{ ft} = 330 \text{ ft} \times \dfrac{1 \text{ yd}}{3 \text{ ft}}$

$= \dfrac{330}{3} \times 1 \text{ yd}$

$= 110 \times 1 \text{ yd}$
$= 110 \text{ yd}$

**27.** $3520 \text{ yd} = 3520 \text{ yd} \times \dfrac{3 \text{ ft}}{1 \text{ yd}} \times \dfrac{1 \text{ mi}}{5280 \text{ ft}}$

$= \dfrac{10{,}560}{5280} \times 1 \text{ mi}$

$= 2 \times 1 \text{ mi}$
$= 2 \text{ mi}$

**29.** $100 \text{ yd} = 100 \times 1 \text{ yd}$
$= 100 \times 3 \text{ ft}$
$= 300 \text{ ft}$

**31.** $63{,}360 \text{ in.} = 63{,}360 \text{ in.} \times \dfrac{1 \text{ ft}}{12 \text{ in.}} \times \dfrac{1 \text{ mi}}{5280 \text{ ft}}$

$= \dfrac{63{,}360}{63{,}360} \times 1 \text{ mi}$

$= 1 \times 1 \text{ mi}$
$= 1 \text{ mi}$

**33.** a) 1 km = _____ m

Think:  To go from km to m in the table is a move of 3 places to the right.  Thus, we move the decimal point 3 places to the right.  This requires writing three additional zeros.

1   1.000.

1 km = 1000 m

b) 1 m = _____ km

Think:  To go from m to km in the table is a move of 3 places to the left.  Thus, we move the decimal point 3 places to the left.  This requires writing two additional zeros.

1   0.001.

1 m = 0.001 km

**35.** a) 1 dam = _____ m

Think:  To go from dam to m in the table is a move of 1 place to the right.  Thus, we move the decimal point 1 place to the right.  This requires writing an additional zero.

1   1.0.

1 dam = 10 m

b) 1 m = _____ dam

Think:  To go from m to dam in the table is a move of 1 place to the left.  Thus, we move the decimal point 1 place to the left.

1   0.1.

1 m = 0.1 dam

**37.** a) 1 cm = _____ m

Think:  To go from cm to m in the table is a move of 2 places to the left.  Thus, we move the decimal point 2 places to the left.  This requires writing an additional zero.

1   0.01.

1 cm = 0.01 m

b) 1 m = _____ cm

Think:  To go from m to cm in the table is a move of 2 places to the right.  Thus, we move the decimal point 2 places to the right.  This requires writing two additional zeros.

1   1.00.

1 m = 100 cm

**39.** 6.7 km = _____ m

Think:  To go from km to m in the table is a move of 3 places to the right.  Thus, we move the decimal point 3 places to the right.  This requires writing two additional zeros.

6.7   6.700.

6.7 km = 6700 m

**41.** 98 cm = _____ m

Think:  To go from cm to m in the table is a move of 2 places to the left.  Thus, we move the decimal point 2 places to the left.

98   0.98.

98 cm = 0.98 m

**43.** 8921 m = _____ km

Think:  To go from m to km in the table is a move of 3 places to the left.  Thus, we move the decimal point 3 places to the left.

8921   8.921.

8921 m = 8.921 km

**45.** 56.66 m = _____ km

Think:  To go from m to km in the table is a move of 3 places to the left.  Thus, we move the decimal point 3 places to the left.  This requires writing an additional zero.

56.66   0.056.66

56.66 m = 0.05666 km

**47.** 5666 m = _____ cm

Think:  To go from m to cm in the table is a move of 2 places to the right.  Thus, we move the decimal point 2 places to the right.  This requires writing two additional zeros.

5666   5666.00.

5666 m = 566,600 cm

**49.** 477 cm = _____ m

Think:  To go from cm to m in the table is a move of 2 places to the left.  Thus, we move the decimal point 2 places to the left.

477   4.77.

477 cm = 4.77 m

**51.** 6.88 m = _____ cm

Think:  To go from m to cm in the table is a move of 2 places to the right.  Thus, we move the decimal point 2 places to the right.

6.88   6.88.

6.88 m = 688 cm

**53.** 1 mm = _____ cm

Think:  To go from mm to cm in the table is a move of 1 place to the left.  Thus, we move the decimal point 1 place to the left.

1   0.1.

1 mm = 0.1 cm

**55.** 1 km = _____ cm

Think: To go from km to cm in the table is a move of 5 places to the right. Thus, we move the decimal point 5 places to the right. This requires writing five additional zeros.

1   1.00000.

1 km = 100,000 cm

**57.** 14.2 cm = _____ mm

Think: To go from cm to mm in the table is a move of 1 place to the right. Thus, we move the decimal point 1 place to the right.

14.2   14.2.

14.2 cm = 142 mm

**59.** 8.2 mm = _____ cm

Think: To go from mm to cm in the table is a move of 1 place to the left. Thus, we move the decimal point 1 place to the left.

8.2   0.8.2

8.2 mm = 0.82 cm

**61.** 4500 mm = _____ cm

Think: To go from mm to cm in the table is a move of 1 place to the left. Thus, we move the decimal point 1 place to the left.

4500   450.0.

4500 mm = 450 cm

**63.** 0.024 mm = _____ m

Think: To go from mm to m in the table is a move of 3 places to the left. Thus, we move the decimal point 3 places to the left. This requires writing three additional zeros.

0.024   0.000.024

0.024 mm = 0.000024 m

**65.** 6.88 m = _____ dam

Think: To go from m to dam in the table is a move of 1 place to the left. Thus, we move the decimal point 1 place to the left.

6.88   0.6.88

6.88 m = 0.688 dam

**67.** 2.3 dam = _____ dm

Think: To go from dam to dm in the table is a move of 2 places to the right. Thus, we move the decimal point 2 places to the right. This requires writing an additional zero.

2.3   2.30.

2.3 dam = 230 dm

**69.** $10 \text{ km} \approx 10 \text{ km} \times \dfrac{0.621 \text{ mi}}{1 \text{ km}} \approx 6.21 \text{ mi}$

**71.** $14 \text{ in.} \approx 14 \text{ in.} \times \dfrac{2.54 \text{ cm}}{1 \text{ in.}} \approx 35.56 \text{ cm}$

**73.** $65 \text{ mph} = 65 \dfrac{\text{mi}}{\text{hr}} = 65 \times \dfrac{1 \text{ mi}}{\text{hr}} \approx 65 \times \dfrac{1.609 \text{ km}}{\text{hr}} = 104.585 \text{ km/h}$

**75.** $330 \text{ ft} \approx 330 \text{ ft} \times \dfrac{1 \text{ m}}{3.3 \text{ ft}} \approx \dfrac{330}{3.3} \times \dfrac{\text{ft}}{\text{ft}} \times 1 \text{ m} \approx 100 \text{ m}$

**77.** $180 \text{ cm} \approx 180 \text{ cm} \times \dfrac{1 \text{ in.}}{2.54 \text{ cm}} \approx \dfrac{180}{2.54} \text{ in.} \approx 70.866 \text{ in.}$

**79.** $36 \text{ yd} = 36 \cdot 3 \text{ ft} = 108 \text{ ft} \approx 108 \text{ ft} \times \dfrac{1 \text{ m}}{3.3 \text{ ft}} \approx \dfrac{108}{3.3} \text{ m} \approx 32.727 \text{ m}$

**81.**
$$-7x - 9x = 24$$
$$-16x = 24 \qquad \text{Collecting like terms}$$
$$\dfrac{-16x}{-16} = \dfrac{24}{-16} \qquad \text{Dividing by } -16 \text{ on both sides}$$
$$x = \dfrac{3 \cdot 8}{-2 \cdot 8} = \dfrac{3}{-2} \cdot \dfrac{8}{8}$$
$$x = \dfrac{3}{-2}, \text{ or } -\dfrac{3}{2}$$

**83.** Let $c$ represent the cost of 7 calculators. We translate to a proportion.

$$\text{Number} \rightarrow \quad \dfrac{3}{43.50} = \dfrac{7}{c} \quad \leftarrow \text{Number}$$
$$\text{Cost} \rightarrow \qquad \qquad \qquad \leftarrow \text{Cost}$$

Solve: $3 \cdot c = 43.50 \cdot 7 \qquad$ Equating cross-products

$c = \dfrac{43.50 \cdot 7}{3} \qquad$ Dividing by 3 on both sides

$c = \dfrac{304.50}{3} \qquad$ Multiplying

$c = 101.50 \qquad$ Dividing

Seven calculators would cost $101.50.

**85.** a) Multiply by 100 to move the decimal point two places to the right.

0.47.

b) Write a percent symbol: 47%

Thus, 0.47 = 47%.

**87.** ◈

**89.** ◈

**91.** $10 \text{ km} \approx 10 \text{ km} \cdot \dfrac{1000 \text{ m}}{1 \text{ km}} \cdot \dfrac{39.37 \text{ in.}}{1 \text{ m}} \approx 393,700 \text{ in.}$

$\left(\text{The result is } 393,465.6 \text{ in. if the conversion is done using } \dfrac{0.621 \text{ mi}}{1 \text{ km}}.\right)$

**93.**    $\dfrac{100 \text{ m}}{9.86 \text{ sec}}$

$\approx \dfrac{100 \text{ m}}{9.86 \text{ sec}} \cdot \dfrac{60 \text{ sec}}{1 \text{ min}} \cdot \dfrac{60 \text{ min}}{1 \text{ hr}} \cdot \dfrac{3.3 \text{ ft}}{1 \text{ m}} \cdot \dfrac{1 \text{ mi}}{5280 \text{ ft}}$

$\approx \dfrac{1,188,000}{52,060.8} \dfrac{\text{mi}}{\text{hr}}$

$\approx 22.8 \dfrac{\text{mi}}{\text{hr}},$ or 22.8 mph

(If we first convert 100 m to 0.1 km and then use the conversion factor $\dfrac{1 \text{ mi}}{1.609 \text{ km}}$, the result is about 22.7 mph.)

**95.** Since 1 in. is larger than 1 cm (1 in. $\approx$ 2.54 cm), we have 59 in. > 59 cm.

**97.** Since 1 km $\approx$ 0.621 mi, then 7 km $\approx 7 \times 0.6$ mi $\approx 4.2$ mi so 7 km < 6 mi.

**99.** Since 1 ft $\approx$ 0.303 m, then 24 ft $\approx 24 \times 0.3$ m $\approx 7.2$ m, so 24 ft > 6 m.

## Exercise Set 9.2

**1.**   $A = b \cdot h$      Area of a parallelogram
$= 8 \text{ cm} \cdot 4 \text{ cm}$     Substituting 8 cm for $b$ and 4 cm for $h$
$= 32 \text{ cm}^2$

**3.**   $A = \dfrac{1}{2} \cdot h \cdot (a + b)$      Area of a trapezoid

$= \dfrac{1}{2} \cdot 8 \text{ ft} \cdot (6 + 20) \text{ ft}$    Substituting 8 ft for $h$, 6 ft for $a$, and 20 ft for $b$

$= \dfrac{8 \cdot 26}{2} \text{ ft}^2$

$= \dfrac{2 \cdot 4 \cdot 26}{1 \cdot 2} \text{ ft}^2$

$= 104 \text{ ft}^2$

**5.**   $A = b \cdot h$      Area of a parallelogram
$= 8 \text{ m} \cdot 8 \text{ m}$     Substituting 8 m for $b$ and 8 m for $h$
$= 64 \text{ m}^2$

**7.**   $A = b \cdot h$      Area of a parallelogram
$= 2.3 \text{ cm} \cdot 3.5 \text{ cm}$    Substituting 2.3 cm for $b$ and 3.5 cm for $h$
$= 8.05 \text{ cm}^2$

**9.**   $A = \dfrac{1}{2} \cdot h \cdot (a + b)$      Area of a trapezoid

$= \dfrac{1}{2} \cdot 9 \text{ mi} \cdot (13 + 19) \text{ mi}$    Substituting 9 mi for $h$, 13 mi for $a$, and 19 mi for $b$

$= \dfrac{9 \cdot 32}{2} \text{ mi}^2$

$= \dfrac{9 \cdot 2 \cdot 16}{1 \cdot 2} \text{ mi}^2$

$= 144 \text{ mi}^2$

**11.**   $A = b \cdot h$      Area of a parallelogram

$= 12\dfrac{1}{4} \text{ ft} \cdot 4\dfrac{1}{2} \text{ ft}$    Substituting $12\dfrac{1}{4}$ ft for $b$ and $4\dfrac{1}{2}$ ft for $h$

$= \dfrac{49}{4} \cdot \dfrac{9}{2} \cdot \text{ ft}^2$

$= \dfrac{441}{8} \text{ ft}^2$

$= 55\dfrac{1}{8} \text{ ft}^2$

**13.**   $A = \dfrac{1}{2} \cdot h \cdot (a + b)$      Area of a trapezoid

$= \dfrac{1}{2} \cdot 7 \text{ m} \cdot (9 + 5) \text{ m}$    Substituting 7 m for $h$, 9 m for $a$, and 5 m for $b$

$= \dfrac{7 \cdot 14}{2} \text{ m}^2$

$= \dfrac{7 \cdot 2 \cdot 7}{1 \cdot 2} \text{ m}^2$

$= 49 \text{ m}^2$

**15.**   $A = b \cdot h$      Area of a parallelogram
$= 12 \text{ cm} \cdot 9 \text{ cm}$    Substituting 12 cm for $b$ and 9 cm for $h$
$= 108 \text{ cm}^2$

**17.**   $A = \dfrac{1}{2} \cdot h \cdot (a + b)$      Area of a trapezoid

$= \dfrac{1}{2} \cdot 8 \text{ yd} \cdot (9.1 + 7.9) \text{ yd}$    Substituting 8 yd for $h$, 9.1 yd for $a$, and 7.9 yd for $b$

$= \dfrac{8 \cdot 17}{2} \text{ yd}^2$

$= \dfrac{2 \cdot 4 \cdot 17}{1 \cdot 2} \text{ yd}^2$

$= 68 \text{ yd}^2$

**19.**   $d = 2 \cdot r$
$= 2 \cdot 7 \text{ cm} = 14 \text{ cm}$

**21.**   $d = 2 \cdot r$

$= 2 \cdot \dfrac{3}{4} \text{ in.} = \dfrac{6}{4} \text{ in.} = \dfrac{3}{2} \text{ in., or } 1\dfrac{1}{2} \text{ in.}$

**23.**   $r = \dfrac{d}{2}$

$= \dfrac{32 \text{ ft}}{2} = 16 \text{ ft}$

**25.**   $r = \dfrac{d}{2}$

$= \dfrac{1.4 \text{ cm}}{2} = 0.7 \text{ cm}$

**27.**   $C = 2 \cdot \pi \cdot r$

$\approx 2 \cdot \dfrac{22}{7} \cdot 7 \text{ cm} \approx \dfrac{2 \cdot 22 \cdot 7}{7} \text{ cm} \approx 44 \text{ cm}$

**29.**   $C = 2 \cdot \pi \cdot r$

$\approx 2 \cdot \dfrac{22}{7} \cdot \dfrac{3}{4} \text{ in.} \approx \dfrac{2 \cdot 22 \cdot 3}{7 \cdot 4} \text{ in.} \approx \dfrac{132}{28} \text{ in.} \approx \dfrac{33}{7} \text{ in.,}$

or $4\dfrac{5}{7}$ in.

**31.** $C = \pi \cdot d$
$\approx 3.14 \cdot 32 \text{ ft} \approx 100.48 \text{ ft}$

**33.** $C = \pi \cdot d$
$\approx 3.14 \cdot 1.4 \text{ cm} \approx 4.396 \text{ cm}$

**35.** $A = \pi \cdot r \cdot r$
$\approx \frac{22}{7} \cdot 7 \text{ cm} \cdot 7 \text{ cm} \approx \frac{22}{7} \cdot 49 \text{ cm}^2 \approx 154 \text{ cm}^2$

**37.** $A = \pi \cdot r \cdot r$
$\approx \frac{22}{7} \cdot \frac{3}{4} \text{ in.} \cdot \frac{3}{4} \text{ in.} \approx \frac{22 \cdot 3 \cdot 3}{7 \cdot 4 \cdot 4} \text{ in}^2 \approx \frac{99}{56} \text{ in}^2,$
or $1\frac{43}{56} \text{ in}^2$

**39.** $A = \pi \cdot r \cdot r$
$\approx 3.14 \cdot 16 \text{ ft} \cdot 16 \text{ ft} \qquad (r = \frac{d}{2}; r = \frac{32 \text{ ft}}{2} = 16 \text{ ft})$
$\approx 3.14 \cdot 256 \text{ ft}^2$
$\approx 803.84 \text{ ft}^2$

**41.** $A = \pi \cdot r \cdot r$
$\approx 3.14 \cdot 0.7 \text{ cm} \cdot 0.7 \text{ cm}$
$\qquad (r = \frac{d}{2}; r = \frac{1.4 \text{ cm}}{2} = 0.7 \text{ cm})$
$\approx 3.14 \cdot 0.49 \text{ cm}^2 \approx 1.5386 \text{ cm}^2$

**43.** $d = 2 \cdot r$
$= 2 \cdot 1 \text{ cm} = 2 \text{ cm}$
The diameter is 2 cm.
$C = \pi \cdot d$
$\approx 3.14 \cdot 2 \text{ cm} \approx 6.28 \text{ cm}$
The circumference is about 6.28 cm.
$A = \pi \cdot r \cdot r$
$\approx 3.14 \cdot 1 \text{ cm} \cdot 1 \text{ cm}$
$\approx 3.14 \text{ cm}^2$
The area is about 3.14 cm$^2$.

**45.** $A = \pi \cdot r \cdot r$
$\approx 3.14 \cdot 220 \text{ mi} \cdot 220 \text{ mi} \approx 151,976 \text{ mi}^2$
The broadcast area is about 151,976 mi$^2$.

**47.** $C = \pi \cdot d$
$\approx 3.14 \cdot 2.5 \text{ cm} \approx 7.85 \text{ cm}$
The circumference of a quarter is about 7.85 cm.
In order to find the area, we first find the radius.
$r = \frac{d}{2}$
$= \frac{2.5 \text{ cm}}{2} = 1.25 \text{ cm}$
$A = \pi \cdot r \cdot r$
$\approx 3.14 \cdot 1.25 \text{ cm} \cdot 1.25 \text{ cm} \approx 4.90625 \text{ cm}^2$
The area of a quarter is about 4.90625 cm$^2$.

**49.** The tree's circumference is 47.1 in.
$C = \pi \cdot d$
$47.1 \approx 3.14 \cdot d$
$\frac{47.1}{3.14} \approx d$
$15 \approx d$
The tree's diameter is about 15 in.

**51.** The shortest distance is composed of two semicircular lengths and two straight 85.56 yd segments. The two semicircles are equivalent to a single circle with an 85.56 yd diameter. First we find the circumference of the circle.
$C = \pi \cdot d$
$\approx 3.14 \cdot 85.56 \text{ yd} \approx 268.6584 \text{ yd}$
Then the total distance is 268.6584 yd + 85.56 yd + 85.5 yd = 439.7784 yd.

**53.** The perimeter consists of the circumferences of three semicircles, each with diameter 8 ft, and one side of a square of length 8 ft. We first find the circumference of one semicircle. This is one-half the circumference of a circle with diameter 8 ft:
$\frac{1}{2} \cdot \pi \cdot d \approx \frac{1}{2} \cdot 3.14 \cdot 8 \text{ ft} = 12.56 \text{ ft}$
Then we multiply by 3:
$3 \cdot (12.56 \text{ ft}) = 37.68 \text{ ft}$
Finally we add the circumferences of the semicircles and the length of the side of the square:
$37.68 \text{ ft} + 8 \text{ ft} = 45.68 \text{ ft}$
The perimeter is 45.68 ft.

**55.** The perimeter consists of three-fourths of the perimeter of a square with side of length 10 yd and the circumference of a semicircle with diameter 10 yd. First we find three-fourths of the perimeter of the square:
$\frac{3}{4} \cdot 4 \cdot s = \frac{3}{4} \cdot 4 \cdot 10 \text{ yd} = 30 \text{ yd}$
Then we find one-half of the circumference of a circle with diameter 10 yd:
$\frac{1}{2} \cdot \pi \cdot d \approx \frac{1}{2} \cdot 3.14 \cdot 10 \text{ yd} = 15.7 \text{ yd}$
Then we add:
$30 \text{ yd} + 15.7 \text{ yd} = 45.7 \text{ yd}$
The perimeter is 45.7 yd.

**57.** The shaded region consists of a circle of radius 8 m, with two circles each of diameter 8 m, removed. First we find the area of the large circle:
$A = \pi \cdot r \cdot r \approx 3.14 \cdot 8 \text{ m} \cdot 8 \text{ m} = 200.96 \text{ m}^2$
Then we find the area of one of the small circles:
The radius is $\frac{8 \text{ m}}{2} = 4 \text{ m}$.
$A = \pi \cdot r \cdot r \approx 3.14 \cdot 4 \text{ m} \cdot 4 \text{ m} = 50.24 \text{ m}^2$
We multiply this area by 2 to find the area of the two small circles:
$2 \cdot 50.24 \text{ m}^2 = 100.48 \text{ m}^2$
Finally we subtract to find the area of the shaded region:
$200.96 \text{ m}^2 - 100.48 \text{ m}^2 = 100.48 \text{ m}^2$
The area of the shaded region is 100.48 m$^2$.

**59.** The shaded region consists of one-half of a circle with diameter 2.8 cm and a triangle with base 2.8 cm and height 2.8 cm. First we find the area of the semicircle. The radius is $\dfrac{2.8 \text{ cm}}{2} = 1.4$ cm.

$$A = \frac{1}{2} \cdot \pi \cdot r \cdot r \approx \frac{1}{2} \cdot 3.14 \cdot 1.4 \text{ cm} \cdot 1.4 \text{ cm} = 3.0772 \text{ cm}^2$$

Then we find the area of the triangle:

$$A = \frac{1}{2} \cdot b \cdot h = \frac{1}{2} \cdot 2.8 \text{ cm} \cdot 2.8 \text{ cm} = 3.92 \text{ cm}^2$$

Finally we add to find the area of the shaded region:

$$3.0772 \text{ cm}^2 + 3.92 \text{ cm}^2 = 6.9972 \text{ cm}^2$$

The area of the shaded region is 6.9972 cm$^2$.

**61.** $9.25\% = \dfrac{9.25}{100}$

$$= \frac{9.25}{100} \cdot \frac{100}{100}$$

$$= \frac{925}{10,000}$$

$$= \frac{25 \cdot 37}{25 \cdot 400}$$

$$= \frac{25}{25} \cdot \frac{37}{400}$$

$$= \frac{37}{400}$$

**63.** a) First find decimal notation by division.

```
        1 .3 7 5
   8 | 1 1 .0 0 0
        8
        ‾‾‾
        3 0
        2 4
        ‾‾‾
          6 0
          5 6
          ‾‾‾
            4 0
            4 0
            ‾‾‾
              0
```

$\dfrac{11}{8} = 1.375$

b) Convert the decimal notation to percent notation. Move the decimal point two places to the right and write a % symbol.

1.37.5

$\llcorner\!\uparrow$

$\dfrac{11}{8} = 137.5\%$

**65.** $\dfrac{5}{4} = \dfrac{5}{4} \cdot \dfrac{25}{25} = \dfrac{125}{100} = 125\%$

**67.** ◈

**69.** ◈

**71.** Area of each side: $A = \dfrac{1}{2} \cdot h \cdot (a + b) =$

$\dfrac{1}{2} \cdot 9$ cm $\cdot (14 + 21)$ cm $= 157.5$ cm$^2$

Area of all 4 sides: $4 \cdot 157.5$ cm$^2 = 630$ cm$^2$

Area of the top: $A = s \cdot s = 14$ cm $\cdot 14$ cm $= 196$ cm$^2$

Area of the bottom: $A = s \cdot s = 21$ cm $\cdot 21$ cm $= 441$ cm$^2$

Total area: $630$ cm$^2 + 196$ cm$^2 + 441$ cm$^2 = 1267$ cm$^2$

**73.** We first convert 500 mi to inches.

$$500 \text{ mi} = 500 \times 1 \text{ mi} = 500 \times 5280 \text{ ft}$$
$$= 500 \times 5280 \times 1 \text{ ft} = 500 \times 5280 \times 12 \text{ in.}$$
$$= 31,680,000 \text{ in.}$$

Next we find the circumference of the tires.

$$C = 2 \cdot \pi \cdot r$$
$$C \approx 2 \cdot \frac{22}{7} \cdot 14 \text{ in.} \approx 88 \text{ in.}$$

Then solve $88 \cdot n = 31,680,000$.

$$n = 360,000$$

Thus, about 360,000 revolutions of each tire occurred.

**75.** The height of the stack of tennis balls is three times the diameter of one ball, or $3 \cdot d$.

The circumference of the can is about the circumference of one ball, or $\pi \cdot d$.

The circumference of the can is greater than the height of the stack of balls, because $\pi > 3$.

---

## Exercise Set 9.3

---

**1.** $4$ yd$^2 = 4 \cdot 9$ ft$^2$    Substituting 9 ft$^2$ for 1 yd$^2$
   $= 36$ ft$^2$

(Had we preferred to use canceling, we could have multiplied 4 yd$^2$ by $\dfrac{9 \text{ ft}^2}{1 \text{ yd}^2}$.)

**3.** $7$ ft$^2 = 7 \cdot 144$ in$^2$    Substituting 144 in$^2$ for 1 ft$^2$

   $= 1008$ in$^2$

(Had we preferred to use canceling, we could have multiplied 7 ft$^2$ by $\dfrac{144 \text{ in}^2}{1 \text{ ft}^2}$.)

**5.** $432$ in$^2 = 432$ $\cancel{\text{in}^2} \times \dfrac{1 \text{ ft}^2}{144 \ \cancel{\text{in}^2}}$

   $= \dfrac{432}{144} \times \text{ ft}^2$

   $= 3$ ft$^2$

**7.** $22$ yd$^2 = 22 \times 9$ ft$^2$     Substituting 9 ft$^2$ for 1 yd$^2$

   $= 198$ ft$^2$

**9.** $44$ yd$^2 = 44 \cdot 9$ ft$^2$     Substituting 9 ft$^2$ for 1 yd$^2$

   $= 396$ ft$^2$

**11.** $20 \text{ mi}^2 = 20 \cdot 640 \text{ acres}$     Substituting 640 acres
                             for 1 mi$^2$
         $= 12,800 \text{ acres}$

**13.** $69 \text{ ft}^2 = 69 \cancel{\text{ft}^2} \times \dfrac{1 \text{ yd}^2}{9 \cancel{\text{ft}^2}}$

$\phantom{69 \text{ ft}^2} = \dfrac{69}{9} \times \text{ yd}^2$

$\phantom{69 \text{ ft}^2} = \dfrac{23}{3} \text{ yd}^2, \text{ or } 7\dfrac{2}{3} \text{ yd}^2$

**15.** $720 \text{ in}^2 = 720 \cancel{\text{in}^2} \times \dfrac{1 \text{ ft}^2}{144 \cancel{\text{in}^2}}$

$\phantom{720 \text{ in}^2} = \dfrac{720}{144} \times \text{ ft}^2$

$\phantom{720 \text{ in}^2} = 5 \text{ ft}^2$

**17.** $1 \text{ in}^2 = 1 \cancel{\text{in}^2} \times \dfrac{1 \text{ ft}^2}{144 \cancel{\text{in}^2}}$

$\phantom{1 \text{ in}^2} = \dfrac{1}{144} \times \text{ ft}^2$

$\phantom{1 \text{ in}^2} = \dfrac{1}{144} \text{ ft}^2$

**19.** $1 \text{ acre} = 1 \cancel{\text{acre}} \cdot \dfrac{1 \text{ mi}^2}{640 \cancel{\text{acres}}}$

$\phantom{1 \text{ acre}} = \dfrac{1}{640} \cdot \text{ mi}^2$

$\phantom{1 \text{ acre}} = \dfrac{1}{640} \text{ mi}^2, \text{ or } 0.0015625 \text{ mi}^2$

**21.** $17 \text{ km}^2 = \underline{\phantom{xxxx}} \text{ m}^2$

Think: A kilometer is 1000 times as big as a meter, so 1 km$^2$ is 1,000,000 times as big as 1 m$^2$. We shift the decimal point six places to the right.

    17    17.000000.
           └─────↑

$17 \text{ km}^2 = 17,000,000 \text{ m}^2$

**23.** $6.31 \text{ m}^2 = \underline{\phantom{xxxx}} \text{ cm}^2$

Think: A meter is 100 times as big as a centimeter, so 1 m$^2$ is 10,000 times as big as 1 cm$^2$. We shift the decimal point 4 places to the right.

    6.31    6.3100.
           └──↑

$6.31 \text{ m}^2 = 63,100 \text{ cm}^2$

**25.** $2345.6 \text{ mm}^2 = \underline{\phantom{xxxx}} \text{ cm}^2$

Think: To convert from mm to cm, we shift the decimal point one place to the left. To convert from mm$^2$ to cm$^2$, we shift the decimal point two places to the left.

    2345.6    23.45.6
              ↑──┘

$2345.6 \text{ mm}^2 = 23.456 \text{ cm}^2$

**27.** $349 \text{ cm}^2 = \underline{\phantom{xxxx}} \text{ m}^2$

Think: To convert from cm to m, we shift the decimal point two places to the left. To convert from cm$^2$ to m$^2$, we shift the decimal point four places to the left.

    349    0.0349.
          ↑──┘

$349 \text{ cm}^2 = 0.0349 \text{ m}^2$

**29.** $250,000 \text{ mm}^2 = \underline{\phantom{xxxx}} \text{ cm}^2$

Think: To convert from mm to cm, we shift the decimal point one place to the left. To convert from mm$^2$ to cm$^2$, we shift the decimal point two places to the left.

    250,000    2500.00.
                ↑──┘

$250,000 \text{ mm}^2 = 2500 \text{ cm}^2$

**31.** $472,800 \text{ m}^2 = \underline{\phantom{xxxx}} \text{ km}^2$

Think: To convert from m to km, we shift the decimal point three places to the left. To convert from m$^2$ to km$^2$, we shift the decimal point six places to the left.

    472,800    0.472800.
              ↑────┘

$472,800 \text{ m}^2 = 0.4728 \text{ km}^2$

**33.** First we convert 3 in. to feet.

$3 \text{ in.} = 3 \cancel{\text{in.}} \times \dfrac{1 \text{ ft}}{12 \cancel{\text{in.}}}$

$\phantom{3 \text{ in.}} = \dfrac{3}{12} \times \text{ ft}$

$\phantom{3 \text{ in.}} = \dfrac{1}{4} \text{ ft}$

Then we find the area using the formula for the area of a rectangle.

$A = l \cdot w$

$\phantom{A} = 8 \text{ ft} \cdot \dfrac{1}{4} \text{ ft}$

$\phantom{A} = \dfrac{8}{4} \text{ ft}^2$

$\phantom{A} = 2 \text{ ft}^2$

**35.** We convert 4 in. and 7 yd to feet.

$4 \text{ in.} = 4 \cancel{\text{in.}} \times \dfrac{1 \text{ ft}}{12 \cancel{\text{in.}}}$

$\phantom{4 \text{ in.}} = \dfrac{4}{12} \times \text{ ft}$

$\phantom{4 \text{ in.}} = \dfrac{1}{3} \text{ ft}$

$7 \text{ yd} = 7 \cdot 3 \text{ ft} = 21 \text{ ft}$

Then we find the area using the formula for the area of a trapezoid.

$A = \dfrac{1}{2} \cdot h \cdot (a + b)$

$\phantom{A} = \dfrac{1}{2} \cdot \dfrac{1}{3} \text{ ft} \cdot (5 + 21) \text{ ft}$

$\phantom{A} = \dfrac{26}{2} \text{ ft}^2$

$\phantom{A} = \dfrac{13}{3} \text{ ft}^2, \text{ or } 4\dfrac{1}{3} \text{ ft}^2$

**37.** The area of the small triangle is
$$A = \frac{1}{2} \cdot b \cdot h$$
$$= \frac{1}{2} \cdot 1 \text{ cm} \cdot 1 \text{ cm}$$
$$= \frac{1}{2} \text{ cm}^2.$$
Then the total area of the 6 small triangles is
$$6 \cdot \frac{1}{2} \text{ cm}^2 = 3 \text{ cm}^2.$$
The area of the large triangle is
$$A = \frac{1}{2} \cdot b \cdot h$$
$$= \frac{1}{2} \cdot 6 \text{ cm} \cdot 6 \text{ cm}$$
$$= 18 \text{ cm}^2.$$
The total area is $3 \text{ cm}^2 + 18 \text{ cm}^2 = 21 \text{ cm}^2$.

**39.** We first find the interest for 1 year:
$$5\% \times 700 = 0.05 \times 700 = 35$$
Then we multiply that amount by $\frac{1}{2}$:
$$\frac{1}{2} \times 35 = \frac{35}{2} = 17.5$$
The interest for $\frac{1}{2}$ yr is \$17.50.

**41.** We first find the interest for 1 year:
$$8.9\% \times 1200 = 0.089 \times 1200 = 106.8$$
Then we multiply that amount by $\frac{30}{365}$:
$$\frac{30}{365} \times 106.8 = \frac{3204}{365} \approx 8.78$$
The interest for 30 days is \$8.78.

**43.** ◈

**45.** ◈

**47.** $1 \text{ m}^2 = 1 \times 1 \text{ m} \times 1 \text{ m}$
$\approx 1 \times 3.3 \text{ ft} \times 3.3 \text{ ft}$
$\approx 1 \times 3.3 \times 3.3 \times \text{ ft} \times \text{ ft}$
$\approx 10.89 \text{ ft}^2$

**49.** $2 \text{ yd}^2 = 2 \times 1 \text{ yd} \times 1 \text{ yd}$
$\approx 2 \times 3 \text{ ft} \times 3 \text{ ft} \times \dfrac{1 \text{ m}}{3.3 \text{ ft}} \times \dfrac{1 \text{ m}}{3.3 \text{ ft}}$
$\approx \dfrac{2 \times 3 \times 3}{3.3 \times 3.3} \times \text{ m} \times \text{ m}$
$\approx 1.65 \text{ m}^2$

**51.** $20,175 \text{ ft}^2 = 20,175 \times 1 \text{ ft} \times 1 \text{ ft}$
$\approx 20,175 \times 1 \text{ ft} \times 1 \text{ ft} \times$
$\dfrac{1 \text{ m}}{3.3 \text{ ft}} \times \dfrac{1 \text{ m}}{3.3 \text{ ft}}$
$\approx \dfrac{20,175}{3.3 \times 3.3} \times \text{ m} \times \text{ m}$
$\approx 1852.6 \text{ m}^2$

**53.** Area of floor: $A = l \cdot w = 9 \text{ ft} \cdot 12 \text{ ft} = 108 \text{ ft}^2$
Convert 108 ft² to yd²:
$$108 \text{ ft}^2 = 108 \text{ ft}^2 \times \frac{1 \text{ yd}^2}{9 \text{ ft}^2} = 12 \text{ yd}^2$$
Cost of carpeting: $12 \cdot \$8.45 = \$101.40$
Perimeter of floor:
$$P = 2 \cdot (l + w) = 2 \cdot (9 \text{ ft} + 12 \text{ ft}) = 42 \text{ ft}$$
Perimeter of floor less width of doorway:
$$42 \text{ ft} - 3 \text{ ft} = 39 \text{ ft}$$
Cost of moulding: $39 \cdot \$0.87 = \$33.93$
Total cost of materials: $\$101.40 + \$33.93 = \$135.33$

## Exercise Set 9.4

**1.** The angle can be named in six different ways:
angle $GHI$, angle $IHG$, angle $H$, $\angle GHI$, $\angle IHG$, or $\angle H$.

**3.** Place the △ of the protractor at the vertex of the angle, and line up one of the sides at 0°. We choose the horizontal side. Since 0° is on the inside scale, we check where the other side of the angle crosses the inside scale. It crosses at 10°. Thus, the measure of the angle is 10°.

**5.** Place the △ of the protractor at the vertex of the angle, point $B$. Line up one of the sides at 0°. We choose the side that contains point $A$. Since 0° is on the outside scale, we check where the other side crosses the outside scale. It crosses at 180°. Thus, the measure of the angle is 180°.

**7.** Place the △ of the protractor at the vertex of the angle, and line up one of the sides at 0°. We choose the horizontal side. Since 0° is on the inside scale, we check where the other side crosses the inside scale. It crosses at 90°. Thus, the measure of the angle is 90°.

**9.** Every circle graph contains a total of 360°.
10% of the circle is a 0.1(360°), or 36° angle;
26% of the circle is a 0.26(360°), or 93.6° angle;
35% of the circle is a 0.35(360°), or 126° angle;
29% of the circle is a 0.29(360°), or 104.4° angle.

To draw a 36° angle, draw a horizontal segment and use a protractor to mark off the angle. From that mark, draw a segment to complete the angle. From that segment draw a 93.6° angle and, from the segment that completes that angle, draw a 126° angle. To confirm that the remainder of the circle is 104.4°, we measure it using a protractor.

**Sporting Goods Purchases**

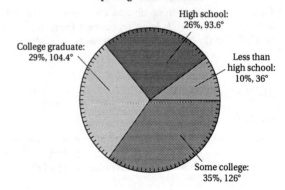

High school: 26%, 93.6°
College graduate: 29%, 104.4°
Less than high school: 10%, 36°
Some college: 35%, 126°

**11.** Using a protractor, we find that the measure of the angle in Exercise 1 is 148°. Since its measure is greater than 90°and less than 180°, it is an obtuse angle.

**13.** The measure of the angle in Exercise 3 is 10°. Since its measure is greater than 0° and less than 90°, it is an acute angle.

**15.** The measure of the angle in Exercise 5 is 180°. It is a straight angle.

**17.** The measure of the angle in Exercise 7 is 90°. It is a right angle.

**19.** The measure of the angle in Margin Exercise 1 is 30°. Since its measure is greater than 0° and less than 90°, it is an acute angle.

**21.** The measure of the angle in Margin Exercise 3 is 126°. Since its measure is greater than 90° and less than 180°, it is an obtuse angle.

**23.** Two angles are complementary if the sum of their measures is 90°.
$$90° - 11° = 79°.$$
The measure of a complement is 79°.

**25.** Two angles are complementary if the sum of their measures is 90°.
$$90° - 67° = 23°.$$
The measure of a complement is 23°.

**27.** Two angles are complementary if the sum of their measures is 90°.
$$90° - 58° = 32°.$$
The measure of a complement is 32°.

**29.** Two angles are complementary if the sum of their measures is 90°.
$$90° - 29° = 61°.$$
The measure of a complement is 61°.

**31.** Two angles are supplementary if the sum of their measures is 180°.
$$180° - 3° = 177°.$$
The measure of a supplement is 177°.

**33.** Two angles are supplementary if the sum of their measures is 180°.
$$180° - 139° = 41°.$$
The measure of a supplement is 41°.

**35.** Two angles are supplementary if the sum of their measures is 180°.
$$180° - 85° = 95°.$$
The measure of a supplement is 95°.

**37.** Two angles are supplementary if the sum of their measures is 180°.
$$180° - 102° = 78°.$$
The measure of a supplement is 78°.

**39.** Replace the % symbol with $\times 0.01$: $56.1 \times 0.01$

Multiply to move the decimal point two places to the left:
$$0.56.1$$

Thus, $56.1\% = 0.561$.

**41.**
$$3.1x + 4.3 = x + 9.55$$
$$3.1x + 4.3 - x = x + 9.55 - x \quad \text{Subtracting } x \text{ on both sides}$$
$$2.1x + 4.3 = 9.55$$
$$2.1x + 4.3 - 4.3 = 9.55 - 4.3 \quad \text{Subtracting 4.3 on both sides}$$
$$2.1x = 5.25$$
$$\frac{2.1x}{2.1} = \frac{5.25}{2.1} \quad \text{Dividing by 2.1 on both sides}$$
$$x = 2.5$$
The solution is 2.5.

**43.** $-9.7 + 3.8$

The difference of the absolute values is 5.9. Since $-9.7$ has the larger absolute value, the answer is negative.
$$-9.7 + 3.8 = -5.9$$

**45.** ◈

**47.** ◈

**49.** Find $m\angle 1$:
$$m\angle 1 + m\angle 2 + m\angle 3 = 180°$$
$$m\angle 1 + 42.17° + 81.9° = 180°$$
$$m\angle 1 + 124.07° = 180°$$
$$m\angle 1 = 55.93°$$
Similarly, $m\angle 2 + m\angle 3 + m\angle 4 = 180°$, so $m\angle 4 = 55.93°$.

Find $m\angle 5$:
$$m\angle 3 + m\angle 4 + m\angle 5 = 180°$$
$$81.9° + 55.93° + m\angle 5 = 180°$$
$$137.83° + m\angle 5 = 180°$$
$$m\angle 5 = 42.17°$$

Find $m\angle 6$:
$$m\angle 5 + m\angle 6 + m\angle 1 = 180°$$
$$42.17° + m\angle 6 + 55.93° = 180°$$
$$98.1° + m\angle 6 = 180°$$
$$m\angle 6 = 81.9°$$

## Exercise Set 9.5

**1.** The square roots of 16 are 4 and $-4$, because $4^2 = 16$ and $(-4)^2 = 16$.

**3.** The square roots of 121 are 11 and $-11$, because $11^2 = 121$ and $(-11)^2 = 121$.

**5.** The square roots of 169 are 13 and $-13$, because $13^2 = 169$ and $(-13)^2 = 169$.

**7.** The square roots of 6400 are 80 and $-80$, because $80^2 = 6400$ and $(-80)^2 = 6400$.

**9.** $\sqrt{49} = 7$

The square root of 49 is 7, because $7^2 = 49$ and 7 is positive.

**11.** $\sqrt{81} = 9$

The square root of 81 is 9, because $9^2 = 81$ and 9 is positive.

**13.** $\sqrt{225} = 15$

The square root of 225 is 15, because $15^2 = 225$ and 15 is positive.

**15.** $\sqrt{625} = 25$

The square root of 625 is 25 because $25^2 = 625$ and 25 is positive.

**17.** $\sqrt{400} = 20$

The square root of 400 is 20, because $20^2 = 400$ and 20 is positive.

**19.** $\sqrt{10,000} = 100$

The square root of 10,000 is 100 because $100^2 = 10,000$ and 100 is positive.

**21.** $\sqrt{48} \approx 6.928$

**23.** $\sqrt{8} \approx 2.828$

**25.** $\sqrt{3} \approx 1.732$

**27.** $\sqrt{12} \approx 3.464$

**29.** $\sqrt{19} \approx 4.359$

**31.** $\sqrt{110} \approx 10.488$

**33.**
$$a^2 + b^2 = c^2 \quad \text{Pythagorean equation}$$
$$3^2 + 5^2 = c^2 \quad \text{Substituting}$$
$$9 + 25 = c^2$$
$$34 = c^2$$
$$\sqrt{34} = c \quad \text{Exact answer}$$
$$5.831 \approx c \quad \text{Approximation}$$

**35.**
$$a^2 + b^2 = c^2 \quad \text{Pythagorean equation}$$
$$7^2 + 7^2 = c^2 \quad \text{Substituting}$$
$$49 + 49 = c^2$$
$$98 = c^2$$
$$\sqrt{98} = c \quad \text{Exact answer}$$
$$9.899 \approx c \quad \text{Approximation}$$

**37.**
$$a^2 + b^2 = c^2$$
$$a^2 + 12^2 = 13^2$$
$$a^2 + 144 = 169$$
$$a^2 + 144 - 144 = 169 - 144$$
$$a^2 = 169 - 144$$
$$a^2 = 25$$
$$a = 5$$

**39.**
$$a^2 + b^2 = c^2$$
$$6^2 + b^2 = 10^2$$
$$36 + b^2 = 100$$
$$36 + b^2 - 36 = 100 - 36$$
$$b^2 = 100 - 36$$
$$b^2 = 64$$
$$b = 8$$

**41.**
$$a^2 + b^2 = c^2$$
$$10^2 + 24^2 = c^2$$
$$100 + 576 = c^2$$
$$676 = c^2$$
$$26 = c$$

**43.**
$$a^2 + b^2 = c^2$$
$$9^2 + b^2 = 15^2$$
$$81 + b^2 = 225$$
$$81 + b^2 - 81 = 225 - 81$$
$$b^2 = 225 - 81$$
$$b^2 = 144$$
$$b = 12$$

**45.**
$$a^2 + b^2 = c^2$$
$$1^2 + b^2 = 32^2$$
$$1 + b^2 = 1024$$
$$1 + b^2 - 1 = 1024 - 1$$
$$b^2 = 1024 - 1$$
$$b^2 = 1023$$
$$b = \sqrt{1023} \quad \text{Exact answer}$$
$$b \approx 31.984 \quad \text{Approximation}$$

**47.**
$$a^2 + b^2 = c^2$$
$$4^2 + 3^2 = c^2$$
$$16 + 9 = c^2$$
$$25 = c^2$$
$$5 = c$$

**49.** *Familiarize.* We first make a drawing. In it we see a right triangle. We let $s = $ the length of the string of lights.

*Translate.* We substitute 8 for $a$, 12 for $b$, and $s$ for $c$ in the Pythagorean equation.

$$a^2 + b^2 = c^2$$
$$8^2 + 12^2 = s^2$$

*Solve.* We solve the equation for $w$.

$$64 + 144 = s^2$$
$$208 = s^2$$
$$\sqrt{208} = s \quad \text{Exact answer}$$
$$14.4 \approx s \quad \text{Approximation}$$

**Check.** $8^2 + 12^2 = 64 + 144 = 208 = (\sqrt{208})^2$

**State.** The length of the string of lights is $\sqrt{208}$ ft, or about 14.4 ft.

**51. Familiarize.** We refer to the drawing in the text. We let $d =$ the distance from home plate to second base.

**Translate.** We substitute 90 for $a$, 90 for $b$, and $d$ for $c$ in the Pythagorean equation.

$$a^2 + b^2 = c^2$$
$$90^2 + 90^2 = d^2$$

**Solve.** We solve the equation for $d$.

$$8100 + 8100 = d^2$$
$$16,200 = d^2$$
$$\sqrt{16,200} = d$$
$$127.3 \approx d$$

**Check.** $90^2 + 90^2 = 8100 + 8100 = 16,200 = (\sqrt{16,200})^2$

**State.** The distance from home plate to second base is $\sqrt{16,200}$ ft, or about 127.3 ft.

**53. Familiarize.** We refer to the drawing in the text.

**Translate.** We substitute in the Pythagorean equation.

$$a^2 + b^2 = c^2$$
$$20^2 + h^2 = 30^2$$

**Solve.** We solve the equation for $h$.

$$400 + h^2 = 900$$
$$400 + h^2 - 400 = 900 - 400$$
$$h^2 = 900 - 400$$
$$h^2 = 500$$
$$h = \sqrt{500}$$
$$h \approx 22.4$$

**Check.** $20^2 + (\sqrt{500})^2 = 400 + 500 = 900 = 30^2$

**State.** The height of the tree is $\sqrt{500}$ ft, or about 22.4 ft.

**55. Familiarize.** We refer to the drawing in the text. We let $h =$ the plane's horizontal distance from the airport.

**Translate.** We substitute 4100 for $a$, $h$ for $b$, and 15,100 for $c$ in the Pythagorean equation.

$$a^2 + b^2 = c^2$$
$$4100^2 + h^2 = 15,100^2$$

**Solve.** We solve the equation for $h$.

$$16,810,000 + h^2 = 228,010,000$$
$$h^2 = 228,010,000 - 16,810,000$$
$$h^2 = 211,200,000$$
$$h = \sqrt{211,200,000}$$
$$h \approx 14,532.7$$

**Check.** $4100^2 + (\sqrt{211,200,000})^2 = 16,810,000 + 211,200,000 = 228,010,000 = 15,100^2$

**State.** The plane's horizontal distance from the airport is $\sqrt{211,200,000}$ ft, or about 14,532.7 ft.

**57. Familiarize.** Let $f =$ the amount the family spends for food.

**Translate.** We rephrase the question and translate.

What is 26% of $1800?
$f$ = 26% × 1800

**Solve.** Convert 26% to decimal notation and multiply.

$$f = 26\% \times 1800 = 0.26 \times 1800 = 468$$

**Check.** The answer seems reasonable since we are finding about one-fourth of 2000, which is 500. We can also repeat the calculation. The answer checks.

**State.** The family spends $468 for food.

**59. Familiarize.** This is a two-step problem. First we find the amount of the increase. Let $a =$ the amount by which the population increases.

**Translate.** We rephrase the question and translate.

What is 4% of 180,000?
$a$ = 4% × 180,000

**Solve.** Convert 4% to decimal notation and multiply.

$$a = 4\% \times 180,000 = 0.04 \times 180,000 = 7200$$

Now we add 7200 to the former population to find the new population.

$$180,000 + 7200 = 187,200$$

**Check.** We can do a partial check by estimating. The old population is approximately 200,000 and 4% of 200,000 is $0.04 \times 200,000$, or 8000. The new population would be about $180,000 + 8000$, or 188,000. Since 188,000 is close to 187,200, we have a partial check. We can also repeat the calculations. The answer checks.

**State.** The population will be 187,200.

**61. Familiarize.** Let $s =$ the number of students who are seniors.

**Translate.** We rephrase the question and translate.

What is 17.5% of 1850?
$s$ = 17.5% × 1850

**Solve.** Convert 17.5% to decimal notation and multiply.

$$s = 17.5\% \times 1850 = 0.175 \times 1850 = 323.75 \approx 324$$

**Check.** We repeat the calculation. The answer checks.

**State.** About 324 students are seniors.

**63.** ◈

**65.** ◈

**67.** To find the areas we must first use the Pythagorean equation to find the height of each triangle and then use the formula for the area of a triangle.

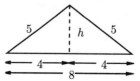

$a^2 + b^2 = c^2$       Pythagorean equation

$4^2 + h^2 = 5^2$      Substituting

$16 + h^2 = 25$

$h^2 = 25 - 16 = 9$

$h = 3$

$A = \dfrac{1}{2} \cdot b \cdot h$     Area of a triangle

$A = \dfrac{1}{2} \cdot 8 \cdot 3$     Substituting

$A = 12$

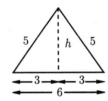

$a^2 + b^2 = c^2$       Pythagorean equation

$3^2 + h^2 = 5^2$      Substituting

$9 + h^2 = 25$

$h^2 = 25 - 9 = 16$

$h = 4$

$A = \dfrac{1}{2} \cdot b \cdot h$ Area of a triangle

$A = \dfrac{1}{2} \cdot 6 \cdot 4$ Substituting

$A = 12$

The areas of the triangles are the same (12 square units).

**69.** We let $w$ = the width of a 42-in. screen and $l$ = its length. We consider the following similar triangles.

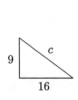 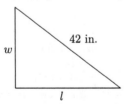

First we use the Pythagorean equation to find the length $c$.

$a^2 + b^2 = c^2$

$16^2 + 9^2 = c^2$

$256 + 81 = c^2$

$337 = c^2$

$18.358 \approx c$

Now we use proportions to find $w$ and $l$.

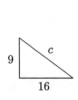 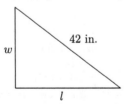

$\dfrac{42}{18.358} = \dfrac{w}{9}$

$42 \cdot 9 = 18.358 \cdot w$

$\dfrac{42 \cdot 9}{18.358} = w$

$\dfrac{378}{18.358} = w$

$20.6 \approx w$

The length of the screen is 20.6 in.

$\dfrac{42}{18.358} = \dfrac{l}{16}$

$42 \cdot 16 = 18.358 \cdot l$

$\dfrac{42 \cdot 16}{18.358} = l$

$\dfrac{672}{18.358} = l$

$36.6 \approx l$

The width of the screen is 36.6 in.

## Exercise Set 9.6

**1.** $V = l \cdot w \cdot h$

$V = 10 \text{ cm} \cdot 5 \text{ cm} \cdot 5 \text{ cm}$

$V = 50 \cdot 5 \text{ cm}^3$

$V = 250 \text{ cm}^3$

**3.** $V = l \cdot w \cdot h$

$V = 9 \text{ in.} \cdot 3 \text{ in.} \cdot 5 \text{ in.}$

$V = 27 \cdot 5 \text{ in}^3$

$V = 135 \text{ in}^3$

**5.** $V = l \cdot w \cdot h$

$V = 10 \text{ m} \cdot 5 \text{ m} \cdot 1.5 \text{ m}$

$V = 50 \cdot 1.5 \text{ m}^3$

$V = 75 \text{ m}^3$

**7.** $V = l \cdot w \cdot h$

$V = 6\dfrac{1}{2} \text{ yd} \cdot 5\dfrac{1}{2} \text{ yd} \cdot 10 \text{ yd}$

$V = \dfrac{13}{2} \cdot \dfrac{11}{2} \cdot 10 \text{ yd}^3$

$V = \dfrac{1430}{4} \text{ yd}^3 = \dfrac{2 \cdot 715}{2 \cdot 2} \text{ yd}^3$

$V = \dfrac{715}{2} \text{ yd}^3$

$V = 357\dfrac{1}{2} \text{ yd}^3$

**9.** $V = Bh = \pi \cdot r^2 \cdot h$

$\approx 3.14 \times 10 \text{ ft} \times 10 \text{ ft} \times 13 \text{ ft}$

$\approx 4082 \text{ ft}^3$

**11.** $V = Bh = \pi \cdot r^2 \cdot h$

$\approx 3.14 \times 4 \text{ cm} \times 4 \text{ cm} \times 7.5 \text{ cm}$

$\approx 376.8 \text{ cm}^3$

**13.** $V = Bh = \pi \cdot r^2 \cdot h$

$\approx \dfrac{22}{7} \times 210 \text{ yd} \times 210 \text{ yd} \times 300 \text{ yd}$

$\approx 41,580,000 \text{ yd}^3$

**15.** $V = \dfrac{4}{3} \cdot \pi \cdot r^3$

$\approx \dfrac{4}{3} \times 3.14 \times (100 \text{ in.})^3$

$\approx \dfrac{4 \times 3.14 \times 1,000,000 \text{ in}^3}{3}$

$\approx 4,186,666.\overline{6} \text{ in}^3, \text{ or } 4,186,666\dfrac{2}{3} \text{ in}^3$

**17.** $V = \dfrac{4}{3} \cdot \pi \cdot r^3$

$\approx \dfrac{4}{3} \times 3.14 \times (3.1 \text{ m})^3$

$\approx \dfrac{4 \times 3.14 \times 29.791 \text{ m}^3}{3}$

$\approx 124.725 \text{ m}^3$

**19.** $V = \dfrac{4}{3} \cdot \pi \cdot r^3$

$\approx \dfrac{4}{3} \times \dfrac{22}{7} \times (7 \text{ km})^3$

$\approx \dfrac{4 \times 22 \times 343 \text{ km}^3}{3 \times 7}$

$\approx 1437\dfrac{1}{3} \text{ km}^3$

**21.** $1 \text{ L} = 1000 \text{ mL} = 1000 \text{ cm}^3$

These conversion relations appear in the text on page 558.

**23.** $87 \text{ L} = 87 \times (1 \text{ L})$

$= 87 \times (1000 \text{ mL})$

$= 87,000 \text{ mL}$

**25.** $49 \text{ mL} = 49 \times (1 \text{ mL})$

$= 49 \times (0.001 \text{ L})$

$= 0.049 \text{ L}$

**27.** $27.3 \text{ L} = 27.3 \times (1 \text{ L})$

$= 27.3 \times (1000 \text{ cm}^3)$

$= 27,300 \text{ cm}^3$

**29.** $5 \text{ gal} = 5 \times 1 \text{ gal}$

$= 5 \times 4 \text{ qt}$

$= 5 \times 4 \times 1 \text{ qt}$

$= 5 \times 4 \times 2 \text{ pt}$

$= 40 \text{ pt}$

**31.** $10 \text{ qt} = 10 \times 1 \text{ qt}$

$= 10 \times 2 \text{ pt}$

$= 10 \times 2 \times 1 \text{ pt}$

$= 10 \times 2 \times 16 \text{ oz}$

$= 320 \text{ oz}$

**33.** $24 \text{ oz} = 24 \text{ oz} \times \dfrac{1 \text{ cup}}{8 \text{ oz}}$

$= \dfrac{24}{8} \cdot 1 \text{ cup}$

$= 3 \text{ cups}$

**35.** $8 \text{ gal} = 8 \text{ gal} \times \dfrac{4 \text{ qt}}{1 \text{ gal}}$

$= 8 \cdot 4 \text{ qt}$

$= 32 \text{ qt}$

**37.** First we convert 3 gal to quarts:

$3 \text{ gal} = 3 \text{ gal} \times \dfrac{4 \text{ qt}}{1 \text{ gal}}$

$= 3 \cdot 4 \text{ qt}$

$= 12 \text{ qt}$

Next we convert 12 qt to pints:

$12 \text{ qt} = 12 \text{ qt} \times \dfrac{2 \text{ pt}}{1 \text{ qt}}$

$= 12 \cdot 2 \text{ pt}$

$= 24 \text{ pt}$

Finally we convert 24 pt to cups:

$24 \text{ pt} = 24 \text{ pt} \times \dfrac{2 \text{ cups}}{1 \text{ pt}}$

$= 24 \cdot 2 \text{ cups}$

$= 48 \text{ cups}$

**39.** First we convert 15 pt to quarts:

$15 \text{ pt} = 15 \text{ pt} \times \dfrac{1 \text{ qt}}{2 \text{ pt}}$

$= \dfrac{15}{2} \cdot 1 \text{ qt}$

$= 7.5 \text{ qt}$

Then we convert 7.5 qt to gallons:

$7.5 \text{ qt} = 7.5 \text{ qt} \times \dfrac{1 \text{ gal}}{4 \text{ qt}}$

$= \dfrac{7.5}{4} \cdot 1 \text{ gal}$

$= 1.875 \text{ gal}, \text{ or } 1\dfrac{7}{8} \text{ gal}$

**41.** We convert 0.5 L to milliliters:

$0.5 \text{ L} = 0.5 \text{ L} \times \dfrac{1000 \text{ mL}}{1 \text{ L}}$

$= 0.5 \times 1000 \text{ mL}$

$= 500 \text{ mL}$

Dr. Carey ordered 500 mL.

**43.** $0.5 \text{ L} \approx 0.5 \text{ qt} = 0.5 \times 32 \text{ oz} = 16 \text{ oz}$

The 16 oz bottle comes closest to filling the prescription.

**45.** We must find the radius of the rung in order to use the formula for the volume of a circular cylinder.

$r = \dfrac{d}{2} = \dfrac{2 \text{ in.}}{2} = 1 \text{ in.}$

Now we find the volume.

$V = Bh = \pi \cdot r^2 \cdot h$

$\approx 3.14 \times 1 \text{ in.} \cdot 1 \text{ in.} \cdot 18 \text{ in.}$

$\approx 56.52 \text{ in}^3$

**47.** We must find the radius of the log in order to use the formula for the volume of a circular cylinder.

$r = \dfrac{d}{2} = \dfrac{12 \text{ cm}}{2} = 6 \text{ cm}$

Now we find the volume.

$$V = Bh = \pi \cdot r^2 \cdot h$$
$$\approx 3.14 \times 6 \text{ cm} \cdot 6 \text{ cm} \cdot 42 \text{ cm}$$
$$\approx 4747.68 \text{ cm}^3$$

**49.** First we find the radius of the can.
$$r = \frac{d}{2} = \frac{6.5 \text{ cm}}{2} = 3.25 \text{ cm}$$

The height of the can is the length of the diameters of 3 tennis balls.
$$h = 3(6.5 \text{ cm}) = 19.5 \text{ cm}$$

Now we find the volume.
$$V = Bh = \pi \cdot r^2 \cdot h$$
$$\approx 3.14 \times 3.25 \text{ cm} \times 3.25 \text{ cm} \times 19.5 \text{ cm}$$
$$\approx 646.74 \text{ cm}^3$$

**51.** A cube is a rectangular solid.
$$V = l \cdot w \cdot h$$
$$= 18 \text{ yd} \cdot 18 \text{ yd} \cdot 18 \text{ yd}$$
$$= 5832 \text{ yd}^3$$

**53.** $V = Bh = \pi \cdot r^2 \cdot h$
$$\approx \frac{22}{7} \cdot 14 \text{ cm} \cdot 14 \text{ cm} \cdot 100 \text{ cm}$$
$$\approx 61,600 \text{ cm}^3$$

Now we convert $61,600 \text{ cm}^3$ to liters.
$$61,600 \text{ cm}^3 = 61,600 \text{ cm}^3 \times \frac{1 \text{ L}}{1000 \text{ cm}^3}$$
$$= \frac{61,600}{1000} \times \frac{\text{cm}^3}{\text{cm}^3} \times 1 \text{ L}$$
$$= 61.6 \text{ L}$$

**55.** Interest $= (\text{Interest for 1 year}) \times \dfrac{1}{2}$
$$= (8\% \times \$600) \times \frac{1}{2}$$
$$= 0.08 \times \$600 \times \frac{1}{2}$$
$$= \$48 \times \frac{1}{2}$$
$$= \$24$$

The interest is \$24.

**57.** Let $c$ represent the cost of 12 pens. We translate to a proportion and solve.

$$\text{Pens} \rightarrow \quad \frac{9}{\$8.01} = \frac{12}{c} \quad \leftarrow \text{Pens}$$
$$\text{Cost} \rightarrow \qquad\qquad\qquad \leftarrow \text{Cost}$$

$$9 \cdot c = \$8.01 \cdot 12 \quad \text{Equating cross-products}$$
$$c = \frac{\$8.01 \cdot 12}{9}$$
$$c = \frac{\$96.12}{9}$$
$$c = \$10.68$$

12 pens would cost \$10.68.

**59.**
$$-5y + 3 = -12y - 4$$
$$-5y + 3 + 12y = -12y - 4 + 12y \quad \text{Adding } 12y \text{ on both sides}$$
$$7y + 3 = -4$$
$$7y + 3 - 3 = -4 - 3 \quad \text{Subtracting 3 on both sides}$$
$$7y = -7$$
$$\frac{7y}{7} = \frac{-7}{7} \quad \text{Dividing by 7 on both sides}$$
$$y = -1$$

The solution is $-1$.

**61.** ◈

**63.** ◈

**65.** Radius of water stream: $\dfrac{2 \text{ cm}}{2} = 1 \text{ cm}$

To convert 30 m to centimeters, think: 1 meter is 100 times as large as 1 centimeter. Thus, we move the decimal point 2 places to the right:

$$30 \text{ m} = 3000 \text{ cm}$$
$$V = Bh = \pi \cdot r^2 \cdot h$$
$$\approx 3.141593 \cdot 1 \text{ cm} \cdot 1 \text{ cm} \cdot 3000 \text{ cm}$$
$$\approx 9424.779 \text{ cm}^3$$

Now we convert $9424.779 \text{ cm}^3$ to liters:
$$9424.779 \text{ cm}^3 = 9424.779 \text{ cm}^3 \cdot \frac{1 \text{ L}}{1000 \text{ cm}^3}$$
$$= 9.424779 \text{ L}$$

There is about 9.424779 L of water in the hose.

**67.** First find the volume of one one-dollar bill in cubic inches:
$$V = l \cdot w \cdot h$$
$$= 6.0625 \text{ in.} \cdot 2.3125 \text{ in.} \cdot 0.0041 \text{ in.}$$
$$\approx 0.05748 \text{ in}^3 \quad \text{Rounding}$$

Then multiply to find the volume of one million one-dollar bills in cubic inches:
$$1,000,000 \cdot 0.05748 \text{ in}^3 = 57,480 \text{ in}^3$$

The volume of one million one-dollar bills is about $57,480 \text{ in}^3$.

**69.** The length of a diagonal of the cube is the length of the diameter of the sphere, 1 m. Visualize a triangle whose hypotenuse is a diagonal of the cube and with one leg a side $s$ of the cube and the other leg a diagonal $c$ of a side of the cube.

We want to find the length of a side $s$ of the cube in order to find the volume. We begin by using the Pythagorean theorem to find $c$.
$$s^2 + s^2 = c^2$$
$$2s^2 = c^2$$
$$\sqrt{2s^2} = c$$

Now use the Pythagorean theorem again to find $s$.

$$c^2 + s^2 = 1^2$$
$$(\sqrt{2s^2})^2 + s^2 = 1 \qquad \text{Subtracting } \sqrt{2s^2} \text{ for } c$$
$$2s^2 + s^2 = 1$$
$$3s^2 = 1$$
$$s^2 = \frac{1}{3}$$
$$s = \sqrt{\frac{1}{3}} \approx 0.577$$

Next we find the volume of the cube.

$$V = l \cdot w \cdot h$$
$$= 0.577 \text{ m} \cdot 0.577 \text{ m} \cdot 0.577 \text{ m}$$
$$= 0.192 \text{ m}^3$$

Find the volume of the sphere. The radius is $\dfrac{1 \text{ m}}{2}$, or 0.5 m.

$$V = \frac{4}{3} \cdot \pi \cdot r^3$$
$$\approx \frac{4}{3} \times 3.14 \times (0.5 \text{ m})^3$$
$$\approx \frac{4 \times 3.14 \times 0.125 \text{ m}^3}{3}$$
$$\approx 0.523 \text{ m}^3$$

Finally we subtract to find how much more volume is in the sphere.

$$0.523 \text{ m}^3 - 0.192 \text{ m}^3 = 0.331 \text{ m}^3$$

There is 0.331 m$^3$ more volume in the sphere.

---

## Exercise Set 9.7

**1.** 1 lb = 16 oz

This conversion relation is given in the text on page 565.

**3.** $6000 \text{ lb} = 6000 \text{ lb} \times \dfrac{1 \text{ T}}{2000 \text{ lb}}$    Writing 1 with tons on the top and pounds on the bottom

$$= \frac{6000}{2000} \text{ T}$$
$$= 3 \text{ T}$$

**5.** $3 \text{ lb} = 3 \times 1 \text{ lb}$
$= 3 \times 16 \text{ oz}$    Substituting 16 oz for 1 lb
$= 48 \text{ oz}$

**7.** $3.5 \text{ T} = 3.5 \times 1 \text{ T}$
$= 3.5 \times 2000 \text{ lb}$    Substituting 2000 lb for 1 T
$= 7000 \text{ lb}$

**9.** $4800 \text{ lb} = 4800 \text{ lb} \times \dfrac{1 \text{ T}}{2000 \text{ lb}}$    Writing 1 with tons on the top and pounds on the bottom

$$= \frac{4800}{2000} \text{ T}$$
$$= 2.4 \text{ T}$$

**11.** $72 \text{ oz} = 72 \text{ oz} \times \dfrac{1 \text{ lb}}{16 \text{ oz}}$

$$= \frac{72}{16} \text{ lb}$$
$$= 4.5 \text{ lb}$$

**13.** 1 kg = _____ g

Think: A kilogram is 1000 times the mass of a gram. Thus, we move the decimal point 3 places to the right.

1   1.000.

1 kg = 1000 g

**15.** 1 g = _____ kg

Think: It takes 1000 grams to have 1 kilogram. Thus, we move the decimal point 3 places to the left.

1   0.001.

1 g = 0.001 kg

**17.** 1 cg = _____ g

Think: It takes 100 centigrams to have 1 gram. Thus, we move the decimal point 2 places to the left.

1   0.01.

1 cg = 0.01 g

**19.** 1 g = _____ mg

Think: A gram is 1000 times the mass of a milligram. Thus, we move the decimal point 3 places to the right.

1   1.000.

1 g = 1000 mg

**21.** 1 g = _____ dg

Think: A gram is 10 times the mass of a decigram. Thus, we move the decimal point 1 place to the right.

1   1.0.

1 g = 10 dg

**23.** Complete: 725 kg = _____ g

Think: A kilogram is 1000 times the mass of a gram. Thus, we move the decimal point 3 places to the right.

725   725.000.

725 kg = 725,000 g

**25.** Complete: 6345 g = _____ kg

Think: It takes 1000 grams to have 1 kilogram. Thus, we move the decimal point 3 places to the left.

6345   6.345.

6345 g = 6.345 kg

**27.** 897 mg = _____ kg

Think: It takes 1,000,000 milligrams to have 1 kilogram. Thus, we move the decimal point 6 places to the left.

897   0.000897.

897 mg = 0.000897 kg

**29.** 7.32 kg = _____ g

Think: A kilogram is 1000 times the mass of a gram. Thus, we move the decimal point 3 places to the right.

7.32   7.320.

7.32 kg = 7320 g

**31.** Complete: 6780 g = _____ kg

Think: It takes 1000 grams to have 1 kilogram. Thus, we move the decimal point 3 places to the left.

6780   6.780.

6780 g = 6.78 kg

**33.** Complete: 69 mg = _____ cg

Think: It takes 10 milligrams to have 1 centigram. Thus, we move the decimal point 1 place to the left.

69   6.9.

69 mg = 6.9 cg

**35.** Complete: 8 kg = _____ cg

Think: A kilogram is 100,000 times the mass of a centigram. Thus, we move the decimal point 5 places to the right.

8   8.00000.

8 kg = 800,000 cg

**37.** 1 t = 1000 kg

This conversion relation is given in the text on page 566.

**39.** Complete: 3.4 cg = _____ dag

Think: It takes 1000 centigrams to have 1 dekagram. Thus, we move the decimal point 3 places to the left.

3.4   0.003.4

3.4 cg = 0.0034 dag

**41.** By laying a straightedge horizontally between the scales on page 567, we see that 178°F ≈ 80°C.

**43.** By laying a straightedge horizontally between the scales on page 567, we see that 140°F ≈ 60°C.

**45.** By laying a straightedge horizontally between the scales on page 567, we see that 68°F ≈ 20°C.

**47.** By laying a straightedge horizontally between the scales on page 567, we see that 10°F ≈ −10°C.

**49.** By laying a straightedge horizontally between the scales on page 567, we see that 86°C ≈ 190°F.

**51.** By laying a straightedge horizontally between the scales on page 567, we see that 58°C ≈ 140°F.

**53.** By laying a straightedge horizontally between the scales on page 567, we see that −10°C ≈ 10°F.

**55.** By laying a straightedge horizontally between the scales on page 567, we see that 5°C ≈ 40°F.

**57.** $F = \dfrac{9}{5} \cdot C + 32$

$F = \dfrac{9}{5} \cdot 25 + 32$

$= 45 + 32$

$= 77$

Thus, 25°C = 77°F.

**59.** $F = \dfrac{9}{5} \cdot C + 32$

$F = \dfrac{9}{5} \cdot 40 + 32$

$= 72 + 32$

$= 104$

Thus, 40°C = 104°F.

**61.** $F = \dfrac{9}{5} \cdot C + 32$

$F = \dfrac{9}{5} \cdot 3000 + 32$

$= 5400 + 32$

$= 5432$

Thus, 3000°C = 5432°F.

**63.** $C = \dfrac{5}{9} \cdot (F - 32)$

$C = \dfrac{5}{9} \cdot (86 - 32)$

$= \dfrac{5}{9} \cdot 54$

$= 30$

Thus, 86°F = 30°C.

**65.** $C = \dfrac{5}{9} \cdot (F - 32)$

$C = \dfrac{5}{9} \cdot (131 - 32)$

$= \dfrac{5}{9} \cdot 99$

$= 55$

Thus, 131°F = 55°C.

**67.** $C = \dfrac{5}{9} \cdot (F - 32)$

$C = \dfrac{5}{9} \cdot (98.6 - 32)$

$= \dfrac{5}{9} \cdot 66.6$

$= 37$

Thus, 98.6°F = 37°C.

**69.** 0.0043

a) Multiply by 100 to move the decimal    0.00.43
point two places to the right.

b) Write a percent symbol.      0.43%

0.0043 = 0.43%

**71.** Let $c$ represent the number of cans that can be bought for $7.45. We translate to a proportion and solve.

$$\text{Cans} \rightarrow \frac{2}{\$1.49} = \frac{c}{\$7.45} \leftarrow \text{Cans} \atop \text{Cost} \rightarrow \quad \leftarrow \text{Cost}$$

$$2 \cdot \$7.45 = \$1.49 \cdot c \quad \text{Equating cross-products}$$

$$\frac{2 \cdot \$7.45}{\$1.49} = c$$

$$\frac{\$14.90}{\$1.49} = c$$

$$10 = c$$

You can buy 10 cans for $7.45.

**73.**
$$9(x - 3) = 4x - 5$$
$$9x - 27 = 4x - 5 \quad \text{Using the distributive law}$$
$$9x - 27 + 27 = 4x - 5 + 27 \quad \text{Adding 27 on both sides}$$
$$9x = 4x + 22$$
$$9x - 4x = 4x + 22 - 4x \quad \text{Subtracting } 4x \text{ on both sides}$$
$$5x = 22$$
$$\frac{5x}{5} = \frac{22}{5} \quad \text{Dividing by 5 on both sides}$$
$$x = \frac{22}{5}$$

The solution is $\frac{22}{5}$.

**75.**

**77.**

**79.** $1 \text{ g} \approx 1 \text{ g} \cdot \frac{1 \text{ lb}}{453.5 \text{ g}} \approx \frac{1}{453.5} \text{ lb} \approx 0.0022 \text{ lb}$

**81. a)** First we convert 18 grams to milligrams. (We could equivalently convert 90 milligrams to grams.) Think: A gram is 1000 times the mass of a milligram. Thus, we move the decimal point 3 places to the right.

18    18.000.

18 g = 18,000 mg

Now we divide to find the number of actuations in one inhaler:

$$18,000 \div 90 = 200$$

There are 200 actuations in one inhaler.

**b)** At 4 actuations per day, in one month (30 days) Myra will need $4 \cdot 30$, or 120 actuations. Then in 4 months she will need $4 \cdot 120$, or 480 actuations.

At 200 actuations per inhaler, Myra will need $480 \div 200$, or 2.4 inhalers. Since 2 entire inhalers and 0.4 of a third are needed, she should take 3 inhalers.

**83.** First we find the volume of a sphere with diameter $5\frac{1}{2}$ cm.

$$r = \frac{d}{2}$$
$$= \left(5\frac{1}{2}\right) \div 2 = \frac{11}{2} \div 2$$
$$= \frac{11}{2} \cdot \frac{1}{2} = \frac{11}{4} \text{ cm}$$
$$V = \frac{4}{3} \cdot \pi \cdot r^3$$
$$\approx \frac{4}{3} \cdot 3.14 \cdot \left(\frac{11}{4} \text{ cm}\right)^3$$
$$\approx \frac{4 \cdot 3.14 \cdot 1331 \text{ cm}^3}{3 \cdot 64}$$
$$\approx 87.07 \text{ cm}^3$$

Now find the volume of a sphere with diameter 4 cm.

$$r = \frac{d}{2} = \frac{4 \text{ cm}}{2} = 2 \text{ cm}$$
$$V = \frac{4}{3} \cdot \pi \cdot r^3$$
$$\approx \frac{4}{3} \cdot 3.14 \cdot (2 \text{ cm})^3$$
$$\approx \frac{4 \cdot 3.14 \cdot 8 \text{ cm}^3}{3}$$
$$\approx 33.49 \text{ cm}^3$$

Next we average the volumes of the two spheres.

$$\frac{87.07 \text{ cm}^3 + 33.49 \text{ cm}^3}{2} = \frac{120.56 \text{ cm}^3}{2} = 60.28 \text{ cm}^3$$

Since 1 cm$^3$ of water has a mass of 1 g, the mass of the egg is about 60 g.

**85.** We solve using a proportion.

$$\text{cephalexin} \rightarrow \frac{250}{5} = \frac{400}{a} \leftarrow \text{cephalexin} \atop \text{liquid} \rightarrow \quad \leftarrow \text{liquid}$$

Solve: $250 \cdot a = 5 \cdot 400$
$$a = \frac{5 \cdot 400}{250}$$
$$a = 8$$

Thus, 8 mL of liquid would be required.

**87.** First we find the volume of the shot put in cubic centimeters.

$$r = \frac{d}{2} = \frac{4.5 \text{ in.}}{2} = 2.25 \text{ in.}$$
$$2.25 \text{ in.} = 2.25 \times 1 \text{ in.} \approx 2.25 \times 2.54 \text{ cm} \approx 5.715 \text{ cm}$$
$$V = \frac{4}{3} \cdot \pi \cdot r^3$$
$$\approx \frac{4}{3} \cdot 3.14 \cdot (5.715 \text{ cm})^3$$
$$\approx 781.476 \text{ cm}^3$$

On page 566 of the text we are told that 1 kg $\approx$ 2.2 lb. Since 1 kg = 1000 g, we know that 1000 g $\approx$ 2.2 lb. We use this fact to convert the weight of the shot put to grams.

$$8.8 \text{ lb} \approx 8.8 \text{ lb} \times \frac{1000 \text{ g}}{2.2 \text{ lb}} \approx \frac{8800}{2.2} \text{ g} = 4000 \text{ g}$$

Finally, we divide to find the mass per cubic centimeter.

$$\frac{4000 \text{ g}}{781.476 \text{ cm}^3} \approx 5.1 \text{ g/cm}^3$$

Thus, the mass per cubic centimeter of the shot put is about 5.1 g/cm$^3$.

# Chapter 10

# Polynomials

---

## Exercise Set 10.1

---

**1.** $(2x + 7) + (-4x + 3) = (2 - 4)x + (7 + 3) = -2x + 10$

**3.** $(-9x + 5) + (x^2 + x - 3) =$
$x^2 + (-9 + 1)x + (5 - 3) = x^2 - 8x + 2$

**5.** $(x^2 - 7) + (x^2 + 7) = (1 + 1)x^2 + (-7 + 7) = 2x^2$

**7.** $(6t^4 + 4t^3 - 1) + (5t^2 - t + 1) =$
$6t^4 + 4t^3 + 5t^2 - t + (-1 + 1) =$
$6t^4 + 4t^3 + 5t^2 - t$

**9.** $(3 + 4x + 6x^2 + 7x^3) + (6 - 4x + 6x^2 - 7x^3) =$
$(3 + 6) + (4 - 4)x + (6 + 6)x^2 + (7 - 7)x^3 =$
$9 + 0x + 12x^2 + 0x^3 = 9 + 12x^2$, or $12x^2 + 9$

**11.** $(9x^8 - 7x^4 + 2x^2 + 5) + (8x^7 + 4x^4 - 2x) =$
$9x^8 + 8x^7 + (-7 + 4)x^4 + 2x^2 - 2x + 5 =$
$9x^8 + 8x^7 - 3x^4 + 2x^2 - 2x + 5$

**13.** $(9t^4 + 6t^3 - t^2 + 3t) + (5t^4 - 2t^3 + t - 7) =$
$(9 + 5)t^4 + (6 - 2)t^3 - t^2 + (3 + 1)t - 7 =$
$14t^4 + 4t^3 - t^2 + 4t - 7$

**15.** $(-5x^4y^3 + 7x^3y^2 - 4xy^2) + (2x^3y^3 - 3x^3y^2 - 5xy) =$
$-5x^4y^3 + 2x^3y^3 + (7 - 3)x^3y^2 - 4xy^2 - 5xy =$
$-5x^4y^3 + 2x^3y^3 + 4x^3y^2 - 4xy^2 - 5xy$

**17.** $(8a^3b^2 + 5a^2b^2 + 6ab^2) + (5a^3b^2 - a^2b^2 - 4a^2b) =$
$(8 + 5)a^3b^2 + (5 - 1)a^2b^2 + 6ab^2 - 4a^2b =$
$13a^3b^2 + 4a^2b^2 + 6ab^2 - 4a^2b$

**19.** $(17.5abc^3 + 4.3a^2bc) + (-4.9a^2bc - 5.2abc) =$
$17.5abc^3 + (4.3 - 4.9)a^2bc - 5.2abc =$
$17.5abc^3 - 0.6a^2bc - 5.2abc$

**21.** Two equivalent expressions for the additive inverse of $-5x$ are
a) $-(-5x)$ and
b) $5x$. (Changing the sign)

**23.** Two equivalent expressions for the additive inverse of $-x^2 + 10x - 2$ are
a) $-(-x^2 + 10x - 2)$ and
b) $x^2 - 10x + 2$. (Changing the sign of every term)

**25.** Two equivalent expressions for the additive inverse of $12x^4 - 3x^3 + 3$ are
a) $-(12x^4 - 3x^3 + 3)$ and
b) $-12x^4 + 3x^3 - 3$. (Changing the sign of every term)

**27.** We change the sign of every term inside parentheses.
$-(3x - 7) = -3x + 7$

**29.** We change the sign of every term inside parentheses.
$-(4x^2 - 3x + 2) = -4x^2 + 3x - 2$

**31.** We change the sign of every term inside parentheses.
$-\left(-4x^4 + 6x^2 + \dfrac{3}{4}x - 8\right) = 4x^4 - 6x^2 - \dfrac{3}{4}x + 8$

**33.** $(3x + 2) - (-4x + 3) = 3x + 2 + 4x - 3$

Changing the sign of every term inside parentheses

$= 7x - 1$

**35.** $(9t^2 + 7t + 5) - (5t^2 + t - 1)$
$= 9t^2 + 7t + 5 - 5t^2 - t + 1$
$= 4t^2 + 6t + 6$

**37.** $(-6x + 2) - (x^2 + x - 3) = -6x + 2 - x^2 - x + 3$
$= -x^2 - 7x + 5$

**39.** $(7a^2 + 5a - 9) - (2a^2 + 7)$
$= 7a^2 + 5a - 9 - 2a^2 - 7$
$= 5a^2 + 5a - 16$

**41.** $(6x^4 + 3x^3 - 1) - (4x^2 - 3x + 3)$
$= 6x^4 + 3x^3 - 1 - 4x^2 + 3x - 3$
$= 6x^4 + 3x^3 - 4x^2 + 3x - 4$

**43.** $(1.2x^3 + 4.5x^2 - 3.8x) - (-3.4x^3 - 4.7x^2 + 23)$
$= 1.2x^3 + 4.5x^2 - 3.8x + 3.4x^3 + 4.7x^2 - 23$
$= 4.6x^3 + 9.2x^2 - 3.8x - 23$

**45.** $\left(\dfrac{5}{8}x^3 - \dfrac{1}{4}x - \dfrac{1}{3}\right) - \left(-\dfrac{1}{8}x^3 + \dfrac{1}{4}x - \dfrac{1}{3}\right)$
$= \dfrac{5}{8}x^3 - \dfrac{1}{4}x - \dfrac{1}{3} + \dfrac{1}{8}x^3 - \dfrac{1}{4}x + \dfrac{1}{3}$
$= \dfrac{6}{8}x^3 - \dfrac{2}{4}x$
$= \dfrac{3}{4}x^3 - \dfrac{1}{2}x$

**47.** $(5x^3y^3 + 8x^2y^2 + 7xy) - (3x^3y^3 - 2x^2y + 3xy)$
$= 5x^3y^3 + 8x^2y^2 + 7xy - 3x^3y^3 + 2x^2y - 3xy$
$= 2x^3y^3 + 8x^2y^2 + 2x^2y + 4xy$

**49.** $-5x + 2 = -5 \cdot 4 + 2 = -20 + 2 = -18$

**51.** $2x^2 - 5x + 7 = 2 \cdot 4^2 - 5 \cdot 4 + 7 = 2 \cdot 16 - 20 + 7 =$
$32 - 20 + 7 = 19$

**53.** $x^3 - 5x^2 + x = 4^3 - 5 \cdot 4^2 + 4 = 64 - 5 \cdot 16 + 4 =$
$64 - 80 + 4 = -12$

**55.** $3x + 5 = 3(-1) + 5 = -3 + 5 = 2$

**57.** $x^2 - 2x + 1 = (-1)^2 - 2(-1) + 1 = 1 + 2 + 1 = 4$

**59.** $-3x^3 + 7x^2 - 3x - 2 =$
$-3(-1)^3 + 7(-1)^2 - 3(-1) - 2 =$
$-3(-1) + 7 \cdot 1 + 3 - 2 = 3 + 7 + 3 - 2 = 11$

**61.** We evaluate the polynomial for $a = 18$ :
$$0.4a^2 - 40a + 1039 = 0.4(18)^2 - 40(18) + 1039$$
$$= 0.4(324) - 40(18) + 1039$$
$$= 129.6 - 720 + 1039$$
$$= 448.6$$

The daily number of accidents involving 18-year-old drivers is 448.6, or about 449.

**63.** We evaluate the polynomial for $t = 8$:
$$16t^2 = 16(8)^2 = 16 \cdot 64 = 1024$$

The cliff is 1024 ft high.

**65.** We evaluate the polynomial for $x = 75$:
$$280x - 0.4x^2 = 280(75) - 0.4(75)^2$$
$$= 280(75) - 0.4(5625)$$
$$= 21,000 - 2250$$
$$= 18,750$$

The total revenue from the sale of 75 stereos is $18,750.

**67.** We evaluate the polynomial for $x = 500$:
$$5000 + 0.6x^2 = 5000 + 0.6(500)^2$$
$$= 5000 + 0.6(250,000)$$
$$= 5000 + 150,000$$
$$= 155,000$$

The total cost of producing 500 stereos is $155,000.

**69.** $\dfrac{7 \text{ servings}}{10 \text{ lb}} = \dfrac{7}{10} \dfrac{\text{servings}}{\text{lb}}$, or $0.7 \dfrac{\text{servings}}{\text{lb}}$

**71.** The sales tax is
$$\underbrace{\text{Sales tax rate}}_{\downarrow} \cdot \underbrace{\text{Purchase price}}_{\downarrow}$$
$$\quad 5\% \qquad \cdot \qquad \$1350,$$
or $0.05 \cdot 1350$, or $67.5$. Thus, the tax is $67.50.

**73.** $A = \pi \cdot r \cdot r$
$\approx 3.14 \cdot 20 \text{ cm} \cdot 20 \text{ cm}$
$\approx 1256 \text{ cm}^2$

**75.** ◈

**77.** ◈

**79.** $2004 - 1985 = 19$, so we evaluate the polynomial for $x = 19$:
$$0.04x^3 - 0.23x^2 + 0.94x - 0.05$$
$$= 0.04(19)^3 - 0.23(19)^2 + 0.94(19) - 0.05$$
$$= 0.04(6859) - 0.23(361) + 0.94(19) - 0.05$$
$$= 274.36 - 83.03 + 17.86 - 0.05$$
$$= 209.14$$

In 2004 it is predicted that there will be about 209 million cellular phones in use.

**81.** $(3x^2 - 4x + 6) - (-2x^2 + 4) + (-5x - 3)$
$= 3x^2 - 4x + 6 + 2x^2 - 4 - 5x - 3$
$= 5x^2 - 9x - 1$

**83.** $(-4 + x^2 + 2x^3) - (-6 - x + 3x^3) - (-x^2 - 5x^3)$
$= -4 + x^2 + 2x^3 + 6 + x - 3x^3 + x^2 + 5x^3$
$= 2 + x + 2x^2 + 4x^3$

**85.** $9t^3 - 2t^3 = 7t^3$, $2t^2 - 2t^2 = 0t^2$, $t - 4t = -3t$, and $-3 + 7 = 4$ so we have $8t^4 + \underline{9t^3} - 2t^3 + \underline{2t^2} - 2t^2 + t - \underline{4t} - 3 + \underline{7} = 8t^4 + 7t^3 - 3t + 4$.

---

## Exercise Set 10.2

**1.** $(5a)(9a) = (5 \cdot 9)(a \cdot a)$
$= 45a^2$

**3.** $(-4x)(15x) = (-4 \cdot 15)(x \cdot x)$
$= -60x^2$

**5.** $(7x^5)(4x^3) = (7 \cdot 4)(x^5 \cdot x^3)$
$= 28x^8$

**7.** $(-0.1x^6)(0.2x^4) = (-0.1 \cdot 0.2)(x^6 \cdot x^4)$
$= -0.02x^{10}$

**9.** $(5x^2y^3)(7x^4y^9) = (5 \cdot 7)(x^2 \cdot x^4)(y^3 \cdot y^9)$
$= 35x^6y^{12}$

**11.** $(4a^3b^4c^2)(3a^5b^4) = (4 \cdot 3)(a^3 \cdot a^5)(b^4 \cdot b^4)(c^2)$
$= 12a^8b^8c^2$

**13.** $(3x^2)(-4x^3)(2x^6) = (3)(-4)(2)(x^2 \cdot x^3 \cdot x^6) = -24x^{11}$

**15.** $3x(-x + 5) = 3x(-x) + 3x \cdot 5$
$= -3x^2 + 15x$

**17.** $-3x(x - 1) = -3x(x) - 3x(-1)$
$= -3x^2 + 3x$

**19.** $x^2(x^3 + 1) = x^2 \cdot x^3 + x^2 \cdot 1$
$= x^5 + x^2$

**21.** $3x(2x^2 - 6x + 1) = 3x \cdot 2x^2 + 3x(-6x) + 3x \cdot 1$
$= 6x^3 - 18x^2 + 3x$

**23.** $4xy(3x^2 + 2y) = 4xy \cdot 3x^2 + 4xy \cdot 2y$
$= 12x^3y + 8xy^2$

**25.** $3a^2b(4a^5b^2 - 3a^2b^2) = 3a^2b \cdot 4a^5b^2 - 3a^2b \cdot 3a^2b^2$
$= 12a^7b^3 - 9a^4b^3$

**27.** $2x + 6 = 2 \cdot x + 2 \cdot 3$
$= 2(x + 3)$

**29.** $7a - 21 = 7 \cdot a - 7 \cdot 3$
$= 7(a - 3)$

**31.** $14x + 21y = 7 \cdot 2x + 7 \cdot 3y$
$= 7(2x + 3y)$

**33.** $9a - 27b + 81 = 9 \cdot a - 9 \cdot 3b + 9 \cdot 9$
$$= 9(a - 3b + 9)$$

**35.** $24 - 6m = 6 \cdot 4 - 6 \cdot m$
$$= 6(4 - m)$$

**37.** $-16 - 8x + 40y = -8 \cdot 2 - 8 \cdot x - 8(-5y)$
$$= -8(2 + x - 5y)$$

**39.** $3x^5 + 3x = 3x \cdot x^4 + 3x \cdot 1$
$$= 3x(x^4 + 1)$$

**41.** $a^3 - 8a^2 = a^2 \cdot a - a^2 \cdot 8$
$$= a^2(a - 8)$$

**43.** $8x^3 - 6x^2 + 2x = 2x \cdot 4x^2 - 2x \cdot 3x + 2x \cdot 1$
$$= 2x(4x^2 - 3x + 1)$$

**45.** $12a^4b^3 + 18a^5b^2 = 6a^4b^2 \cdot 2b + 6a^4b^2 \cdot 3a$
$$= 6a^4b^2(2b + 3a)$$

**47. *Familiarize*.** The tennis court is a rectangle that measures 27 ft by 78 ft. We make a drawing.

27 ft

78 ft

***Translate*.** We substitute in the formula for the area of a rectangle.
$$P = 2 \cdot (l + w) = 2 \cdot (78 \text{ ft} + 27 \text{ ft})$$
***Solve*.** We carry out the calculation.
$$P = 2 \cdot (78 \text{ ft} + 27 \text{ ft})$$
$$= 2 \cdot 105 \text{ ft}$$
$$= 210 \text{ ft}$$
***Check*.** We repeat the calculation.

***State*.** The perimeter of the tennis court is 210 ft.

**49. *Familiarize*.** Let $m$ = the old truck's mileage.

***Translate*.**

| 21 miles per gallon | is | old truck's mileage | plus 20% of | old truck's mileage. |
|---|---|---|---|---|
| ↓ | ↓ | ↓ | ↓ ↓ ↓ | ↓ |
| 21 | = | m | + 20% · | m |

***Solve*.** We convert 20% to decimal notation and solve the equation.
$$21 = m + 0.2m$$
$$21 = 1.2m$$
$$\frac{21}{1.2} = \frac{1.2m}{1.2}$$
$$17.5 = m$$

***Check*.** Note that the new truck's mileage is 100% of the old truck's mileage plus 20% of the old truck's mileage, or 120% of the old truck's mileage. Since $1.2(17.5) = 21$, the answer checks.

***State*.** The old truck got 17.5 miles per gallon.

**51. *Familiarize*.** The number of fish that were not trout is $8 - 3$, or 5. Let $n$ = the percent that were not trout.

***Translate*.**

5 is what percent of 8?
↓ ↓     ↓     ↓ ↓
$5 =$     $n$     $\cdot\ 8$

***Solve*.**
$$5 = n \cdot 8$$
$$\frac{5}{8} = n$$
$$0.625 = n$$
$$62.5\% = n$$

***Check*.** We could solve the same problem as a proportion. Also note that 62.5% of 8 is 5, so the answer checks.

***State*.** 62.5% of the fish were not trout.

**53.** ◈

**55.** ◈

**57.** $703a^{437} + 437a^{703}$
$$= 19 \cdot 37 \cdot a^{437} + 19 \cdot 23 \cdot a^{437} \cdot a^{266}$$
$$= 19a^{473}(37 + 23a^{266})$$

**59.**

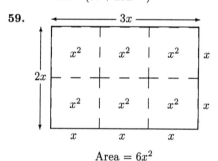

Area $= 6x^2$

---

## Exercise Set 10.3

**1.** $(x + 7)(x + 2) = (x + 7)x + (x + 7)2$
$$= x \cdot x + 7 \cdot x + x \cdot 2 + 7 \cdot 2$$
$$= x^2 + 7x + 2x + 14$$
$$= x^2 + 9x + 14$$

**3.** $(x + 5)(x - 2) = (x + 5)x + (x + 5)(-2)$
$$= x \cdot x + 5 \cdot x + x(-2) + 5(-2)$$
$$= x^2 + 5x - 2x - 10$$
$$= x^2 + 3x - 10$$

**5.** $(x + 6)(x - 2) = x \cdot x - 2 \cdot x + 6 \cdot x + 6(-2)$
$$= x^2 - 2x + 6x - 12$$
$$= x^2 + 4x - 12$$

**7.** $(x - 7)(x - 3) = x \cdot x - 3 \cdot x - 7 \cdot x + (-7)(-3)$
$$= x^2 - 3x - 7x + 21$$
$$= x^2 - 10x + 21$$

**9.**  $(x+6)(x-6) = x^2 - 6 \cdot x + 6 \cdot x + 6(-6)$
$$= x^2 - 6x + 6x - 36$$
$$= x^2 - 36$$

**11.**  $(3+x)(6+2x) = 3 \cdot 6 + 3 \cdot 2x + 6 \cdot x + x \cdot 2x$
$$= 18 + 6x + 6x + 2x^2$$
$$= 18 + 12x + 2x^2$$

**13.**  $(3x-4)(3x-4) = 3x \cdot 3x - 4 \cdot 3x - 4 \cdot 3x + (-4)(-4)$
$$= 9x^2 - 12x - 12x + 16$$
$$= 9x^2 - 24x + 16$$

**15.**  $\left(x - \dfrac{5}{2}\right)\left(x + \dfrac{2}{5}\right) = x \cdot x + \dfrac{2}{5} \cdot x - \dfrac{5}{2} \cdot x - \dfrac{5}{2} \cdot \dfrac{2}{5}$
$$= x^2 + \dfrac{2}{5}x - \dfrac{5}{2}x - \dfrac{5 \cdot 2}{2 \cdot 5}$$
$$= x^2 + \dfrac{4}{10}x - \dfrac{25}{10}x - 1$$
$$= x^2 - \dfrac{21}{10}x - 1$$

**17.**  $(x^2 + x + 1)(x - 1)$
$$= (x^2 + x + 1)x + (x^2 + x + 1)(-1)$$
$$= x^2 \cdot x + x \cdot x + 1 \cdot x + x^2(-1) + x(-1) + 1(-1)$$
$$= x^3 + x^2 + x - x^2 - x - 1$$
$$= x^3 - 1$$

**19.**  $(2x + 1)(2x^2 + 6x + 1)$
$$= 2x(2x^2 + 6x + 1) + 1(2x^2 + 6x + 1)$$
$$= 2x \cdot 2x^2 + 2x \cdot 6x + 2x \cdot 1 + 1 \cdot 2x^2 + 1 \cdot 6x + 1 \cdot 1$$
$$= 4x^3 + 12x^2 + 2x + 2x^2 + 6x + 1$$
$$= 4x^3 + 14x^2 + 8x + 1$$

**21.**  $(y^2 - 3)(3y^2 - 6y + 2)$
$$= y^2(3y^2 - 6y + 2) - 3(3y^2 - 6y + 2)$$
$$= y^2 \cdot 3y^2 + y^2(-6y) + y^2 \cdot 2 - 3 \cdot 3y^2 - 3(-6y) - 3 \cdot 2$$
$$= 3y^4 - 6y^3 + 2y^2 - 9y^2 + 18y - 6$$
$$= 3y^4 - 6y^3 - 7y^2 + 18y - 6$$

**23.**  $(x^3 + x^2)(x^3 + x^2 - x)$
$$= x^3(x^3 + x^2 - x) + x^2(x^3 + x^2 - x)$$
$$= x^3 \cdot x^3 + x^3 \cdot x^2 + x^3(-x) + x^2 \cdot x^3 + x^2 \cdot x^2 + x^2(-x)$$
$$= x^6 + x^5 - x^4 + x^5 + x^4 - x^3$$
$$= x^6 + 2x^5 - x^3$$

**25.**

$$\begin{array}{r} 2t^2 - \ t - 4 \\ 3t^2 + 2t - 1 \\ \hline \end{array}$$

$$\begin{array}{ll} -\ 2t^2 + \ t + 4 & \text{Multiplying by } -1 \\ 4t^3 - \ 2t^2 - 8t & \text{Multiplying by } 2t \\ 6t^4 - 3t^3 - 12t^2 & \text{Multiplying by } 3t^2 \\ \hline 6t^4 + \ t^3 - 16t^2 - 7t + 4 & \end{array}$$

**27.**

$$\begin{array}{ll} x \qquad - x^3 \quad + x^5 & \\ -1 + x^2 \qquad + x^4 & \text{Rewriting in ascending order} \\ \hline x^5 - x^7 + x^9 & \text{Multiplying by } x^4 \\ x^3 - x^5 + x^7 & \text{Multiplying by } x^2 \\ -x + \ x^3 - x^5 & \text{Multiplying by } -1 \\ \hline -x + 2x^3 - x^5 \qquad + x^9 & \end{array}$$

**29.**  *Familiarize.* We label the width of the sidewalk $s$.

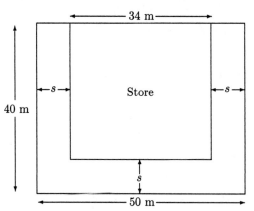

*Translate.* This is a two-step problem. We first find the width of the sidewalk. Then we find the area of the sidewalk.

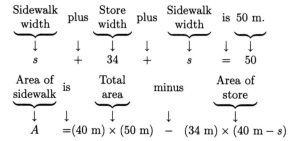

*Solve.* We solve the first equation.

$$s + 34 + s = 50$$
$$2s + 34 = 50 \qquad \text{Collecting like terms}$$
$$2s = 16 \qquad \text{Subtracting 34 on both sides}$$
$$s = 8 \qquad \text{Dividing by 2 on both sides}$$

Thus the width of the sidewalk is 8 m.

Then we solve the second equation.

$$A = (40 \text{ m}) \times (50 \text{ m}) - (34 \text{ m}) \times (40 \text{ m} - s)$$
$$A = (40 \text{ m}) \times (50 \text{ m}) - (34 \text{ m}) \times (40 \text{ m} - 8 \text{ m})$$
$$\qquad\qquad\qquad \text{Substituting 8 m for } s$$
$$A = (40 \text{ m}) \times (50 \text{ m}) - (34 \text{ m}) \times (32 \text{ m})$$
$$A = (40 \times 50 \times \text{ m} \times \text{ m}) - (34 \times 32 \times \text{ m} \times \text{ m})$$
$$A = 2000 \text{ m}^2 - 1088 \text{ m}^2$$
$$A = 912 \text{ m}^2$$

*Check.* We repeat the calculations.

*State.* The area of the sidewalk is 912 m$^2$.

**31.**  *Translate:*

$$\underbrace{\text{What percent}}_{\downarrow \atop n} \text{ of } \underbrace{24 \text{ is } 32?}_{\substack{\downarrow \ \downarrow \ \downarrow \ \downarrow \\ \times \ 24 = \ 32}}$$

*Solve:* We divide by 24 on both sides and convert the result to percent notation.

$$n \times 24 = 32$$
$$\frac{n \times 24}{24} = \frac{32}{24}$$
$$n = 1.\overline{3} = 133.\overline{3}\%, \text{ or } 133\frac{1}{3}\%$$

Thus, 24 is $133.\overline{3}\%$ of 32. The answer is $133.\overline{3}\%$, or $133\frac{1}{3}\%$.

**33.** We rephrase the question and translate.

$$\underbrace{\text{What percent}}_{\downarrow \atop n} \text{ of } 162 \text{ is } 111?$$
$$\qquad\qquad \downarrow \ \downarrow \ \downarrow \ \downarrow$$
$$\qquad\qquad \times \ 162 = 111$$

To solve, we divide on both sides by 162 and convert to percent notation.

$$n \times 162 = 111$$
$$n = \frac{111}{162}$$
$$n \approx 0.685$$
$$n \approx 68.5\%$$

**35.**

**37.**

**39.**
$$(x+2)(x+3) + (x-4)^2$$
$$= (x+2)(x+3) + (x-4)(x-4)$$
$$= x\cdot x + 3\cdot x + 2\cdot x + 2\cdot 3 + x\cdot x - 4\cdot x - 4\cdot x + (-4)(-4)$$
$$= x^2 + 3x + 2x + 6 + x^2 - 4x - 4x + 16$$
$$= 2x^2 - 3x + 22$$

**41.** a) $P = 2\cdot(l+w) = 2\cdot[(m+5)+(m+5)] = 2\cdot(2m+10) = 4m+20$

b) $A = s\cdot s = (m+5)(m+5) =$
$m\cdot m + 5\cdot m + 5\cdot m + 5\cdot 5 =$
$m^2 + 5m + 5m + 25 = m^2 + 10m + 25$

**43.**

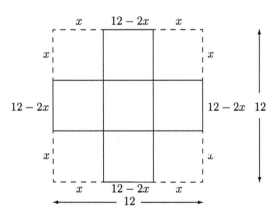

The dimensions of the box are $12-2x$ by $12-2x$ by $x$. The volume is the product of the dimensions (volume = length $\times$ width $\times$ height):

Volume $= (12-2x)(12-2x)x$
$$= (12\cdot 12 - 2x\cdot 12 - 2x\cdot 12 - 2x(-2x))x$$
$$= (144 - 24x - 24x + 4x^2)x$$
$$= (144 - 48x + 4x^2)x$$
$$= 144x - 48x^2 + 4x^3$$

The outside surface area is the sum of the area of the bottom and the areas of the four sides. The dimensions of the bottom are $12-2x$ by $12-2x$, and the dimensions of each side are $x$ by $12-2x$.

$$\begin{aligned} \text{Surface} \atop \text{area} &= {\text{Area of bottom}+ \atop \qquad 4\cdot \text{Area of each side}} \\ &= (12-2x)(12-2x) + 4\cdot x(12-2x) \\ &= 144 - 48x + 4x^2 + 48x - 8x^2 \\ &= 144 - 4x^2 \end{aligned}$$

**45.** $(9x+4)^2 = (9x+4)(9x+4)$
$$= 9x\cdot 9x + 4\cdot 9x + 4\cdot 9x + 4\cdot 4$$
$$= 81x^2 + 36x + 36x + 16$$
$$= 81x^2 + 72x + 16$$

---

## Exercise Set 10.4

**1.** $9^0 = 1 \qquad (b^0 = 1, \text{ for any nonzero number } b.)$

**3.** $3.14^0 = 1 \qquad (b^0 = 1, \text{ for any nonzero number } b.)$

**5.** $(-19.57)^1 = -19.57$

**7.** $(-5.43)^0 = 1$

**9.** $x^0 = 1, \ x \neq 0$

**11.** $\begin{aligned} (3x-17)^0 &= (3\cdot 10 - 17)^0 \quad \text{Substituting} \\ &= (30-17)^0 \qquad \text{Multiplying} \\ &= 13^0 \\ &= 1 \end{aligned}$

**13.** $\begin{aligned} (5x-3)^1 &= (5\cdot 4 - 3)^1 \quad \text{Substituting} \\ &= (20-3)^1 \qquad \text{Multiplying} \\ &= 17^1 \\ &= 17 \end{aligned}$

**15.** $\begin{aligned} (4m-19)^0 &= (4\cdot 3 - 19)^0 \\ &= (12-19)^0 \\ &= (-7)^0 \\ &= 1 \end{aligned}$

**17.** $\begin{aligned} 3x^0 + 4 &= 3(-2)^0 + 4 \\ &= 3\cdot 1 + 4 \\ &= 3 + 4 \\ &= 7 \end{aligned}$

**19.** $\begin{aligned} (3x)^0 + 4 &= [3(-2)]^0 + 4 \\ &= (-6)^0 + 4 \\ &= 1 + 4 \\ &= 5 \end{aligned}$

**21.** $(5 - 3x^0)^1 = (5 - 3 \cdot 19^0)^1$
$\qquad\qquad = (5 - 3 \cdot 1)^1$
$\qquad\qquad = (5 - 3)^1$
$\qquad\qquad = 2^1$
$\qquad\qquad = 2$

**23.** $3^{-2} = \dfrac{1}{3^2} = \dfrac{1}{9}$

**25.** $10^{-4} = \dfrac{1}{10^4} = \dfrac{1}{10,000}$

**27.** $a^{-3} = \dfrac{1}{a^3}$

**29.** $(-5)^{-2} = \dfrac{1}{(-5)^2} = \dfrac{1}{25}$

**31.** $3x^{-7} = 3\left(\dfrac{1}{x^7}\right) = \dfrac{3}{x^7}$

**33.** $\dfrac{x}{y^{-4}} = xy^4$   Instead of dividing by $y^{-4}$, multiply by $y^4$.

**35.** $\dfrac{a^3}{b^{-4}} = a^3 b^4$   Instead of dividing by $b^{-4}$, multiply by $b^4$.

**37.** $-7a^{-9} = -7\left(\dfrac{1}{a^9}\right) = \dfrac{-7}{a^9}$, or $-\dfrac{7}{a^9}$

**39.** $\left(\dfrac{2}{5}\right)^{-2} = \left(\dfrac{5}{2}\right)^2 = \dfrac{5}{2} \cdot \dfrac{5}{2} = \dfrac{25}{4}$

**41.** $\left(\dfrac{5}{a}\right)^{-3} = \left(\dfrac{a}{5}\right)^3 = \dfrac{a}{5} \cdot \dfrac{a}{5} \cdot \dfrac{a}{5} = \dfrac{a^3}{125}$

**43.** $\dfrac{1}{4^3} = 4^{-3}$

**45.** $\dfrac{9}{x^3} = 9x^{-3}$

**47.** $x^{-2} \cdot x = x^{-2+1} = x^{-1} = \dfrac{1}{x}$

**49.** $x^4 \cdot x^{-4} = x^{4+(-4)} = x^0 = 1$, assuming $x \neq 0$

**51.** $x^{-7} \cdot x^{-6} = x^{-7+(-6)} = x^{-13} = \dfrac{1}{x^{13}}$

**53.** $\quad (3a^2 b^{-7})(2ab^9)$
$\quad = 3 \cdot 2 \cdot a^2 \cdot a \cdot b^{-7} \cdot b^9$   Using the commutative and associative laws
$\quad = 6a^{2+1} b^{-7+9}$   Using the product rule
$\quad = 6a^3 b^2$

**55.** $\quad (-2x^{-3} y^8)(3xy^{-2})$
$\quad = -2 \cdot 3 \cdot x^{-3} \cdot x \cdot y^8 \cdot y^{-2}$   Using the commutative and associative laws
$\quad = -6x^{-3+1} y^{8+(-2)}$   Using the product rule
$\quad = -6x^{-2} y^6$
$\quad = -6\left(\dfrac{1}{x^2}\right) y^6$
$\quad = -\dfrac{6y^6}{x^2}$

**57.** $\quad (3a^{-4} bc^2)(2a^{-2} b^{-5} c)$
$\quad = 3 \cdot 2 \cdot a^{-4} \cdot a^{-2} \cdot b \cdot b^{-5} \cdot c^2 \cdot c$
$\quad = 6a^{-4+(-2)} b^{1+(-5)} c^{2+1}$
$\quad = 6a^{-6} b^{-4} c^3$
$\quad = \dfrac{6c^3}{a^6 b^4}$

**59.** $\dfrac{450 \text{ km}}{9 \text{ hr}} = 50\dfrac{\text{km}}{\text{hr}}$, or 50 km/h

**61.** **Familiarize**. Let $n =$ the percent of games the Jets won.

**Translate**. We rephrase and translate.

14 is $\underbrace{\text{what percent}}$ of 16?
$\downarrow \ \downarrow \qquad \downarrow \qquad \ \downarrow \ \downarrow$
$14 = \qquad n \qquad \cdot \ 16$

**Solve**.
$\qquad 14 = n \cdot 16$
$\qquad \dfrac{14}{16} = n$
$\qquad 0.875 = n$
$\qquad 87.5\% = n$

**Check**. We could solve the problem using a proportion. We also note that 87.5% of 16 is 14, so the answer checks.

**State**. The Jets won 87.5% of their games.

**63.** $C = 2 \cdot \pi \cdot r$
$\quad \approx 2 \cdot 3.14 \cdot 5$ cm
$\quad \approx 31.4$ cm

**65.** ◈

**67.** ◈

**69.** Use a calculator.

For $x = -3$:
$\dfrac{5^x}{5^{x+1}} = \dfrac{5^{-3}}{5^{-3+1}} = \dfrac{5^{-3}}{5^{-2}} = 0.2$

For $x = -30$:
$\dfrac{5^x}{5^{x+1}} = \dfrac{5^{-30}}{5^{-30+1}} = \dfrac{5^{-30}}{5^{-29}} = 0.2$

**71.** $(y^{2x})(y^{3x}) = y^{2x+3x} = y^{5x}$

**73.** $\dfrac{a^{6t}(a^{7t})}{a^{9t}} = \dfrac{a^{6t+7t}}{a^{9t}} = \dfrac{a^{13t}}{a^{9t}} = a^{13t} \cdot a^{-9t} =$
$a^{13t+(-9t)} = a^{4t}$